Springer Undergraduate Texts in Philosophy

The Springer Undergraduate Texts in Philosophy offers a series of self-contained textbooks aimed towards the undergraduate level that covers all areas of philosophy ranging from classical philosophy to contemporary topics in the field. The texts will include teaching aids (such as exercises and summaries) and will be aimed mainly towards more advanced undergraduate students of philosophy.

The series publishes:

- All of the philosophical traditions
- Introduction books with a focus on including introduction books for specific topics such as logic, epistemology, German philosophy etc.
- Interdisciplinary introductions - where philosophy overlaps with other scientific or practical areas

This series covers textbooks for all undergraduate levels in philosophy particularly those interested in introductions to specific philosophy topics.

We aim to make a first decision within 1 month of submission. In case of a positive first decision the work will be provisionally contracted: the final decision about publication will depend upon the result of the anonymous peer review of the complete manuscript. We aim to have the complete work peer-reviewed within 3 months of submission.

Proposals should include:

- A short synopsis of the work or the introduction chapter
- The proposed Table of Contents
- CV of the lead author(s)
- List of courses for possible course adoption

The series discourages the submission of manuscripts that are below 65,000 words in length.

For inquiries and submissions of proposals, authors can contact Christi.Lue@springer.com

More information about this series at http://www.springer.com/series/13798

Kevin McCain

The Nature of Scientific Knowledge

An Explanatory Approach

Springer

Kevin McCain
Department of Philosophy
University of Alabama at Birmingham
Birmingham, AL, USA

Springer Undergraduate Texts in Philosophy
ISBN 978-3-319-33403-5 ISBN 978-3-319-33405-9 (eBook)
DOI 10.1007/978-3-319-33405-9

Library of Congress Control Number: 2016944014

Printed on acid-free paper

This Springer imprint is published by Springer Nature
The registered company is Springer International Publishing AG Switzerland

For Quid, just because

Preface

The goal of this book is to provide a comprehensive and accessible introduction to the epistemology of science. To the degree that it is successful, this book introduces readers to epistemology in general as well as the particular nuances of scientific knowledge. The chapters that follow, while far from exhaustive treatments of the various topics, provide readers with a solid introduction to philosophical topics that will be of particular use for those seeking to better understand the nature of scientific knowledge.

My own understanding of the nature of scientific knowledge has greatly benefited from discussions with many colleagues and friends: Marshall Abrams, Jon Altschul, Bryan Appley, John G. Bennett, Mike Bergmann, Mike Bishop, Kenny Boyce, Brandon Carey, Eli Chudnoff, Christopher Cloos, Earl Conee, Brett Coppenger, Andy Cullison, Trent Dougherty, John Dudley, Rich Feldman, Bill Fitzpatrick, Richard Fumerton, Chris Gadsden, Jeff Glick, Alvin Goldman, David Grober-Morrow, Ali Hasan, Sommer Hodson, Kostas Kampourakis, Matt King, Court Lewis, Clayton Littlejohn, Todd Long, Jack Lyons, Peter Markie, Josh May, Matt McGrath, Andrew Moon, Alyssa Ney, Tim Perrine, Kate Phillips, Ted Poston, Jason Rogers, Bill Rowley, Carl Sachs, Greg Stoutenburg, Philip Swenson, Chris Tweedt, Jonathan Vogel, Brad Weslake, Ed Wierenga, Chase Wrenn, Sarah Wright, and several others. Thank you all.

I am particularly grateful to John Dudley, Matt Frise, and Kostas Kampourakis. John and Matt both read and provided helpful comments on significant portions of this book. Kostas provided me sound advice and support at every stage of this project, and it was his encouragement that prompted me to write this book in the first place. Finally, I am deeply indebted to my fiancée, Molly Hill, for the love and support that make this project and many others possible.

In places (particularly, chapters nine and ten) material from the following article is reprinted with kind permission from Springer: "Explanation and the Nature of Scientific Knowledge." *Science & Education*, (2015) 24 (7–8): 827–854. I am grateful to the publishers, journal editor, and anonymous referees for helpful advice concerning this material.

Birmingham, AL, USA Kevin McCain

Contents

About the Author

Kevin McCain is Assistant Professor of Philosophy at the University of Alabama at Birmingham. His research focuses on issues in epistemology and philosophy of science—particularly where these areas intersect. In addition to numerous journal articles, he has published a research monograph on the nature of epistemic justification, *Evidentialism and Epistemic Justification*. He is also currently co-editing (with Ted Poston) a volume on inference to the best explanation and a volume on philosophical skepticism.

Chapter 1
The Importance of Understanding the Nature of Scientific Knowledge

> *"Science without epistemology is—insofar as it is thinkable at all—primitive and muddled"*
>
> (Einstein 1949, p. 683)

Abstract This chapter explains and motivates the importance of understanding the nature of scientific knowledge. The chapter begins by briefly exploring some of the recent science education literature and some of the ways that the literature might benefit from stronger philosophical foundations. Roughly, it will be noted that since scientific knowledge is just a special instance of knowledge, understanding the nature of knowledge in general can provide key insights into the nature of scientific knowledge. These insights into knowledge in general and scientific knowledge in particular seem to hold promise for bolstering the effectiveness of the science education literature on the nature of science. It is because of this that it is important to understand the basics of key debates in epistemology. Also, it is noted that challenges to our general knowledge of the world around us are equally challenges to our scientific knowledge. After briefly explaining the relevance of understanding scientific knowledge this chapter provides an overview of the remaining chapters of the book.

It is clear that the knowledge gained from the sciences has an enormous impact on our lives every single day. No rational person will deny that scientific knowledge is integral to our daily lives and that in many ways it has made our lives better. Scientific knowledge and its applications allow us to cure and treat various diseases, communicate with people all over the world, heat and cool our homes, travel the world, enjoy various entertainments, and so on. Not surprisingly, science, and the knowledge it produces, is often highly valued.

As we do with many things we value, we encourage people to study the sciences to appreciate and benefit from the scientific knowledge already available as well as to help contribute to the production of new scientific knowledge. It is widely held that merely studying the content of the various sciences is not enough by itself to achieve these goals. One should also learn about the nature of science (NOS) itself. That is to say, one should not simply learn the contents of the current state of scientific knowledge, but also learn about the methods that produce such knowledge and the characteristics of scientific knowledge (Kampourakis 2016).

K. McCain, *The Nature of Scientific Knowledge*, Springer Undergraduate Texts in Philosophy, DOI 10.1007/978-3-319-33405-9_1

Emphasis on the importance of understanding NOS is not a new development—understanding NOS has been advocated as a major goal for the study of science since at least the beginning of the last century (Central Association of Science and Mathematics Teachers 1907). This is not simply a goal promoted by the educational systems of a small set of countries—understanding NOS "has been advocated as a critical educational outcome by various science education reform documents worldwide" (Lederman 2007, p. 831). Most nations regard the development of students' understanding of NOS as a primary objective of science education (Eurydice Network 2011; Feng Deng et al. 2011; National Research Council 2012; NGSS Lead States 2013). In fact, the goal of understanding NOS is so widespread that several scholars around the world have argued for its primacy as an educational goal (Lederman 1999, 2007).

This widespread advocating of understanding NOS as a primary goal of science education prompts one question immediately: what exactly is NOS? The answer to this question is the focus of intense debate in the science education literature. According to Norman G. Lederman, and others, NOS can be best understood in terms of what Gürol Irzik and Robert Nola (2011) label the "consensus view".[1] According to this consensus view, the way to best conceptualize NOS is in terms of a fairly small number of general characteristics such as: being based on empirical evidence, tentative, theory-laden, and so on.[2]

Others claim that this consensus view is mistaken. Some argue that the general features that the consensus view uses to characterize NOS are too broad.[3] Others argue, relatedly, that science is simply too heterogeneous to fit the sort of model offered by the consensus view.[4] So, although the consensus view is "the most widely adopted conceptualization of NOS", it is not without its critics (Kampourakis 2016, p. 1).[5]

Not only does the consensus view have critics, it also has rivals. Chief among the rivals to the consensus view is the "Family Resemblance" approach to NOS. Advocates of the Family Resemblance approach maintain that "it is useful to understand NOS not as some list of necessary and sufficient conditions for a practice to be scientific, but rather as something that, following Wittgenstein's terminology, identifies a 'family resemblance' of features that warrant different enterprises being

[1]It is worth noting that Irzik and Nola make it clear that this view is largely the result of work by Lederman along with his various collaborators. See, for example, Abd-El-Khalick (2004), Bell (2004), Cobern and Loving (2001), Flick and Lederman (2004), Hanuscin et al. (2006), Khishfe and Lederman (2006), McComas et al. (1998), McComas and Olson (1998), Osborne et al. (2003), Schwartz and Lederman (2008), Smith and Scharmann (1999), and Ziedlier et al. (2002).

[2]For particularly clear expressions of the consensus view see Lederman (1999), Swartz and Lederman (2008) and Irzik and Nola (2011).

[3]See Allchin (2011), Rudolph (2000), Irzik and Nola (2011).

[4]See Elby and Hammer (2001), Erduran and Dagher (2014), Irzik and Nola (2011, 2014), Matthews (2015), and van Dijk (2011, 2014).

[5]See Abd-El-Khalick (2012), McCain (2015), and Schwartz et al. (2012) for responses to some of these objections for the consensus view.

called scientific" (Matthews 2012, p. 4).[6] The key idea of this approach is that we should "investigate the ways in which each of the sciences are similar or dissimilar, thereby building up from scratch polythetic sets of characteristics for each individual science" instead of trying to come up with necessary and sufficient conditions for NOS (Irzik and Nola 2011, p. 595).

Rather than attempting to adjudicate between the consensus view, the Family Resemblance approach, and other conceptions of NOS, the focus in this book will be to provide a philosophical foundation that can help illuminate this debate over the proper understanding of NOS. By exploring, and becoming clear on, core issues in epistemology and philosophy of science this book will provide tools that will help make this debate (and perhaps others in science education) more tractable.[7]

Getting clear about what "NOS" refers to is very important, and there is serious debate concerning how best to understand NOS. However, our discussion up to this point prompts another key question: why is NOS such an important feature of science education? One answer that springs to mind is simply that the better students understand NOS, the better equipped they will be to become good scientists. After all, it seems plausible that a deeper understanding of the activity you are engaged in will allow you to be better at that activity. The best chess players tend to be those who have a deep understanding of the game of chess. The best carpenters tend to be those who have a deep understanding of construction processes (measuring, leveling, framing, etc.). So, it is not unreasonable to think that the best scientists will tend to be those who have a deep understanding of NOS. More important than the benefit of simply being a better scientist is the benefit of increased scientific literacy. After all, most students will not become professional scientists, but they will all be citizens who may someday need to make decisions about socio-scientific issues. For this reason, of the several other potential benefits of an adequate understanding of NOS that have been identified it seems that 2, 3, and 4 below are the most important:

1. It is necessary for understanding the process of science and it helps with managing technological objects.
2. It is necessary for making informed decisions about socio-scientific issues (e.g. global warming, stem cell research, etc.)
3. It is necessary for fully appreciating the importance of science in contemporary culture.

[6] Also see Irzik and Nola (2011, 2014) and Erduran and Dagher (2014) for defense of this sort of view.

[7] For example, there is ongoing debate concerning what the proper goal of teaching various scientific theories, such as evolution, should be. Some, e.g., Goldman (1999), argue that belief is the primary goal of science education. Others, e.g., Smith and Siegel (2004), argue that belief, while important, should not be the goal of science education. Instead, they maintain that understanding should be the primary goal of science education with belief as a potentially desirable outcome. Getting clearer on such epistemological concepts as 'belief', 'knowledge', and 'understanding' can shed light on this debate over the primary goals of science education.

4. It helps facilitate an understanding of the norms of the scientific community—particularly, the moral commitments of this community and how they are valuable to society as a whole.
5. It aids in the learning of scientific subject matter—the principles and findings of particular sciences.[8]

Not only are these potential benefits intuitively plausible, empirical research has supported the link between increased understanding of NOS and the attainment of these benefits (Feng Deng et al. 2011).

Of course, given that an adequate understanding of NOS is necessary for obtaining a number of these benefits it follows that one simply cannot possess certain benefits if she has an inferior understanding of NOS. That is to say, without a proper understanding of NOS one cannot truly understand the process of science, make well-informed decisions about socio-scientific issues, or fully appreciate the importance science has in our contemporary culture. To give just one illustration of this, Kostas Kampourakis (2014) persuasively argues that one of the reasons why many people are resistant to accepting evolutionary theory (one of the best confirmed theories in the history of science) is that they fail to properly understand NOS. Specifically, Kampourakis argues that many people fail to distinguish scientific knowledge from other things that they *merely* believe, but do not really know, and this failure leads them to side against mountains of evidence by not accepting the truth of evolutionary theory. It is not difficult to imagine how this sort of tendency could lead to many uninformed socio-scientific decisions and other errors.

Now that some of the benefits of properly understanding NOS and the widespread emphasis on increased understanding of NOS as a premier educational goal have been made clear, yet another question clearly presents itself: how well do students and science educators understand NOS? Unfortunately, the answer to this question is "not well at all". Sadly, "the longevity of this educational objective [increased understanding of NOS] has been surpassed only by the longevity of students' inability to articulate the meaning of the phrase 'nature of science' and to delineate the associated characteristics of science" (Lederman and Niess 1997, pp. 1). Numerous studies have shown that students' understandings of NOS are lacking in many areas (Lederman 2007). The findings of these studies are particularly telling because they employ a variety apparatuses for measuring students' understandings of NOS with the same results—students exhibit poor understanding. As Lederman (2007, pp. 838) points out, "the overwhelming conclusion that students did not possess adequate conceptions of the nature of science or scientific reasoning is considered particularly significant when one realizes that a wide variety of assessment instruments were used".

[8]These five benefits are closely based on those put forward by Driver et al. (1996). See Lederman (2007) and Feng Deng et al. (2011) for further discussion.

Even more alarming than the fact that students lack a proper understanding of NOS is the fact that some studies, such as Miller (1963), suggest that many secondary science teachers do not understand NOS as well as students do! These results have been supported by more recent investigations to the point that it is at least safe to say, "science teachers do not possess adequate conceptions of NOS, irrespective of the instrument used to assess understandings" (Lederman 2007, pp. 852). The evidence supports that regardless of how we assess understanding of NOS, "student and teacher understandings are not at the desired levels" (Lederman 2007, pp. 861). This is a most unfortunate situation.

At this point a brief recap is in order. We have seen that there are numerous benefits both to individuals and to society as a whole when citizens possess a proper understanding of NOS. These benefits are so widely recognized that increased understanding of NOS has been almost unanimously held to be a critical educational goal for more than one hundred years. However, despite the recognized benefits of properly understanding NOS and the emphasis on this understanding as an educational objective, students and teachers both fail to possess adequate understanding of NOS. What is to be done?

Fortunately, a number of factors, which seem to aid in increasing understanding of NOS, have been identified. It seems sufficient background in the history and philosophy of science clearly influences teachers' ability to teach science (King 1991). Teachers are not the only ones who benefit from a background in philosophy. There is evidence that students enrolled in a philosophy of science course develop a deeper understanding of NOS than students who are only enrolled in courses on scientific methods (Abd-El-Khalick 2005). Astoundingly, Mekritt Kimball (1967) has noted that undergraduate philosophy majors outscore both science teachers and professional scientists on measures designed to track understanding of NOS. This finding led Kimball to suggest that adding some philosophy courses to undergraduate science curricula may help improve understanding of NOS.[9] Given the results of these studies it is not surprising that an explicitly reflective approach in instruction has been linked to greater success in teaching NOS (Schwartz et al. 2012). After all, philosophy is known for its employment of an explicitly reflective approach to understanding and assessing fundamental problems and theories in many domains. Of course, this is not to suggest that a grasp of the philosophical issues surrounding and interwoven with NOS will guarantee an adequate understanding of NOS. However, the evidence does suggest that some appreciation of philosophy, particularly as it is related to science, can help put one on the path to an adequate understanding of NOS. It is the purpose of this book to facilitate, at least to some degree, the initial steps on this path.

In order to help facilitate a deeper understanding of NOS, the goal of this book is to offer a comprehensive and accessible introduction to the epistemology of science—an introduction that does not presuppose familiarity with philosophy.

[9]Carey and Stauss (1968) also recommended the inclusion of philosophy of science courses in undergraduate science curricula as a means to enhancing understanding of NOS.

That is to say, I will attempt to provide an introduction to epistemology (theory of knowledge) in general as well as the particular nuances of philosophical work on scientific knowledge that will help initiate some increase in understanding of NOS on its own, and more importantly, help provide some of the tools necessary to facilitate even more in-depth investigation of NOS and the surrounding science education debate.

One might wonder, if a deeper understanding of NOS is the primary goal, why are we bothering with epistemology in general in this book? The reason is simple. Scientific knowledge is itself a kind of knowledge, so understanding the nature of knowledge in general can provide key insights into the nature of scientific knowledge. A firm grounding in the theory of knowledge can help provide a strong foundation for deeper appreciation of NOS. Additionally, other important debates in science education revolve around key concepts of general epistemology such as the nature of belief, knowledge, and understanding. The chapters in this book, while far from exhaustive treatments of the various topics, provide a solid introduction to philosophical topics that will be of particular use for science education. By exploring the basics of general epistemology as well as philosophy of science and particular challenges to our scientific knowledge, the various epistemological components of key science education debates will become clearer.

Now, it would be a mistake to think that this book, or any other single work for that matter, will provide easy solutions to debates in science education or a cure-all for the prevailing inadequacies in understanding of NOS. Rather than attempting the impossible, in this book I instead seek to aid in these endeavors by providing an accessible overview of the major components of knowledge in general and scientific knowledge in particular. In addition to explaining the nature of scientific knowledge (from the perspective of contemporary philosophy) I also explore some of the challenges that have been raised for the possibility of our having such knowledge at all. Although in many cases the relevant philosophical issues are explained and various moves in the debate are explored in a neutral manner, I do not always remain neutral. In particular, I do not simply mention challenges to our scientific knowledge in various chapters, I also argue for what I take to be the strongest rebuttals of such challenges. Additionally, I present a picture of knowledge throughout this book that, while being a picture that many scientists, science educators, and philosophers would accept, is not universally accepted. Unfortunately, universal acceptance is another lofty goal that is likely unreachable. So, instead of trying for this, I offer the reader a picture of the nature of scientific knowledge that is consistent with commonsense, is acceptable to scientists, and helps to provide a foundation for understanding NOS and gaining clarity in various science education debates. Even if the reader finds various components of this view of scientific knowledge unsatisfactory, examining the view as well as what can be said in its favor will be instructive both as a way to make clear key ideas in epistemology and philosophy of science and as a way of illuminating how alternative conceptions might be profitably explored.

The remainder of this book is divided into four parts. The seven chapters in the first part are focused on the "traditional account of knowledge". In the following chapter, I introduce the traditional account of knowledge. First, I distinguish between three main kinds of knowledge: acquaintance knowledge, knowledge-how, and propositional knowledge. The nature of each of these kinds of knowledge and their differences are illuminated. It will become clear that scientific knowledge is best understood as a particular variety of propositional knowledge. After clarifying the differences between these kinds of knowledge, I turn to a brief examination of the traditional account of propositional knowledge. This traditional account holds that in order for someone to have knowledge of a particular proposition three conditions must be satisfied: she must believe the proposition, the proposition must be true, and she must have justification for believing the proposition. My discussion of the traditional account of knowledge in this chapter sets the stage for the more in-depth examination of the general features of knowledge that is the focus for the remaining chapters in this part of the book.

Chapter 3 is the first of three in which I explore each of the components of the traditional account of knowledge in detail. In this chapter I investigate the nature of belief. I contrast the idea of *believing in* something with the idea of *believing that* something is true. The first notion of belief is really an expression used to signify trust or faith in something rather than the sort of belief that is a component of knowledge. Once the importance of *believing that* is made clear I briefly examine various accounts of the nature of belief. I explain why we do not need to decide which of these accounts is superior for the purpose of understanding our scientific knowledge. Further, I lay out some of the main distinctions concerning kinds of beliefs and touch on various philosophical issues that arise from the consideration of belief. Fortunately, we do not need to settle all (or even many) of these issues for our purposes. It is sufficient for the present focus to simply have a good grasp of the general nature of belief as well as of some of the various ways an account of belief might be developed. However, consideration of these distinctions and issues helps to deepen our understanding of the nature of belief, and so deepens our understanding of the nature of knowledge.

In Chap. 4 I focus on the nature of truth, the second component of the traditional account of knowledge. Although it is often taken to be obviously clear, the nature of truth is a very complex philosophical issue. I examine both traditional and contemporary theories of truth as well as realist and anti-realist conceptions of truth. Further, I briefly look at some of the major challenges for a successful theory of truth. Ultimately, I argue for a commonsensical, realist conception of truth. This conception of truth is supported by both philosophical argument as well as the recognition of its presupposition in scientific practice.

Next, I turn toward the final component of the traditional account of knowledge: justification. Traditionally, justification has been understood as having good reasons for believing that a particular claim is true. However, there are several important distinctions and debates about how best to understand justification as good reasons. In fact, one major contemporary debate in epistemology concerns whether we should keep with tradition and understand justification in terms of good reasons at all. Internalists say, "yes, one always needs good reasons in order to be justified in her beliefs", but externalists disagree. I explain each of these views of justification in some detail in this chapter. After explaining internalism and externalism about justification, I consider some of the major moves in the debate between these two positions. It becomes clear by chapter five's end that whether internalists or externalists are correct about justification in general, the sort of justification required for scientific knowledge does require good reasons, which I argue are best understood as evidence.

Since Chap. 5 concludes by noting that scientific knowledge requires evidence, it is quite natural that evidence is the focus of Chap. 6. In this chapter I explore two central issues of evidence. The first issue concerns the nature of evidence itself. There are two primary theories of the nature of evidence. The first claims that evidence consists of non-factive mental states and the second claims that evidence consists of propositions. I explain both of these theories and consider some of the major challenges to each theory. The second issue I explore in this chapter is that of what it takes for someone to have an item of information as evidence. Extreme views of evidence possession each have serious problems, however, moderate views face challenges too. After elucidating some of the challenges facing the various views, I argue that there are some promising ways of providing a moderate account of what it takes to have evidence.

In Chap. 7 I clarify the very important distinction between having justification for believing a proposition (propositional justification) and justifiedly believing a proposition (doxastic justification). Since justified belief is a necessary condition of knowledge, it is extremely important to understand what is required to move from merely having propositional justification to having doxastic justification. In this chapter I explore the relation that one's belief has to bear to her propositional justification in order to be doxastically justified—what epistemologists call the "basing relation". Accounts of the basing relation fall into three categories: causal accounts, doxastic accounts, and hybrid accounts. I elucidate the general features of each of these kinds of accounts, as well as the challenges they face, in this chapter.

I begin the final chapter of this part of the book, Chap. 8, with a brief recap of what has been discovered about the traditional account of knowledge throughout the previous chapters. After making the traditional account of knowledge clear, I present a decisive objection to that account of knowledge—what is known as the "Gettier Problem". Roughly, this problem illustrates that it is possible to have a belief that is both justified and true, and yet, contrary to the traditional account, the belief is not an instance of knowledge. In addition to explaining how the Gettier Problem shows that the traditional account of knowledge is incomplete I explore some promising responses to the Gettier Problem. Finally, I conclude by noting that even without an answer to the Gettier Problem we can use the traditional account of knowledge as a framework for understanding scientific knowledge. This is so because the Gettier Problem does not threaten the relevant components of scientific knowledge—belief, evidence, and truth. The Gettier Problem simply gives us reason to think that we should be primarily concerned with the relation between scientific claims and evidence, which holds whether or not we are in a situation where the Gettier Problem arises.

In the second part of the book I transition from focusing on the nature of knowledge in general to focusing on scientific knowledge in particular. In the first chapter of this section, Chap. 9, I examine the nature of scientific explanations as well as their relation to understanding. Roughly, I point out that good explanations are those that provide understanding of particular phenomena. In addition to examining the relationship between explanation and understanding, I also explore how understanding scientific theories is related to understanding phenomena. This examination of explanations and how they lead to understanding is very important since it is common in scientific practice to adopt a particular theory because the explanations that it produces allow us to understand various phenomena. This is also particularly important, if as some claim, understanding is the primary goal of science education.

I build upon the insights of the previous chapter when arguing for a close connection between explanation and evidential support in Chap. 10. That is to say, in this chapter I argue that the degree to which a given body of evidence supports believing that a particular proposition is true depends upon how well that proposition explains the evidence. The upshot of this is a clear conception of when we should accept claims in science that can be extended to an account of justification more generally. Thus, in this chapter I seek to demonstrate that the connection between scientific explanatory practices and the justification for any of our beliefs may be a close one. So, the process of expanding our scientific knowledge may be more closely related to the process(es) of expanding our knowledge in any other domain than one might have expected. This might be thought to have particularly important ramifications for debates surrounding the proper understanding of NOS in the science education literature.

In the next part of the book I respond to challenges to our knowledge. One way to challenge our scientific knowledge is to challenge all of our knowledge of the world around us. I explore the challenge to our knowledge posed by external world skepticism in Chap. 11. During the process of examining and responding to arguments for external world skepticism important insights are revealed. One of the foremost of these insights is that knowledge does not require evidence that makes the believed proposition absolutely certain. Instead, significant, yet fallible, evidence is all that is required for knowledge. Another insight is that the explanatory account of evidential support developed in earlier chapters helps to show that, despite initial appearances, external world skepticism is not a significant threat to our knowledge after all.

Chapter 12 centers on another way of challenging our scientific knowledge: challenging our knowledge of unobserved cases. The skeptic about induction claims that while we might observe many, many instances of black ravens this does not allow us to know that the next raven we observe will be black (or even reasonably believe that it will be, or is likely to be, black). This sort of inductive skepticism poses a major threat to our scientific knowledge as well as our commonsense knowledge of the world around us. In this chapter I argue that, again, the skeptical challenge can be overcome by carefully understanding the sort of explanatory account of evidential support that has been developed in earlier chapters.

In Chap. 13 I examine a more limited, but in some ways more worrisome, threat to our scientific knowledge. This challenge comes from empirical research which suggests we are subject to a number of biases and irrational processes when forming our beliefs. Some take this evidence of human irrationality to undercut our knowledge in general. I argue that the challenge posed by evidence of our irrationality is not a significant threat to our scientific knowledge. By recognizing our tendency to make certain systematic mistakes we can take steps to correct for the effects of such mistakes both as individuals and as groups. Thus, at most what evidence of systematic irrationality does is give us further reason for thinking scientific knowledge should be held tentatively—something that is embraced by most all conceptions of NOS.

This part of the book closes with Chap. 14, in which I discuss one of the major debates in philosophy of science, the debate between realists and anti-realists. Realists maintain that our best-confirmed scientific theories are true, but anti-realists think that we should only accept that our best-confirmed scientific theories are useful in some sense without committing to their truth (or even approximate truth). I present and evaluate the major arguments on both sides of this debate. Throughout the chapter I defend a realist stance, which allows for genuine scientific knowledge. Ultimately, I conclude that in the case of our best-confirmed theories the truth of those theories best explains their predictive success, which gives us justification for believing they are true. Anti-realist arguments do not give us reason to think scientific knowledge is beyond our grasp. While this realist stance may be controversial among science educators and philosophers of science, I believe that it is defensible. Additionally, I believe that exploring the debate between realism and anti-realism by way of defending realism can make the debate itself clearer and shed light on where those of anti-realist leanings might focus their rebuttals.

The final part of this book is devoted to important issues in the social epistemology of science. I begin by exploring how individual knowledge of scientific claims is related to knowledge within a scientific community. In Chap. 15 I explore the nature of social evidence. The most prevalent form of social evidence is testimony. In this chapter I discuss the nature of testimony and how it is that we can gain knowledge from the testimony of others. In this chapter I also examine a second kind of social knowledge, which we gain from the very commonplace, yet philosophically interesting, phenomenon of disagreement. I explicate the epistemic significance of disagreement in science. In particular I discuss how we should respond when we discover that others disagree with us about a scientific claim.

In Chap. 16 I move beyond a study of the individualistic characteristics of scientific knowledge by looking at science as an epistemic system. It becomes clear that the thoroughgoing social nature of science leads to some characteristics which make it particularly well suited for adding to the store of scientific knowledge. In particular, the social nature of science leads to a division of cognitive labor. This division of cognitive labor both makes it so that trust plays an integral role in the generation of scientific knowledge and so that scientific progress is enhanced by the scientific community hedging its bets through scientists pursuing a wide variety of research projects utilizing a variety of methods. Although the individual scientists that make up the scientific community are not without human flaws, various social institutions in science help to make good use out of our baser motivations. I argue that while science may not be perfect, it is an epistemic system which has features that make it tremendously successful at generating knowledge of the world around us.

I conclude the book with Chap. 17. I begin by recapping some of the major insights of the earlier chapters of this book. I also point out how these insights can be used to supplement the science education literature. The result, I hope, is a more philosophically grounded science education literature. Such integration holds promise for strengthening both science education and philosophical approaches to understanding scientific knowledge. The philosophical foundation provided by this book holds promise of clarifying a number of epistemological concepts of great importance to debates in science education. This is not to say that I settle the debate concerning how we should understand NOS, or any other key debate in science education—far from it! Instead, this book provides a philosophical basis from which we can better understand and evaluate the key positions in these debates. It is my hope that this philosophical basis, by being relevant to adjudicating between positions in these debates, can also serve as a springboard for increasing understanding of NOS and other key components of science education. I also discuss some of the major areas where further research would be helpful in this final chapter. Although it is often a bit risky to do so, I make some suggestions as to how some of the needed research might be fruitfully conducted and speculate on what some of the results of such research might be. My goal for this final chapter is not to offer precise predictions of how things will turn out, but rather, to encourage further research to continue on the path to greater understanding of NOS and helpfully gesture to good starting places for such research.

References

Abd-El-Khalick, F. (2004). Over and over and over again: College students' views of nature of science. In L. B. Flick & N. G. Lederman (Eds.), *Scientific inquiry and nature of science* (pp. 389–426). Dordrecht: Kluwer.

Abd-El-Khalick, F. (2005). Developing deeper understandings of nature of science: The impact of a philosophy of science course on preservice teachers' views and instructional planning. *International Journal of Science Education, 27*, 15–42.

Abd-El-Khalick, F. (2012). Examining the sources for our understandings about science: Enduring conflations and critical issues in research on nature of science in science education. *International Journal of Science Education, 34*, 353–374.

Allchin, D. (2011). Evaluating knowledge of the nature of (whole) science. *Science Education, 95*, 518–542.

Bell, R. (2004). Perusing Pandora's box: Exploring the what, when, and how of nature of science. In L. B. Flick & N. G. Lederman (Eds.), *Scientific inquiry and nature of science* (pp. 427–446). Dordrecht: Kluwer.

Carey, R. L., & Stauss, N. G. (1968). An analysis of the understanding of the nature of science by prospective secondary science teachers. *Science Education, 52*, 358–363.

Central Association for Science and Mathematics Teachers. (1907). A consideration of the principles that should determine the courses in biology in secondary schools. *School Science and Mathematics, 7*, 241–247.

Cobern, W., & Loving, C. (2001). Defining "Science" in a multicultural world: Implications for science education. *Science Education, 85*, 50–67.

Driver, R., Leach, J., Millar, R., & Scott, P. (1996). *Young peoples's images of science*. Buckingham: Open University Press.

Einstein, A. (1949). Remarks concerning the essays brought together in this co-operative volume. In P. A. Schilpp (Ed.), *Albert Einstein: Philosopher-scientist* (pp. 665–688). Evanston: The Library of Living Philosophers.

Elby, A., & Hammer, D. (2001). On the substance of a sophisticated epistemology. *Science Education, 85*, 554–567.

Erduran, S., & Dagher, Z. (2014). *Reconceptualizing the nature of science for science education: Scientific knowledge, practices and other family categories*. Dordrecht: Springer.

Eurydice Network. (2011). *Science education in Europe: national policies, practices and research*. Brussels: Education, Audiovisual and Culture Executive Agency.

Feng Deng, D. C., Tsai, C., & Chai, C. S. (2011). Student's views of the nature of science: A critical review of research. *Science Education, 95*, 961–999.

Flick, L. B., & Lederman, N. G. (2004). Introduction. In L. B. Flick & N. G. Lederman (Eds.), *Scientific inquiry and nature of science* (pp. ix–xviii). Dordrecht: Kluwer.

Goldman, A. I. (1999). *Knowledge in a social world*. Oxford: Oxford University Press.

Hanuscin, D. L., Akerson, V. L., & Phillipson-Mower, T. (2006). Integrating nature of science instruction into a physical science content course for preservice elementary teachers: NOS views of teaching assistants. *Science Education, 90*, 912–935.

Irzik, G., & Nola, R. (2011). A family resemblance approach to the nature of science for science education. *Science and Education, 20*, 591–607.

Irzik, G., & Nola, R. (2014). New directions for nature of science research. In M. Matthews (Ed.), *International handbook of research in history, philosophy and science teaching* (pp. 999–1021). Dordrecht: Springer.

Kampourakis, K. (2014). *Understanding evolution*. Cambridge, UK: Cambridge University Press.

Kampourakis, K. (2016). The "general aspects" conceptualization as a pragmatic and effective means to introducing students to nature of science. *Journal of Research in Science Teaching, 53*, 667–682.

Khishfe, R., & Lederman, N. G. (2006). Teaching nature of science within a controversial topic: Integrated versus nonintegrated. *Journal of Research in Science Teaching, 43*, 395–418.

Kimball, M. E. (1967). Understanding the nature of science: A comparison of scientists and science teachers. *Journal of Research in Science Teaching, 5*, 110–120.

King, B. B. (1991). Beginning teachers' knowledge of and attitudes toward history and philosophy of science. *Science Education, 75*, 135–141.

Lederman, N. G. (1999). Teachers' understanding of the nature of science and classroom practice: Factors that facilitate or impede the relationship. *Journal of Research in Science Teaching, 36*, 916–929.

Lederman, N. G. (2007). Nature of science: Past, present, and future. In S. K. Abell & N. G. Lederman (Eds.), *Handbook of research on science education* (pp. 831–879). Mahwah: Erlbaum.

Lederman, N. G., & Niess, M. L. (1997). The nature of science: Naturally? *School Science and Mathematics, 97*, 1–2.

Matthews, M. (2012). Changing the focus: From nature of science (NOS) to features of science (FOS). In M. S. Khine (Ed.), *Advances in nature of science research* (pp. 3–26). Dordrecht: Springer.

Matthews, M. (2015). *Science teaching: The contribution of history and philosophy of science* (2nd ed.). New York: Routledge.

McCain, K. (2015). Explanation and the nature of scientific knowledge. *Science & Education, 24*, 827–854.

McComas, W. F., & Olson, J. K. (1998). The nature of science in international science education standards documents. In W. F. McComas (Ed.), *The nature of science in science education: Rationales and strategies* (pp. 41–52). Hingham: Kluwer.
McComas, W. F., Clough, M. P., & Almazroa, H. (1998). The role and character of the nature of science in science education. In W. F. McComas (Ed.), *The nature of science in science education: Rationales and strategies* (pp. 3–40). Hingham: Kluwer.
Miller, P. E. (1963). A comparison of the abilities of secondary teachers and students of biology to understand science. *Iowa Academy of Science, 70*, 510–513.
National Research Council. (2012). *A framework for K-12 science education: Practices, crosscutting concepts, and core ideas*. Committee on a Conceptual Framework for New K-12 Science Education Standards. Board on Science Education, Division of Behavioral and Social Sciences and Education. Washington, DC: The National Academies Press.
NGSS Lead States. (2013). *Next generation science standards: For states, by states*. Washington, DC: The National Academies Press.
Osborne, J., Collins, S., Ratcliffe, M., Millar, R., & Duschl, R. (2003). What "ideas-about-science" should be taught in school science? A Delphi study of the expert community. *Journal of Research in Science Education, 40*, 692–720.
Rudolph, J. L. (2000). Reconsidering the 'nature of science' as a curriculum component. *Journal of Curriculum Studies, 32*, 403–419.
Schwartz, R., & Lederman, N. G. (2008). What scientists say: Scientists' views of nature of science and relation to science context. *International Journal of Science Education, 30*, 727–771.
Schwartz, R., Lederman, N. G., & Abd-El-Khalick, F. (2012). A series of misrepresentations: A response to Allchin's whole approach to assessing nature of science understandings. *Science Education, 96*, 685–692.
Smith, M. U., & Scharmann, L. C. (1999). Defining versus describing the nature of science: A pragmatic analysis for classroom teachers and science educators. *Science Education, 83*, 493–509.
Smith, M. U., & Siegel, H. (2004). Knowing, believing, and understanding: What goals for science education? *Science & Education, 13*, 553–582.
van Dijk, E. M. (2011). Portraying real science in science communication. *Science Education, 95*, 1086–1100.
van Dijk, E. M. (2014). Understanding the heterogeneous nature of science: A comprehensive notion of PCK for scientific literacy. *Science Education, 98*, 397–411.
Ziedler, D. N., Walker, K. A., & Ackett, W. A. (2002). Tangled up in views: Beliefs in the nature of science and responses to socioscientific dilemmas. *Science Education, 86*, 343–367.

Part I
General Features of Knowledge

Chapter 2
The Traditional Account of Knowledge

Abstract This chapter introduces the traditional account of knowledge. First, three main kinds of knowledge are distinguished: acquaintance knowledge, knowledge-how, and propositional knowledge. The nature of each of these kinds of knowledge and their differences from one another are illuminated. It is also made clear that scientific knowledge is best understood as a particular variety of propositional knowledge. After clarifying the differences between these kinds of knowledge the chapter turns to a brief examination of the traditional account of propositional knowledge. This traditional account holds that in order for one to have knowledge of a particular proposition three conditions must be satisfied: the proposition must be true, one must believe the proposition, and one must have justification for believing the proposition. The discussion of the traditional account of knowledge in this chapter sets the stage for the more in-depth examination of the general features of knowledge that is the focus of the remaining chapters in this section of the book.

We know many things. For example, you know that you are reading this book. You know that the sun is more massive than Earth. You know that dogs are animals. You know that electrons have negative charge. You may know how to swim or how to ride a bicycle. You know your friends. You know many other things too. But, what exactly does it mean to say that you *know* these things? What is it that the various things you know have in common? As we saw in the previous chapter, discovering the answers to these questions is vital to truly understanding scientific knowledge. In this chapter we will begin our exploration of the nature of knowledge in general.

2.1 Kinds of Knowledge

Above we noted many things that you know. You have knowledge that various claims are true, but you also have knowledge of how to do things and knowledge of people. We asked what these things have in common, but this may have been a misleading question. Perhaps there are different kinds of knowledge. In fact, epistemologists (philosophers who specialize in the study of knowledge and related concepts) tend to think that there are at least three distinct basic kinds of knowledge

K. McCain, *The Nature of Scientific Knowledge*, Springer Undergraduate Texts in Philosophy, DOI 10.1007/978-3-319-33405-9_2

that are very different from one another. These distinct kinds of knowledge are *acquaintance knowledge*, *knowledge-how*, and *propositional knowledge*. We will briefly examine the first two kinds of knowledge before turning to our primary focus, propositional knowledge.

Acquaintance knowledge is the sort of knowledge you have of your friends and other things with which you are familiar. In order to appreciate the nature of this kind of knowledge, consider the following comparison: your knowledge of your best friend and your knowledge of President Obama. If you are best friends with the president of the United States, congratulations—instead of this example, compare your knowledge of him with your knowledge of some famous person who is not a personal friend of yours. Let us assume for the sake of illustration that your best friend and President Obama are not the same person. You may still know a lot about President Obama. For instance, you might know that his full name is Barack Hussein Obama II. You might know that he is the 44th president of the United States. You might know many other facts about him. Perhaps you have even read several books about him. However, despite your knowledge of all of these facts about President Obama we would not say that you *know him* in the sense that you know your friends. To express this thought we might say something like "you know all about President Obama, but you do not know him personally". When it comes to your best friend you probably know a lot about her as well, but you also know *her*. The difference here is that while you only know various facts about President Obama, you know a number of facts about your best friend, and you also have acquaintance knowledge of her. So, knowledge of facts is different from acquaintance knowledge.

Knowledge-how is different from both acquaintance knowledge and knowledge of facts. It is the sort of knowledge that you have when you have a particular ability or skill. You may know how to ride a bicycle or know how to swim. This sort of knowledge seems very different than your knowledge of your best friend. Knowing how to do something is not the same as having acquaintance knowledge of something. Also, it seems that your knowledge of how to ride a bicycle is very different than your knowledge of facts such as those you know about President Obama. So, we seem to have three distinct kinds of knowledge. The sort of knowledge you have of people you are familiar with, the sort of knowledge that is your skills or abilities, and the sort of knowledge you have of facts.

Let us take a closer look at the third kind of knowledge, the sort of knowledge you have of facts. Consider again your knowledge of President Obama. This knowledge consists of knowing facts about the President. You know that Barack Obama is the current President of the United States. You know that he was a senator in Illinois. And so on. This is all propositional knowledge. That is, it is knowledge that you have of particular propositions.[1] Propositions are what our declarative sentences mean, but they are not identical to those declarative sentences. For instance, consider

[1]The term "facts" as we have been using it so far is simply another word for "true propositions". Also, at times we will speak of "claims" being true or false. As we will use the term, a "claim" is simply a proposition that is asserted or accepted as true by someone.

the English sentence "The ball is red", the Spanish sentence "La pelota es de color rojo", and the German sentence "Der Ball ist rot". Obviously, these sentences are very different; they are composed of different words and different orderings of letters. They also sound very different than one another when read aloud. Despite their many differences, each of these sentences means the same thing. Namely, they each mean that *the ball is red*. How is it that these very different sentences can have the same meaning? It is because they each express the same proposition. It is propositions that are the objects of propositional knowledge. So, when you know that the sun is more massive than Earth you know a particular proposition to be true.[2] Specifically, you know that the proposition we express with the English sentence "The sun is more massive than Earth" is true.

Propositional knowledge seems very different than both acquaintance knowledge and knowledge-how. For example, someone who has never met your best friend may know a number of facts about her. This person may have read an extensive file about your friend and her entire life. However, if he has never met your friend, we would not say that he has acquaintance knowledge of her. He only has propositional knowledge of facts about your friend. Similarly, someone might have read several articles on how to ride a bicycle. He may have read so many articles that he can tell you all the finer points of riding a bicycle like a professional—perhaps he can even describe to you how to ride a bicycle better than a champion cyclist. Nevertheless, if he has never ridden a bicycle himself and lacks the ability to do so, we would say that although he has propositional knowledge of bicycle riding, he does not *know how* to ride a bicycle. He knows a lot about bicycle riding, but not how to actually ride one. So, propositional knowledge seems to be different from knowledge-how too.[3]

When it comes to scientific knowledge it is clear that the relevant kind of knowledge is propositional knowledge. After all, scientific claims are not the sort of thing with which you can have acquaintance knowledge. You might really like a particular scientific truth, but you are not acquainted with it in this sense. General Relativity is not one of your personal friends! Also, although knowledge-how clearly plays a significant role in scientific practice in that you might need to know how to run experiments or know how to proceed in a lab, our scientific knowledge

[2] As we will see in chapter four there is some dispute about whether the primary bearers of truth are propositions or sentences. Although there is dispute about this point, it will not make a difference to our understanding of the nature of knowledge. In light of this, we will follow the majority in holding that propositions are the primary bearers of truth and that when you know something such as that the sun is bigger than Earth what you know is that a particular proposition is true.

[3] Though the dominant view is that knowledge-how is not propositional knowledge, this distinction has become a somewhat controversial philosophical issue in recent years. Stanley (2011) and Stanley and Williamson (2001) have argued that knowledge-how is really just a form of propositional knowledge. Adams (2009), Devitt (2011), Lewis (1990), Poston (2009), and Ryle (1949) have argued in support of the traditional distinction. Contemporary psychology seems to be on the side of the dominant philosophical view because psychologists, such as Schacter et al. (2000) and Squire (1987), distinguish between "procedural" knowledge (knowledge-how) and "semantic"/"declarative" knowledge (propositional knowledge).

is not knowledge-how. Your knowledge of atomic theory, for instance, is not your possessing a particular skill. Instead, it is knowledge that particular propositions are true. Your knowledge of atomic theory includes things like your knowledge that electrons have negative charge, protons have positive charge, etc. So, throughout our discussion of scientific knowledge (and knowledge in general) we will be focused on propositional knowledge . Lucky for us, this is the kind of knowledge that has received by far the most attention from epistemologists.

2.2 The Traditional Account of Propositional Knowledge

The traditional account of propositional knowledge is that knowledge *just is* justified true belief.[4] So, in order for you to know that some proposition, *p* (that the sun is more massive than Earth, say), you must believe that *p*. *p* also must be true—the sun must actually be more massive than Earth. And, your believing that *p* must be justified. Let us take a brief look at each of these conditions and why they are so intuitively plausible that they have formed the foundation of the traditional account of knowledge for over 2000 years (at least since the time of the ancient Greek philosopher Plato).

You cannot know a proposition which you fail to believe. For instance, if you do not believe that the sun is more massive than the Earth, we would not say that you know it is. Now, we have to be careful because sometimes people do talk in a way which makes it seem like knowing does not require believing. In some cases a person might say something like the following: "I do not believe that the sun is more massive than the Earth—I *know* it is!" At first, we might be tempted to take such expressions as evidence that you do not need to, perhaps even cannot, believe that *p* in order to know that *p*. This would be a mistake though. It is much more plausible that when someone says "I do not believe that *p*, I know it" what she is really doing is expressing that she does not *merely* believe that *p*. After all, this person will exhibit the same behaviors as someone who believes that *p*. She will be willing to assert that *p* is true, if she is asked about it. She will be willing to treat *p* as true in her reasoning. And so on. To further illustrate this point consider that someone might exclaim, "That is not a house, it is a mansion!"[5] Does the fact that someone might appropriately say this mean that mansions are not houses? Surely not. What is going on here is that the speaker is emphasizing that the house in question is not *merely* a house, it is a very particular kind of house—a mansion. Similarly, the most plausible explanation of assertions of "I do not believe that *p*, I know it" is that the speaker wants to emphasize that *p* is not something that she *merely* believes. Instead, *p* is

[4]Since our focus will be on propositional knowledge, from this point on, unless otherwise noted "knowledge" will mean "propositional knowledge" and "the traditional account of knowledge" will mean "the traditional account of propositional knowledge".

[5]This illustration is from Lehrer (1974).

something that she has a better connection to than mere belief—she believes that *p* for very good reasons. This is why it is widely accepted among philosophers that knowledge requires belief.[6]

It is even more widely accepted (perhaps almost universally so) that you can only know propositions that are true. Again, we have to be careful because sometimes people say things which might lead us to doubt this obvious truth. For example, people sometimes say things like "In the past people knew that the earth was flat" or when a sports team loses "I just knew that the team was going to win". As we know, the earth was not flat in the past, and it is not flat now. So, if people at one time really did *know* that the earth was flat, knowledge does not require the known proposition to be true. But, is this the best way to understand what is being expressed by these sorts of claims? Intuitively, the answer is "no". A much better explanation is that when someone says, "In the past people knew that the earth was flat" what she really means is that in the past many people *believed* that the earth was flat or, perhaps even more accurately, in the past many people *thought* they knew that the earth was flat. Similarly, when someone says, "I just knew the team was going to win" what she really means is that she was very confident that the team would win or she *thought* she knew they would win. In neither case should we claim that a false proposition was known to be true! One final consideration to help illuminate this fact: imagine someone claims to know that the sun has less mass than Earth. Would we say that he knows something that is false or would we simply say that he is mistaken? We would correctly say the latter—he does not know that the sun has less mass than Earth because this claim is simply not true.

As we have seen, it is widely accepted by philosophers, and intuitively clear, that in order to know some proposition, it must be true and you must believe it. However, knowledge requires more than simply having a true belief. To see this, consider the following case:

> Jim makes a wild guess that the winning numbers in tonight's lottery drawing will be 7, 29, 40, 18, 3, 13, and 8. On the basis of this guess Jim believes that these will be the winning numbers.

Assume for the sake of illustration that as luck would have it Jim's guess happens to be true—these really are the winning lottery numbers. Does Jim *know* that these are the winning numbers before the drawing is held? Surely not. After all, Jim has absolutely no reason to believe that these specific numbers will be drawn. He is simply guessing, and he is aware of the fact that he is guessing. Not only does Jim fail to know that these are the winning numbers, it is unreasonable for him to believe that they are. Given Jim's lack of information about which numbers will be drawn,

[6]Of course, like most things in philosophy, this is not universally accepted. A few philosophers are tempted to claim that there may be rare cases where one can know that *p* without believing that *p*. See, for example, Lewis (1996) and Radford (1990). For strong opposition to this view see, in particular, Armstrong (1969), Lehrer (1974), and Rose and Schaffer (2013). Even if such rare cases of knowledge without belief do occur, which is highly doubtful, we can safely set aside the issue for our purposes because scientific knowledge does not arise in these rare cases.

the rational thing for him to do is to refrain from believing that any particular set of numbers are the winning ones—in other words, Jim should suspend judgment concerning the winning numbers. Jim does have a true belief in this case, but it is clear that he fails to have knowledge. In light of this, we can clearly see that something more than true belief must be needed for knowledge.

In the *Meno*, Plato claims that the something beyond true belief that is required for knowledge is something which tethers one's belief to the truth of the proposition believed. Today we refer to this "tether" as "justification". So, in order to know that *p* you must have justification which supports your believing that *p* is true.[7] As we will see in chapter five, there is much dispute over exactly what justification is and what is required for you to have justification for believing that *p*. At this point, it will suffice to say that your having justification for *p* means you have good reason to believe that *p* is true, and you must have justification in order to know that *p*.

Putting these pieces together we get the traditional account of knowledge:

Someone, S, knows that *p* if and only if[8]:

1. S believes that *p*
2. *p* is true
3. S is justified in believing that *p*

Of course, there are several things that need to be spelled out in this account of knowledge, and there are various complications that we have yet to consider. We will spell out these details and consider some of the complications in the chapters that follow.

2.3 Conclusion

In this chapter we have taken our first steps toward building a philosophical foundation for understanding NOS. We have begun our examination of the nature of knowledge. We distinguished between three kinds of knowledge and recognized that the relevant kind of knowledge for our purposes is propositional knowledge. We have also seen some of the basics of the traditional account of knowledge. In the next three chapters we will devote our time to examining the three key components of the traditional account of knowledge. In the chapter that immediately follows, chapter three, we will discuss what it means for someone to *believe* a proposition. In chapter four we will consider what it takes for a proposition to be *true*. In chapter five we

[7]Although some philosophers wish to draw technical distinctions between them, we will use the terms "rational", "reasonable", and "justified" interchangeably. So, to say that a belief is justified is the same as saying that it is rational or reasonable. Similarly, to say that S is justified in believing that *p* is the same as saying that it is rational or reasonable for S to believe that *p*.

[8]Since the traditional account of knowledge is in terms of "if and only if", it follows that according to this account, anyone who has knowledge satisfies (1)–(3); and, anyone who satisfies (1)–(3) has knowledge.

will explore the nature of *justification*. Afterward, we will turn toward some of the various complications for this account in the remaining chapters of this section of the book. Our goal throughout will be to increase our understanding of knowledge in general. By doing this we will have gone a long way toward understanding scientific knowledge in particular. Additionally, we will have provided ourselves with a solid foundation from which to continue our exploration of NOS and other debated issues in science education.

References

Adams, M. P. (2009). Empirical evidence and the knowledge-that/knowledge-how distinction. *Synthese, 170*, 97–114.

Armstrong, D. M. (1969). Does knowledge entail belief? *Proceedings of the Aristotelian Society, 70*, 21–36.

Devitt, M. (2011). Methodology and the nature of know how. *Journal of Philosophy, 108*, 205–218.

Lehrer, K. (1974). *Knowledge*. Oxford: Clarendon Press.

Lewis, D. (1990). What experience teaches. In W. Lycan (Ed.), *Mind and cognition* (pp. 29–57). Malden: Blackwell.

Lewis, D. (1996). Elusive knowledge. *Australasian Journal of Philosophy, 74*, 549–567.

Poston, T. (2009). Know-how to be Gettiered? *Philosophy and Phenomenological Research, 79*, 743–747.

Radford, C. (1990). Belief, acceptance, and knowledge. *Mind, 99*, 609–617.

Rose, D., & Schaffer, J. (2013). Knowledge entails dispositional belief. *Philosophical Studies, 166*, 19–50.

Ryle, G. (1949). *The concept of mind*. Chicago: University of Chicago Press.

Schacter, D. L., Wagner, A. D., & Buckner, R. L. (2000). Memory systems of 1999. In E. Tulving & F. I. M. Craik (Eds.), *The Oxford handbook of memory* (pp. 627–643). Oxford: Oxford University Press.

Squire, L. R. (1987). *Memory and brain*. New York: Oxford University Press.

Stanley, J. (2011). *Know how*. New York: Oxford University Press.

Stanley, J., & Williamson, T. (2001). Knowing how. *Journal of Philosophy, 98*, 411–444.

Chapter 3
Belief

Abstract This chapter explores the nature of belief. An often-overlooked distinction between *believing in* something and *believing that* something is true is explored. It is made clear that the first notion of belief is really an expression used to signify trust or faith in something rather than the sort of belief that is a component of knowledge. Once the importance of believing that is made clear the various major accounts of the nature of belief are briefly examined. The chapter concludes by making it clear that the dispositional and representationalist accounts of belief are the best options. In light of this it follows that the best way to understand the belief component of knowledge is either in terms of having certain dispositions or in terms of having certain mental representations of information. Further, it is made clear that we do not need to decide which of these accounts is superior for the purpose of understanding scientific knowledge.

The first component of the traditional account of knowledge is *belief*. As we noted in the previous chapter, in order to know that *p* you must believe that *p*. At first glance this seems rather simple. We are all familiar with the role that beliefs and desires play in explaining behavior. We might explain why John is eating the last cookie by citing the facts that he had a *desire* for something sweet and that he *believes* that the cookie is the only easily available sweet thing to eat. In saying that John believes that the cookie is the only easily available sweet thing to eat we are saying something about how he views the world. John takes it that the proposition <the cookie is the only easily available sweet thing to eat> accurately describes the world.[1] More abstractly, S's believing that *p* is her taking *p* to correctly describe the world—her taking *p* to be true.[2] For instance, your believing that the sun is more massive than Earth means that you think the world is such that the sun has more mass than Earth. In simplest terms this is what it means to believe that the sun is more massive than Earth. However, there are many important distinctions and subtleties about belief that are worth exploring. Our goal in this chapter is to draw these distinctions and

[1]Here, and throughout the remaining chapters, we will follow the common convention of using "<*p*>" to signify the content of a proposition.

[2]Unless otherwise noted we will use the term "world", as it is commonly used among philosophers, to refer to the entire universe (or multiverse, depending on the correct physics).

K. McCain, *The Nature of Scientific Knowledge*, Springer Undergraduate Texts in Philosophy, DOI 10.1007/978-3-319-33405-9_3

explore these subtleties in order to better understand belief as well as the knowledge of which it is a component.

3.1 Belief in Versus Belief That

An important distinction to make clear immediately is the distinction between *believing in* something and *believing that* something is true.[3] We believe that many things are true. For example, you probably believe that you are reading this page, that dogs are animals, that $1 + 1 = 2$, that water is H_2O, that the sun is more massive than Earth, that eye-glasses help some people see better, and a whole host of other things. You might also believe in various things. You might believe in God or you might believe in democracy or you might believe in the Loch Ness Monster or other things.

In many cases saying you believe in something is really just a way of expressing that you *believe that* various things are true. When we say that Sam believes in the Loch Ness monster we typically just mean that Sam believes that the Loch Ness monster exists. However, in some instances saying you believe in something means more than just that you believe that various things are true. It often also means that you have faith in certain things. For example, Sally's believing in democracy, of course, means that she believes that various things are true. Sally believes that many issues are best settled in a democratic fashion, she believes that people should be allowed to voice their opinions and have a say in their own governance, and so on. But, believing in democracy also seems to mean something more than just believing that various things are true. It also seems to mean that Sally has faith in democratic processes, so that she will exhibit certain kinds of behaviors: she will encourage others to vote, she will advocate for democratic governments in countries in addition to her own, and so on.[4] Sally not only has beliefs about democracy, she values and esteems democracy. So, it seems that believing in something may often include things beyond beliefs that particular propositions are true.

Knowledge requires belief that, but not belief in. The totalitarian dictator who adamantly opposes democracy can still know various things about democracy. The reason for this is that while the dictator may not *believe in* democracy, he can still *believe that* various things concerning democracy are true. The dictator may believe that democracy is a form of government where each citizen of a certain age gets to participate in the governance of the State, for instance. He may

[3]See Price (1965).

[4]Plantinga (2000) similarly distinguishes between *believing that* God exists from *believing in* God. According to Plantinga, the former state simply consists in believing that a particular sort of being exists. However, he maintains that believing in God includes much more: loving God, trusting God, and seeing God as beautiful and glorious.

also have various false beliefs concerning democracy such as the belief that his people are better off under his brutal rule than they would be if they could have a hand in their own governance. The dictator can have all of these beliefs without believing in democracy. Further, it is plausible that in some cases the dictator's beliefs concerning democracy amount to knowledge (according to the traditional account) because they are true, and he is justified in holding them. So, the relevant notion of belief for our discussion is *belief that*, not the perhaps richer notion of *belief in*.[5]

3.2 What Is a Belief?

We have seen that believing in something seems to, at least in some cases, be a more complex attitude (or set of attitudes) than believing that something is true. This is a useful distinction, but it does not tell us a lot about what a belief is. It does not really illuminate what it means for S to believe that *p*, which is what we need to know in order to more fully understand the traditional account of knowledge. As we mentioned above, when you believe a proposition you take that proposition to accurately describe the world. Your belief that the sun is more massive than Earth is your taking the proposition <the sun is more massive than Earth> to be true—you have a particular attitude toward that proposition. This holds for other beliefs as well; they are attitudes that one has toward a proposition. This is the first step to getting clear about the nature of belief itself. Beliefs are propositional attitudes. That is to say, beliefs are mental states that consist of having some attitude toward a proposition or the state of affairs described by the proposition.

Of course, noting that beliefs are propositional attitudes is far from sufficient for accurately characterizing the nature of belief. Not just any attitude toward a proposition is a belief because there are many propositional attitudes in addition to beliefs. Hopes, fears, and desires are all propositional attitudes, and there are other propositional attitudes besides these. These mental states are all alike in that they consist of one's having a particular attitude toward a proposition. When you hope that you make it home before it starts raining you have a particular attitude toward the proposition <I make it home before it starts raining>. Namely, when you have this sort of hope you *want* it to be the case that the proposition is true. When you fear that there is life on Mars you have a particular attitude toward the proposition <there is life on Mars>. You are *apprehensive* about the truth of that proposition. Similarly, when you believe that the sun is more massive than Earth you have a particular attitude toward the proposition <the sun is more massive than Earth>.

[5]As we noted in the first chapter, this distinction may be relevant to issues in science education such as the issue of the proper goals for science education. It also seems to be particularly relevant to the issue of how we should understand data concerning the acceptance rates of various scientific theories (Kampourakis and McCain 2016).

In the case of belief the attitude you have toward the proposition in question is one of taking that proposition to accurately describe the world, i.e. *you take the proposition to be true*.[6]

As mentioned above, belief is a propositional attitude wherein you take a particular proposition to be true. This is a particular mental state that you have. It is important to note that this is a mental *state*, not simply a mental *event*. Obviously, there is the event of your coming to believe that *p*. There may be the event of your no longer believing that *p* such as when you forget that *p*, or when you change your mind and come to think that *p* is not true, or even when you come to doubt *p* to the point that you are no longer sure whether *p* or not *p* is true. But, the belief that *p* itself is a state of your mind that endures the whole time that you have that belief. To see this, consider the following sort of situation:

> Sara is presented with a large and diverse sampling of data in support of the claim that the sun is more massive than Earth in her science class. In light of all of the data that she has, Sara comes to believe that the sun is more massive than the Earth. As the instructor moves on to other basic facts about our solar system, Sara stores this belief in her memory in the way that properly functioning humans usually do. The next day Sara correctly answers a test question concerning which is more massive the sun or Earth. Many years after this exam Sara continues to believe that the sun is more massive than Earth.

Quite plausibly, the event of Sara's coming to believe that the sun is more massive than Earth occurs when she is presented with the data in support of this proposition during her class. However, it seems that when Sara correctly answers the test question the next day she is relying on this same belief. Further, it seems that so long as there is no point at which she has forgotten that the sun is more massive than Earth she has the same belief throughout this example. Sara's belief does not just occur when she consciously thinks that the sun is more massive than Earth; it is present even when she is not thinking about the proposition. This is why it seems true to say that between the time when Sara first comes to believe and her recalling that the sun is more massive than Earth during the test she believes the same proposition. It seems that her instructor would speak truly if she were to say to another instructor before the exam, "Sara believes that the sun is more massive than Earth". Thus, it makes sense to think that Sara's belief that the sun is more massive than Earth is a continuing mental state that she has—one that consists of her having a particular attitude toward this proposition.[7]

---

[6]For now we will simply stick with an intuitive understanding of what it means to say that a proposition is *true*. In the next chapter we will dive into the deep waters of the nature of truth.

[7]See Armstrong (1973), Lycan (1988), and Moser (1989) for further explication and arguments in support of thinking that beliefs are mental states rather than events.

3.3 Philosophical Theories of Belief

At this point we have seen that beliefs are a particular kind of mental state. They are propositional attitudes consisting of taking a particular proposition to be true. It will be worth briefly exploring some of the major philosophical theories of the nature of this particular kind of mental state to deepen our understanding of belief.

3.3.1 Representationalism

Representationalism is the most prominent theory of belief.[8] According to this theory, a belief is a mental state that involves having a particular representation present (either consciously or stored) in one's mind. As we have noted, a belief is a propositional attitude. Accordingly, representationalism holds that the relevant representation is a mental representation of a proposition. So, Sasha's believing that all dogs are animals consists of her having a representation of the content <all dogs are animals> stored in her mind.

Representationalists disagree about the exact nature of these representations though. Some representationalists hold that the mental representations which constitute beliefs are sentences in a "language of thought". The idea here is that much like a computer has its own code language our minds have their own language of thought. This language of thought is similar to our ordinary spoken languages so that contents in the language of thought are related to one another in ways analogous to how the propositional contents of sentences in our ordinary language are related to one another.[9] Other representationalists hold that beliefs are like maps rather than sentences in a language of thought. The idea here is that we represent the contents of the multiple propositions that we believe in the way that a map represents multiple facts about a particular geographical region.[10] Yet other representationalists maintain that the way to understand the representations is as part of a representational system that has the function of tracking various properties in the world around us. The thought here is that our mental representations form a system that co-varies with our environment in certain ways.[11] Fortunately, we do not need to settle the exact nature of these representations for our purposes.

Although we do not need to take a stance on the intricacies of mental representations to understand the representationalist theory of belief, we do need to clear up one further point. However one comes down on the nature of the representations that constitute beliefs for representationalists, merely representing propositional

---

[8] See, for example, Cummins (1996), Dretske (1988), Fodor (1975), and Millikan (1993).

[9] The most prominent supporter of this version of representationalism is Fodor (1975).

[10] Armstrong (1973) and Ramsey (1931) both accept this version of representationalism.

[11] This version of representationalism is put forward by Dretske (1988) and Millikan (1993).

content in one's mind is not sufficient for belief. After all, Sasha's desiring that all dogs are animals involves her having a representation of the content <all dogs are animals> in her mind—the very same propositional content that she would represent by believing that all dogs are animals. So, it is not sufficient to simply have a representation of the content <all dogs are animals> in her mind, Sasha must have this content in a *particular way* in order to believe that all dogs are animals. D.M. Armstrong (1973, p. 3) claims that the representational content must be present in one's mind in such a way that "it is something *by which we steer*." So, Sasha must have the content <all dogs are animals> in such a way that it plays the appropriate role in her mental life and her behaviors. For instance, part of what it means for her to believe that all dogs are animals is for her to have the content <all dogs are animals> represented in her mind in such a way that when she comes to know that Fido is a dog she is apt to infer that Fido is an animal. Similarly, if Sasha learns that Astro is not an animal, then she will be apt to infer that Astro is not a dog either. So, representationalism says that in order for one to have the belief that *p* she must a) have a representation of the content *p* in her mind and b) have the content *p* in her mind in such a way that it plays the appropriate causal role in her cognitive life.

3.3.2 Dispositionalism

Although representationalism is the dominant view of the nature of belief in contemporary philosophy of mind, it will be worth saying a bit about some of its major competitors. The primary rival of representationalism is dispositionalism. Dispositionalism about the nature of belief denies that having a particular representation of a belief's propositional content in one's mind is required for having that belief. According to this theory, to have a belief that *p* just is to have various dispositions toward *p*. Traditional dispositionalists take the requisite dispositions to be dispositions to exhibit observable behaviors.[12] According to traditional dispositionalists, Sasha's believing that all dogs are animals consists of her having various *behavioral dispositions*.[13] For instance, Sasha is *disposed to assent* to the claim that all dogs are animals in the right kinds of circumstances, she is *disposed to act surprised* when someone else denies that all dogs are animals, she is *disposed to leave* Fido the dog at home when she goes somewhere that forbids non-human animals from entering, and so on. In general, the thought is that when one believes that *p* she will be *disposed to act* as if *p* is true.

Unfortunately, there are at least two fairly decisive objections to traditional dispositionalism. The first objection is that the traditional dispositionalist's apparent goal of reducing belief entirely to facts about observable behavior seems to be

[12]The discussion here follows Schwitzgebel (2014) in referring to the three primary strands of dispositionalism as "traditional dispositionalism", "liberal dispositionalism", and "interpretationism".
[13]Braithwaite (1932–1933) and Marcus (1990) defend traditional dispositionalism.

impossible.[14] The reason for this is that such a reduction would have to reduce a belief such as Sasha's belief that all dogs are animals to observable behaviors without referring to other beliefs or any other mental states for that matter. It seems that such a reduction cannot be done because how someone who believes that *p* will behave in various circumstances will depend on the other beliefs, desires, and so on that she has. So, in addition to the empirical fact that traditional dispositionalists have not provided a single complete reduction of a belief (or any other mental state) there are good conceptual reasons to think that such a reduction simply cannot be completed.

The second objection to traditional dispositionalism is that there are cases where belief does not seem to be closely connected with observable behavior. For instance, there are cases where one might not want to display any of the outward behaviors that may normally be associated with a particular belief. A person living under a totalitarian regime may hold beliefs that lead her to disagree with the regime's policies. In such a case she may fail to display any behavior that would suggest she has such beliefs out of concern for her own safety. Additionally, there are cases where a belief may have very little connection to any practical concerns, and so, is unlikely to manifest in outward behaviors. Simon may believe that the number of stars in the Milky Way galaxy is either even or odd. It does not seem that this belief will result in his behaving in one way rather than another in many (if any) situations. If there are beliefs that are not connected to observable behaviors in the way that traditional dispositionalism maintains, then there is a serious problem for the view.

These objections have led some to abandon traditional dispositionalism in favor of one of two alternatives. The first alternative is liberal dispositionalism.[15] This version of dispositionalism abandons the traditional dispositionalist's project of reducing beliefs to observable behavioral dispositions. Liberal dispositionalists can thereby avail themselves to other mental states when specifying the dispositions that constitute a particular belief. This allows liberal dispositionalists to avoid, or at least mitigate to some degree, both of the objections to traditional dispositionalism. Loosening up the restriction of not referring to other mental states at least makes it more likely that the liberal dispositionalist can account for a particular belief in terms of behavioral dispositions. Also, this loosening seems to mitigate the worry that some beliefs are not closely linked to observable behavior. For instance, liberal dispositionalism allows that the relevant dispositions for believing that the totalitarian regime's policies are unjust may only manifest themselves in ways that are internal to the believer—she might feel certain ways when considering the policies or she may think various things about the regime to herself. Such inward, non-observable behaviors are sufficient for belief according to liberal dispositionalism.

---

[14]See Chisholm (1957).

[15]Audi (1972), Price (1969), and Schwitzgebel (2002) all hold views of this sort.

The second alternative to emerge from the failure of traditional dispositionalism is interpretationism.[16] Interpretationism is similar to traditional dispositionalism in that it focuses on observable behavior. However, whereas traditional dispositionalism holds that beliefs are constituted by various behavioral dispositions of the believer, interpretationism holds that beliefs are constituted by our belief attributing practices. Consequently, whether Sasha counts as believing that all dogs are animals depends upon whether we (as a community) would interpret her behavior in such a way that we would ascribe the belief that all dogs are animals to her.

One obvious problem with interpretationism, which seems to be embraced by its advocates as a strength of the view, is that given this theory whether S believes that *p* is often indeterminate. It may be the case that under some interpretations Sasha believes that all dogs are animals, but under others she does not because some people would attribute this belief to Sasha while others would not. According to interpretationism, there may not be any fact of the matter concerning whether Sasha really believes that all dogs are animals—belief is always relative to an interpretive schema. Embracing such a relativistic stance seems to be warranted only if a strong case can be made in support of interpretationism over the other theories of belief. It does not seem that such a case has been made, nor does it seem likely that it will be.[17]

3.3.3 Eliminativism

A final theory of belief that is worth briefly mentioning is eliminativism.[18] As the name suggests, eliminativism is a theory that eliminates beliefs—according to eliminativism beliefs do not exist. Eliminativists maintain that our talk of beliefs is simply a hold over from our "folk psychology". They claim that when neuroscience and psychology progress to a sufficient level of sophistication this folk psychology

---

[16]The primary supporters of this sort of view are Davidson (1984) and Dennett (1978).

[17]It is worth noting that representationalism, traditional dispositionalism, liberal dispositionalism, and interpretationism all seem to be compatible with a broader view about mental states known as "functionalism". According to functionalism, the actual and potential causal relations to behavior, other mental states, and sensory stimuli are what make something the particular type of mental state that it is. Very roughly, a mental state counts as the type of mental state that it is because of how it functions in one's cognitive system (mind), not because of how it is realized in particular creatures. For example, the mental state of experiencing pain can be realized by the activity of different kinds of neurons in different species or even by something other than neurons in creatures like Martians. Nonetheless, so long as the activity of whatever Martians have in their heads, say, leads to the same sorts of causal relations in the Martian's cognitive system as the activity of the neurons in a human's does in her cognitive system each is experiencing pain. See Armstrong (1968), Block (1991), Fodor (1968), Lewis (1972), and Putnam (1975) for important discussions of functionalism.

[18]Churchland (1981) and Stich (1983) endorse this sort of view, though Stich seems to have taken a more moderate position in his later work.

will be done away with, and the same will occur with the beliefs that are part of it. In essence, eliminativists think that the word "belief" is similar to scientific terms that we have discarded as our knowledge of the world has increased. Hence, "belief" is on a par with "aether" or "phlogiston"—it is an empty term, which refers to nothing in reality. Eliminativists hold that once we see that there is nothing that is referred to by "belief" we will stop using the term except perhaps to illustrate how failed theories are sometimes apt to make use of empty terms.

The primary case for eliminativism seems to rest on the idea that our scientifically sophisticated neuroscience and psychology will one day abandon terms like "belief" and prove that there is no good use for it. However, this prediction has not been born out in the many years since eliminativists began seriously pushing this view, nor does it seem likely to be born out any time in the near future (if it ever will). A cursory glance at the literature in neuroscience and psychology reveals that the term "belief" is quite ubiquitous. "Belief" is not just used to describe some bit of folk psychology; it appears in this literature as a label for a particular mental state that we have aspirations to understand more fully. In light of this (and our own introspective experience), it does not seem that there are good reasons for thinking that beliefs do not exist.

Of course, a brief exploration of theories of belief such as the one we have conducted in this section is not sufficient for establishing definitively which theory is correct. Fortunately, for our purposes we do not need to decide among the major contenders (other than perhaps denying eliminativism, which seems reasonable to do). Instead, we have accomplished our present goal by gaining a grasp of the most prominent theories of the nature of belief.

3.4 Kinds of Beliefs

Now that we have a handle on some of the main theories of the nature of belief it is time to draw some very important distinctions between various kinds of beliefs. In many cases it will be helpful to assume a particular theory of the nature of belief in drawing these various distinctions. In light of this we will take representationalism to be the correct theory of the nature of belief and explicate the distinctions that follow from within a representationalist framework. There are two good reasons for proceeding this way. The first reason is simply that by assuming a particular theory of the nature of belief the discussion that follows can be made less cumbersome and, after all, representationalism is a good choice for the theory to assume because it is the dominant theory of the nature of belief. Also, representationalism provides a straightforward theory for us to work with throughout our subsequent discussions. The second reason is that nothing of significance will hang on our choice of representationalism as our working theory of the nature of belief. If it turns out that one of the other theories, or perhaps some as yet unspecified theory, is correct, the discussion of these important distinctions can be simply re-conceptualized in terms of the correct theory.

3.4.1 Explicit Belief Versus Implicit Belief

Recall from above that according to representationalism having a belief requires having a mental representation of the propositional content of that belief. In order for Sasha to believe that all dogs are animals she has to have a mental representation of the content <all dogs are animals>. When Sasha has such a mental representation of this propositional content in her mind (either consciously or stored in memory) she *explicitly* believes that <all dogs are animals>. So, our earlier discussion of representationalism (and the other theories as well) was in terms of explicit belief—beliefs whose content one has explicit representations of in her mind at the present time.

In contrast to explicit beliefs there are also *implicit*, or tacit, beliefs. Unlike explicit beliefs, when one has an implicit belief she does not have a mental representation of the propositional content of that belief currently in her mind. Plausibly, there are three primary ways of having an implicit belief (Harman 1986). One way is for those beliefs to be easily inferable from things you explicitly believe because they are logically entailed by your explicit beliefs. For example, Molly explicitly believes that there are exactly eight songs on a particular album—so, she has a mental representation of the content <there are exactly eight songs on the album> in her mind in the appropriate way. She can easily infer from this belief that there are less than 20 songs on the album. She can also easily infer that there are less than 21 songs on the album; there are less than 22 songs on the album, and so on. Although Molly can easily infer the truth of these additional propositions because they are implied by her explicit belief that there are exactly 8 songs on the album, she does not at present have mental representations of the content of these propositions in her mind. In this case it seems that Molly believes these propositions about the number of songs on the album, but she only does so implicitly.

A second way of having implicit beliefs is for those beliefs to be easily inferred from your explicit beliefs even though they are not entailed by those explicit beliefs. For example, it is plausible that you implicitly believe that tigers do not wear top hats in the jungle.[19] Notice, this is different than Molly's belief about the number of songs on the album being less than 20. In Molly's case her implicit belief is directly implied by her explicit beliefs. However, in this case it is likely that your explicit beliefs do not directly entail this proposition about tigers. Yet, your explicit beliefs do make it so that you could easily infer tigers do not wear top hats in the jungle. Thus, it seems that one might have implicit beliefs by virtue of their being easily inferred from one's explicit beliefs even when they are not logically implied by those explicit beliefs.

A third way of having implicit beliefs is that they may be implicit in some of your explicit beliefs. The idea here is that when Molly explicitly believes that

[19]See Dennett (1978) for similar examples.

there are exactly eight songs on the album she may implicitly believe that she is justified/rational in believing that there are exactly eight songs on the album (Harman 1986). The idea is that by explicitly believing that there are exactly eight songs on the album Molly may commit herself to believing that she is justified/rational in having that belief. Since Molly commits herself in this way, it is plausible that she implicitly believes that she is justified/rational in believing that there are exactly eight songs on the album.

It seems that there are a variety of ways in which one might implicitly believe that *p*. Importantly, in each of these ways when one has an implicit belief she does not have a mental representation of the content of that belief in her mind whereas such a mental representation is required for explicit belief. Also, a plausible restriction on whether one has an implicit belief is that she must have the potential to make that belief explicit. That is to say, it seems that Molly does not even implicitly believe that she is justified/rational in believing that there are exactly eight songs on the album if she cannot represent the content <my belief that there are exactly eight songs on the album is justified/rational>.

3.4.2 *Occurrent Belief Versus Dispositional Belief Versus Disposition to Believe*

The next set of distinct kinds of belief concerns one's consciousness of the belief in a significant way. We must be careful to recognize that the distinctions drawn in this section are different from the explicit belief/implicit belief distinction of the previous section. The first kind of belief to discuss in this section is *occurrent belief*. Occurrent beliefs are those that one is currently thinking of or those that are "in some other way currently operative in guiding what one is thinking or doing" (Harman 1986, p. 14). For example, when Sasha is currently thinking to herself that all dogs are animals her belief that all dogs are animals is occurrent. Similarly, when Jack is jogging up a hill and jumps over a hole in the ground it is plausible that his belief that there is a hole in the ground is occurrent. Since occurrent beliefs are either currently thought of, or stored in one's mind and currently guiding her thought/behavior, they must be explicit beliefs. Therefore, all occurrent beliefs are explicit; however, not all explicit beliefs are occurrent.

Some explicit beliefs may be *dispositional*. An explicit dispositional belief is an explicit belief that is currently stored in one's mind, which is not presently occurrent but is potentially so. This is worth unpacking a little. Consider a situation where Sasha has formed the explicit belief that all dogs are animals, and she stores this belief in her memory in the normal way. So, she has a mental representation of the content <all dogs are animals>, and when prompted in the appropriate way she will recall this belief in the way we typically do when we remember things that we believe. When Sasha is not currently thinking that all dogs are animals, but her belief is stored in her memory waiting to be recalled, her belief that all dogs are animals is dispositional. It is dispositional because it is a belief that she has, which is not currently conscious but which she can potentially bring to consciousness. Her belief in this situation is also an explicit belief because she has the appropriate mental representation of the content of this belief stored in her mind. Although some dispositional beliefs are explicit, not all are. In fact, any implicit belief will be dispositional.

At this point a brief summary is in order. All occurrent beliefs are explicit beliefs, but not all explicit beliefs are occurrent beliefs. An explicit belief can fail to be occurrent when the sort of mental representation required for explicit belief is merely stored in one's memory. However, when one has an occurrent belief she has a mental representation of the content of the belief before her mind—in other words, her belief is explicit. All implicit beliefs are dispositional because they are beliefs that have not yet been brought to consciousness, and so, are only potentially conscious. Not all dispositional beliefs are implicit though. Some dispositional beliefs are ones that were previously occurrent and are now mental representations stored in memory. So, there are both explicit dispositional beliefs and implicit dispositional beliefs.

One final distinction needs to be drawn in this section—the distinction between *dispositional beliefs* and *dispositions to believe*. The former are beliefs that one actually has, whether explicit or implicit. The latter are not beliefs that one actually has, but dispositions that one has to form certain beliefs. Here is a simple way to understand the difference between the two. When Theresa is sleeping on a train she has the disposition to believe that the man who just sat down across from her is wearing a hat. She has such a disposition because if she were to wake up and look across the aisle, she would believe that the man across from her is wearing a hat.[20] Yet, currently Theresa does not even have this belief dispositionally—she simply has not formed this belief either explicitly or implicitly at this time. So, in this case Theresa has the disposition to believe that the man across from her is wearing a hat, but she does not have the dispositional (nor occurrent) belief that the man across from her is wearing a hat.

With the distinctions from this section in hand we now have a clearer understanding of the sort of belief that is relevant for understanding scientific knowledge. Typically, when we are discussing the sorts of things that we know from the sciences the relevant sense of belief is explicit and quite often occurrent. We will see more about this in later chapters, but for now it is worth noting that we currently have a grasp of some of the most important distinctions concerning belief. This will help us to deepen our understanding of the nature of knowledge.[21]

[20]This is similar to Audi's (1994) discussion of dispositions to believe.

[21]Of course, there are other important distinctions between kinds of beliefs such as the *de re/de dicto* distinction (see Moser (1989) and Quine (1956) for more on this). However, these distinctions are not as central to our purposes so we will not spend time discussing them.

3.5 The Tripartite View Versus Degrees of Belief

It is obvious that we believe some things more strongly than other things. You likely believe that George Washington was the first president of the United States. This belief is something that you probably hold quite firmly—you are quite certain that it is true. Although you believe this strongly, you probably even more strongly believe that you are reading a book. It is plausible that your belief that you exist is held even more strongly than your belief that you are reading a book. After all, you could be mistaken about reading the book—perhaps you are hallucinating or perhaps you are the victim of some massive deception.[22] However, even if you are hallucinating or deceived in various ways, you must still exist. This is something that Descartes pointed out quite well in his *Meditations*. So, it seems that you have stronger grounds for believing that you exist than you do for believing that you are reading a book or for believing that George Washington was the first president of the United States. Given that you have stronger grounds for believing that you exist than you do for these other beliefs, it is plausible that you believe that you exist more strongly than you do these other beliefs.

Recognition of the fact that we believe some things more strongly than others has led some to suggest that we should understand belief not as an all or nothing state, but as a matter of degree. Often the idea is that for every proposition we have considered (sometimes this is understood to apply to every proposition, whether we have considered the proposition or not) we have a degree of belief in the truth of that proposition that falls somewhere between 1 and 0. Under this "standard" way of understanding degrees of belief when you have a degree of belief 1 that *p* you are absolutely certain that *p* is true, when you have a degree of belief 0 that *p* you are absolutely certain that *p* is false, and for anything in between 1 and 0 you believe *p* to a greater or lesser degree. Typically, degree of belief .5 is taken to be the point at which you are completely unsure about *p*'s truth—you believe it is true to the same degree that you believe that it is false; you simply are not swayed one way or the other concerning *p*.

An obvious question that arises when considering degrees of belief is: how are we to understand what these degrees of belief mean? How, for example, should we understand the claim that Alex believes that *p* to degree .65? A common way to analyze degrees of belief is in terms of betting behavior. So, to say that Alex believes that *p* to degree .65 is to say that he would (or should) be willing to bet $0.65 on a wager where he gets nothing if *p* is false and he gets $1 if *p* is true. Of course, there are obvious problems with understanding degrees of belief in terms of betting behavior—some people may simply be more averse to betting than others, there may be situations in which it is improper to bet on the truth of a particular proposition, and so on.[23]

[22]Such skeptical possibilities will be discussed much more fully in chapter eleven.

[23]See Jeffrey (1983) and Skyrms (2000) for more on the degreed approach to belief.

An additional problem for the degree of belief view is that it is simply not clear how it tracks our commonsense tripartite view of belief. According to this commonsense view, when it comes to a particular proposition we can take three attitudes when it comes to belief. We can believe the proposition, we can disbelieve the proposition (believe it is false), or we can suspend judgment concerning the truth of the proposition. In some cases it seems clear that a certain degree of believing amounts to belief on the tripartite view. For instance, degree of belief 1 is clearly believing on the tripartite view, and degree of belief 0 is clearly disbelieving on this view. What about degrees of belief that fall between 1 and 0 though? Does S's having degree of belief .57 in *p* really count as S believing that *p* is true? It is not clear what the correct answers to such questions are.

Fortunately, we do not need to answer such vexing questions here. For our purposes we can work with the tripartite view instead of worrying about a degreed approach. Focusing on the tripartite view has two advantages over focusing on the degree of belief approach. First of all, the traditional account of knowledge is in terms of the tripartite view, so our use of the tripartite view of belief directly links up to the traditional account of knowledge, which is what we are seeking to better understand. Second, the tripartite view is a simpler view to work with—both in terms of actually employing the theory and in terms of bypassing the difficult questions that we just saw arise for the degree of belief approach. This is not to say that our subsequent discussion cannot be translated into the degree of belief approach. With some work, it can be. However, such a translation would, for our purposes, be a considerable amount of work without a corresponding payoff.

3.6 Belief Versus Acceptance

A final distinction that is worth discussing in this chapter is the distinction between *belief* and *acceptance*. Throughout this chapter we have discussed belief in some detail, however, there is a related notion that is of particular importance to our future discussions: acceptance. Gilbert Harman (1986) helpfully distinguishes between full acceptance and acceptance as a working hypothesis. Full acceptance *just is* belief. So, when S fully accepts that *p* she believes that *p*.

Acceptance as a working hypothesis is much more tentative. When S accepts *p* as a working hypothesis she is taking *p* as true to see where it goes. As Harman (1986, p. 46) says, she is "trying it out" in an effort to see what further discoveries or implications acceptance of *p* will lead to. This sort of acceptance as a working hypothesis is a fairly common practice in science (van Fraassen 1980). A researcher may accept *p* as a working hypothesis and then go about constructing experiments and conducting research on various things some of which may yield more conclusive evidence concerning the truth or falsity of *p*.

It is important to recognize the attitude of acceptance as a working hypothesis as distinct from full acceptance/belief. In many cases the reason why scientific knowledge is tentative is that part of the hypotheses that constitute our current scientific theories are simply accepted as working hypotheses—they are not fully accepted/believed to be true.[24] Additionally, it is plausible that a misunderstanding of when a theory is accepted as a working hypothesis and when it is fully accepted/believed as well as when the evidence supports one or the other of these attitudes may help contribute to misunderstandings of scientific theories. For example, one common and sorely misguided objection to evolutionary theory is that it is "just a theory". It may be that some who press this objection are confused about the difference between a theory that is fully accepted (and supported by sufficient evidence to make its acceptance beyond a reasonable doubt) and a theory that is merely accepted as a working hypothesis. In light of this, clarity on the distinction between full acceptance and acceptance as a working hypothesis, as well as the mountains of evidence for evolutionary theory, may be quite helpful in clearing up this particular misguided objection to evolutionary theory.[25]

3.7 Conclusion

In this chapter we have explored the nature of belief as well as several very important distinctions between various kinds of beliefs. This information provides a strong philosophical foundation for understanding the first component of the traditional account of knowledge. As we noted, S's believing that *p* consists of her having a mental representation of *p* and her being guided by that representation—for her to take *p* as true. At this point it is natural to seek further clarity on what exactly we are thinking when we take *p* as true. In particular, it is time to examine what it means for *p* to be true. This second component of the traditional account of knowledge and fundamental philosophical issue will be the focus of the chapter that follows.

References

Armstrong, D. M. (1968). *A materialist theory of the mind*. New York: Routledge.
Armstrong, D. M. (1973). *Belief, truth and knowledge*. London: Cambridge University Press.
Audi, R. (1972). The concept of believing. *Personalist, 53*, 43–62.
Audi, R. (1994). Dispositional beliefs and dispositions to believe. *Nous, 28*, 419–434.
Block, N. (1991). Troubles with functionalism. In D. M. Rosenthal (Ed.), *The nature of mind* (pp. 211–228). New York: Oxford University Press.
Braithwaite, R. B. (1932–1933). The nature of believing. *Proceedings of the Aristotelian Society, 33*, 129–146.

[24]Of course, this is not the only source of the tentativeness of scientific knowledge as we will see later.

[25]For further discussion of the "just a theory" objection to evolutionary theory and detailed explanation of why it is such a misguided objection see McCain and Weslake (2013) and Kampourakis (2014).

Chisholm, R. M. (1957). *Perceiving*. Ithaca: Cornell University Press.
Churchland, P. M. (1981). Eliminative materialism and the propositional attitudes. *Journal of Philosophy, 78*, 67–90.
Cummins, R. (1996). *Representations, targets, and attitudes*. Cambridge, MA: MIT Press.
Davidson, D. (1984). *Inquiries into truth and interpretation*. Oxford: Clarendon Press.
Dennett, D. C. (1978). *Brainstorms*. Cambridge, MA: MIT Press.
Dretske, F. (1988). *Explaining behavior*. Cambridge, MA: MIT Press.
Fodor, J. (1968). *Psychological explanation*. New York: Random House.
Fodor, J. (1975). *The language of thought*. New York: Cromwell.
Harman, G. (1986). *Change in view*. Cambridge, MA: MIT Press.
Jeffrey, R. C. (1983). *The logic of decision* (2nd ed.). Chicago: University of Chicago Press.
Kampourakis, K. (2014). *Understanding evolution*. Cambridge, UK: Cambridge University Press.
Kampourakis, K., & McCain, K. (2016). Belief *in* or *about* evolution? *BioScience, 66*, 187–188.
Lewis, D. (1972). Psychophysical and theoretical identifications. *Australasian Journal of Philosophy, 50*, 249–258.
Lycan, W. G. (1988). *Judgement and justification*. Cambridge, UK: Cambridge University Press.
Marcus, R. B. (1990). Some revisionary proposals about belief and believing. *Philosophy and Phenomenological Research, 50*, 132–153.
McCain, K., & Weslake, B. (2013). Evolutionary theory and the epistemology of science. In K. Kampourakis (Ed.), *The philosophy of biology: A companion for educators* (pp. 101–119). Dordrecht: Springer.
Millikan, R. G. (1993). *White queen psychology and other essays for Alice*. Cambridge, MA: MIT Press.
Moser, P. K. (1989). *Knowledge and evidence*. Cambridge, UK: Cambridge University Press.
Plantinga, A. (2000). *Warranted Christian belief*. New York: Oxford University Press.
Price, H. H. (1965). Belief 'in' and belief 'that'. *Religious Studies, 1*, 5–27.
Price, H. H. (1969). *Belief*. London: Allen & Unwin.
Putnam, H. (1975). *Philosophical papers, vol. 2: Mind, language, and reality*. Cambridge, UK: Cambridge University Press.
Quine, W. V. O. (1956). Quantifiers and propositional attitudes. *Journal of Philosophy, 53*, 177–186.
Ramsey, F. P. (1931). *The foundations of mathematics, and other logical essays*. New York: Routledge.
Schwitzgebel, E. (2002). A phenomenal, dispositional account of belief. *Nous, 36*, 249–275.
Schwitzgebel, E. (2014). Belief. In E. N. Zalta (Ed.), *The Stanford encyclopedia of philosophy* (Spring 2014 Edition). http://plato.stanford.edu/archives/spr2014/entries/belief/
Skyrms, B. (2000). *Choice and chance* (4th ed.). Belmont: Wadsworth.
Stich, S. P. (1983). *From folk psychology to cognitive science*. Cambridge, MA: MIT Press.
van Fraassen, B. C. (1980). *The scientific image*. Oxford: Oxford University Press.

Chapter 4
Truth

Abstract This chapter explores another component of the traditional account of knowledge in detail. Although it is often taken as clear, the nature of truth is a complex philosophical issue and worth careful consideration. This chapter examines both traditional and contemporary theories of truth as well as realist and anti-realist conceptions of truth. Further, it briefly looks at some of the major challenges for a successful theory of truth. Ultimately, this chapter puts forward an argument for a commonsensical, realist conception of truth. This conception of truth is supported by both philosophical argument as well as recognition of its presupposition in science. While neither of these considerations is decisive, together they do provide a strong case for accepting this realist conception of truth. At the very least these considerations make it is clear that working with this realist conception of truth to further understand the traditional account of knowledge is perfectly acceptable.

The second component of the traditional account of knowledge is *truth*. We have noted that it is possible to know that *p* only when *p* is true. But, what does it mean for *p* to be true? In some sense we already have a good grasp on at least a partial answer to this question—*p* is true when it accurately describes the way the world is. As Aristotle (1941, *Metaphysics*, 1011b25) famously said:

> To say of what is that it is not, or of what is not that it is, is false, while to say of what is that it is, and of what is not that it is not, is true.

Not only do we have this basic grasp of what it means to be true, we can easily list a large number of truths: "$1 + 1 = 2$", "all dogs are animals", "circles are not squares", "George Washington was the first president of the United States", "sugar typically dissolves in water", and so on. Nonetheless, it seems that there is more to understanding truth than simply being able to report various truths. Knowing things that are true does not amount to knowing what *truth* is. It seems clear that we might know that various propositions are true without knowing the nature of truth, i.e. without knowing what it is that all true propositions have in common. So, the question we are presented with, which will be the focus of this chapter, is: What is the nature of truth?

K. McCain, *The Nature of Scientific Knowledge*, Springer Undergraduate Texts in Philosophy, DOI 10.1007/978-3-319-33405-9_4

4.1 Preliminaries

Before considering various answers to the question as to the nature of truth it will be useful to make a few preliminary points. First, it is worth emphasizing that a satisfactory answer to our question will not simply amount to a listing of examples of true propositions. Our question concerns the nature of truth itself, not simply knowledge of the truths in some particular domain of inquiry. For example, listing all of the known truths of science will not itself provide an explanation of the nature of truth. Instead, a satisfactory answer to our question—what is the nature of truth?—will explain what it is that all true claims share and all false claims lack (Wrenn 2014). Exploration of various candidates for this sort of answer will be our primary focus in this chapter.

A second preliminary point worth making is that there is a live debate about what constitutes the primary bearers of truth. *Truth bearers* are simply the sorts of things that can be true or false. A rock is not a truth bearer, but a proposition is. Although there are perhaps many candidates for truth bearers, only two are serious contenders for being the *primary* truth bearers (Burgess and Burgess 2011; Wrenn 2014). All other truth bearers are truth bearers only insofar as they are related to the primary truth bearers in a particular way. For example, the two leading contenders for primary truth bearers are propositions and sentences. One might hold that an utterance, such as "All dogs are animals", is true because it expresses a true proposition or because it expresses a true sentence. In this way utterances can be truth bearers, but they are not primary truth bearers.

In order to better understand the debate concerning primary truth bearers, recall the distinction between propositions and sentences from chapter two. As we noted in that chapter, propositions are what our declarative sentences mean, but they are not identical to those declarative sentences. The English sentence "The ball is red", the Spanish sentence "La pelota es de color rojo", and the German sentence "Der Ball ist rot" are all very different sentences. Yet, each of these sentences means the same thing because they all express the same proposition. Most philosophers hold that it is propositions, which are true or false, and sentences are only true or false insofar as they express true or false propositions.[1] Although the debate about which truth bearers are primary is interesting, it is quite intricate, and it is not central to our discussion in this chapter or our subsequent discussions. So, rather than trying to settle this issue we will simply go with the majority of philosophers and assume that propositions are the primary truth bearers.[2]

The final preliminary is a very important distinction. For the most part this distinction will not affect our discussion of theories of the nature of truth. Nevertheless, the distinction is still very important to make because it is relevant to our later

[1]Two notable exceptions are Field (2001) and Quine (1992). Both Field and Quine argue that sentences, rather than propositions, are the primary bearers of truth.

[2]See Soames (1999) for an example of the sorts of reasons that lead the majority of philosophers to claim that propositions rather than sentences are the primary bearers of truth.

discussions of knowledge in general and scientific knowledge in particular. This is the distinction between *necessary* and *contingent* truths. Necessary truths are truths that could not have been false. That is, the denial of a necessary truth describes an impossible situation. For example, <squares have four sides> is a necessary truth. It is impossible for there to be a square that has more or less than four sides. Similarly, <bachelors are married males> is a necessary falsehood—there is no possible situation in which this proposition is true. Although there are many (an infinite number in fact) necessary truths, it is contingent truths that we are most interested in when conducting science. Contingent truths are propositions that in fact are true, but which could have been false. For example, it is true that there is life on Earth, but it could have been that life never arose on this planet. Hence, <there is life on Earth> is true, but only contingently so. It is true that objects near the surface of the Earth fall at a rate of approximately 9.8 m/s^2. However, this truth, like other laws of nature, is contingent. It is possible that in a different world gravitation would have behaved differently than it does in our world.[3] The same holds for other laws of nature.[4] This is not to say that laws of nature do not hold universally in this world or that they are apt to change as time progresses. Rather, the point is simply that laws of nature are contingent truths, not necessary truths.

4.2 Truth and Objectivity

Now that we have cleared up some preliminaries it is time to explore the philosophical debate about the nature of truth. There are three categories of views on the nature of truth: realism, relativism, and anti-realism.[5] One can understand the debate between these three kinds of theories of the nature of truth as hinging on the answers to a pair of questions. As Chase Wrenn (2014, p. 25) explains "First, is anything true irrespective of what anyone believes? If so, then relativism is incorrect. Second, is there anything that is true even though there is no way, in principle, for anyone to know that it is? If so, then anti-realism is incorrect." Roughly, relativism claims that truth depends upon what people believe, anti-realism claims that truth

[3]Recall, as we noted in the previous chapter, "world" refers to the entire universe.

[4]This is something widely accepted by philosophers of science, but like many things in life it is not universally so. Some philosophers hold that laws of nature are necessary truths. For example, see Bird (2007), Fales (1990), and Shoemaker (1998). Fortunately, we do not need to enter into the debate concerning the necessity of natural laws for our purposes.

[5]It is a bit unfortunate that two of the categories are called "realism" and "anti-realism" because there are several debates in philosophy in which various views are labeled "realism" or "anti-realism". As we will see in chapter fourteen a major debate in philosophy of science occurs between realists and anti-realists. Realism and anti-realism in the debate in philosophy of science are different positions than realism and anti-realism about truth. Given that the names "realism" and "anti-realism" are well entrenched in both debates, we will simply follow the literature and stick with these terms. The context of the discussion will make it clear whether realism/anti-realism about truth or realism/anti-realism about science is intended.

depends upon what people can come to know, and realism denies both of these claims. Let us take a closer look at each of these kinds of theories of the nature of truth beginning with realism.

4.2.1 Realism

Realism captures our commonsense view of the nature of truth. It is the view that the world is the way that it is independently of what we think. On a realist view truth is objective and independent of what we can or cannot know. This picture of the nature of truth emerges from consideration of two facts about our experiences (Wrenn 2014). First, there are times when we mistakenly believe things that are not true. In the very distant past many people believed the Earth was flat. They were wrong. You have likely believed something that turned out false. Perhaps you believed (or even currently believe) that Humphrey Bogart's character in *Casablanca* says, "Play it again, Sam". After all, this is one of the more widely misquoted lines in cinema—Bogart does not say, "Play it again, Sam" in the movie though. If you believed that he did, you were mistaken. Unfortunately, we can easily think up numerous further examples like these. It is a fact that we sometimes mistakenly believe false things. Second, we sometimes discover previously unknown truths. For example, when Alexander Fleming made his Nobel Prize winning discovery of penicillin he uncovered a previously unknown truth—this particular antibiotic exists. It is quite possible that you will someday discover a hitherto unknown truth. There are many such truths out there waiting to be discovered.

These facts about our common experiences seem to clearly indicate how we should answer Wrenn's two questions. First, there are truths that exist independently of our beliefs. Denying this would commit us to thinking that when most people thought that the Earth was flat it really was! Note, this is not to say that they meant something different by "flat" than we do; it is to say that since they believed the Earth was flat, it was. Obviously, this is wrong. The mere fact that we believe something does not make it so. Even if everyone on the planet convinces herself or himself to believe that the Earth is flat, that will not change the shape of our planet one bit. It will remain roughly spherical, not flat. Second, there are things that are true whether or not we can ever know that they are. Right now there either is or is not an enormous teddy bear floating ten trillion light years away from our galaxy. This is a fact that we will never be in a position to confirm or disconfirm—we simply will not be able to check to see whether the teddy bear is there or not. Nonetheless, it seems clear that there is a truth of the matter here. If there is no teddy bear there, then it is true that there is not a teddy bear floating ten trillion light years from our galaxy. If there is one, then it is false that there is not a teddy bear floating ten trillion light years from our galaxy. As we noted above, realism claims that truth exists independently of what we know and believe. Realism is correct. The world around us is mind independent.[6]

Although realism seems to be our commonsense view of the nature of truth, its insistence on a mind-independent world does open the way for a particular challenge: skepticism. Since realism holds that the world is independent of what we can know and what we do believe, one might think that we will never be in a position to have objective knowledge about the world. After all, our only contact with the world comes to us from our experiences and perhaps our intuitions. Given realism, how we take the world to be on the basis of our experiences *could be* very different from how the world actually is. It is just this point that the skeptic claims leads to our lacking knowledge of truths about the world around us. Roughly, since our experiences could be misleading, skeptics claim that we cannot know that the world is the way that those experiences make it seem to be. Although this is a significant challenge, it is one that philosophers have been considering and working out replies to for a long time. Furthermore, it is one that we will see (in chapter eleven) that it is reasonable to think can be overcome. For now we will set aside exactly how the skeptical challenge can be met.

Given its strong grounding in facts about our experiences and its fit with commonsense, realism is the default view when it comes to the nature of truth. This is as it should be. In order for us to abandon a realist picture, whichever one we ultimately settle on, we would need strong reasons to prefer a relativist or anti-realist picture instead. As we will see, such reasons are not likely to be had.

4.2.2 *Relativism*

The first rival to realist views of the nature of truth is relativism. Relativism denies the realist's claim that the world is a certain way regardless of what we can know or do believe about it. According to relativism, there is no way the world is objectively. In other words, truth is relative to what people believe. So, it is possible that one and the same proposition can be true for me, and yet, false for you. Of course, there are cases where it is clear that there is no objective fact of the matter. For example, you might like chocolate ice cream best while I like cookies & cream ice cream best. What is the truth concerning the best ice cream? Clearly, this is merely a matter of opinion. Chocolate ice cream is best for you and cookies & cream is best for me.

[6]It is worth noting that these claims hold even if one accepts a Kuhnian view of scientific change. When a paradigm shift occurs things that were once true do not suddenly become false. Likewise, things that were false do not suddenly become true. Instead, the scientific community adopts a new framework from which to organize the truths that we have come to know and to determine how best to uncover new truths. At most, things that were once *thought* to be true or false are no longer accepted or rejected under the new paradigm. The truth itself does not change. For more on the Kuhnian view of scientific change see Kuhn (1962).

Similarly, you might think that a particular joke is funny while I think that it is not funny. What is the truth concerning the joke: is it true that it is funny? Again, this seems to be just a matter of opinion. <The joke is funny> is not simply true or false. It is true or false relative to particular people. It is true for you, but not for me.[7] Relativism is the view that all propositions are like <the joke is funny> in the sense that they are only true or false relative to people. Consequently, even things like <1 + 1 = 2> and <Earth is not flat> are only true relative to certain people.

There are two main kinds of relativism: subjectivism and what we will call "group relativism". Subjectivism is the view that truth is relative to individual people. So, what is true and what is false depends on what the individual person believes and disbelieves. According to subjectivism, if James believes that $1 + 1 = 3$ and disbelieves that the sun is more massive than Earth, then <1 + 1 = 3> is true for him and <the sun is more massive than Earth> is false for him. If James has no opinion about whether the Earth is flat, then subjectivism holds that <Earth is flat> has no truth-value for James—it is neither true nor false.

Group relativism differs from subjectivism because it does not claim truth is relative to individuals. Instead, group relativism holds that truth is relative to groups. Hence, if the relevant group is one's nation, then what is true will depend on what the majority of people who belong to that nation believe. An example will help illustrate group relativism and how it differs from subjectivism:

> Uma is a citizen of the United States. The majority of citizens in the United States believe that Earth is not flat. Uma does not herself accept this; she believes that Earth is flat.

On group relativism, since the majority of people in the United States believe that Earth is not flat, it is true for Uma that Earth is not flat. This is so regardless of the fact that Uma disbelieves this. It is easy to see how subjectivism and group relativism can come apart in this case. Group relativism claims that <Earth is not flat> is true for Uma, but subjectivism claims that <Earth is flat> is true for Uma because she believes it to be so.

Both versions of relativism are plagued by serious difficulties. It will be instructive to consider some of the major problems with these kinds of theories. First of all, subjectivism entails that we are never mistaken about anything. If you believe that *p*, then *p* is true for you. It does not matter what proposition we are discussing! If you believe that pigs fly, then <pigs fly> is true for you. If you later change your mind about this, then you are still correct because now <pigs fly> is false. Clearly, this is not right. If we can be sure of anything about the nature of truth, it is that sometimes we mistakenly think that false propositions are true, and sometimes we mistakenly think that true propositions are false.

Now, one might think that group relativism avoids this problem because it does not make truth relative to whatever individual people believe. It faces a very similar problem of its own though—groups can never be wrong. If the relevant group is

[7]The joke example is borrowed from Wrenn (2014).

one's nation, then whatever the majority of people in the nation believe is true for everyone in that nation. This is obviously false. If it were the case that the majority of people in the United States were to believe that the moon is made of cheese, group relativism would say that it is true that the moon is made of cheese for everyone who is a citizen of the United States. This is ridiculous.

Notice how when speaking of group relativism the claims were always qualified by "if X is the relevant group". This suggests a second problem for group relativism—determining the relevant group. Each of us belongs to many groups. For example, Uma is a citizen of the United States, but she is also a member of the human race. What are we to say when the majority of people in the United States believe that *p*, but the majority of humans believe that *not p*? What is true concerning *p* for Uma? Is *p* both true and not true for her at the same time, or does one group take precedence? It does not seem that group relativism has the tools needed to answer these questions. This appears to be a very serious problem for this sort of view because without a way of determining which group is the relevant group, group relativism cannot tell us what truth is.

A final problem for both relativist views can be traced back to Plato's *Theaetetus*. In that dialogue Plato argues that relativism about truth is self-defeating. Here is a version of Plato's argument applied to relativism in general, and so, applicable to both subjectivism and group relativism:

1. Assume (for the sake of argument) that relativism is true.
2. Either relativism is only true relative to those who believe it (either individuals or groups) or it is true regardless of what anyone believes about it.
3. If relativism is true regardless of what anyone believes about it, then relativism is false (because relativism claims that all truths are relative, but the truth about relativism would not itself be relative).
4. If relativism is only true relative to those who believe it, then it is false for those who disbelieve it (such as realists and anti-realists).
5. Therefore, either relativism is false simpliciter, or it is false for anyone who denies relativism.

This argument is sometimes referred to as the "peritrope" because it is a turning of the table. That is, it turns relativism against itself.[8] This argument provides reason for thinking that relativism, whether it is of the subjectivist or group variety, is a self-defeating view of the nature of truth.

In light of these objections to both subjectivism and group relativism, it is reasonable to think that neither variety of relativism provides a sufficiently plausible alternative to warrant abandoning the realist view of the nature of truth.

[8]For further contemporary discussion of this argument see Swoyer (2014) and Wrenn (2014).

4.2.3 Anti-realism

The second rival to realist theories of truth is anti-realism. Anti-realism differs from relativism in that it does not hold that truth depends on what people believe nor that truth is relative to individuals or groups. However, anti-realism also differs from realism because it denies that the world is mind independent. Instead, anti-realism is the view that truth is closely tied to what it is possible for us to know. According to anti-realism, the only things that are true or false are those that we could come to know are true or false. Thus, if there is no way for us to ever know whether *p* is true or not, anti-realism says that *p* has no truth-value at all—it is neither true nor false. For example, there is no way we can come to know whether there is an enormous teddy bear floating ten trillion light years away from our galaxy. Ten trillion light years is simply too far for us to ever observe; it is outside of our observable universe. In light of this, anti-realism commits one to claiming <there is an enormous teddy bear floating ten trillion light years away from our galaxy> is neither true nor false.

Two of the more prominent versions of anti-realism are the coherence theory of truth and the pragmatic theory of truth. The coherence theory of truth is the view that when *p* is true it is part of a coherent, and sufficiently comprehensive, set of beliefs.[9] The pragmatic theory of truth as first developed by C.S. Peirce (1878/1982) holds that *p* is true when we *would* believe that *p* without doubt after an ideal investigation (an investigation where we have examined everything to the point that there is nothing more for us to learn on the topic). A somewhat different version, developed by William James (1907/2000), holds that *p* is true when believing it is useful to us. That is to say, *p* is true when we succeed at our endeavors by acting on the basis of accepting *p*.[10] Although both coherence and pragmatic theories of truth are interesting and face serious problems unique to their own idiosyncrasies, we will not examine them individually.[11] The reason for this is that as anti-realist views of truth both coherence and pragmatic theories share two very formidable problems that result from the general anti-realist picture. Examining these two problems is sufficient for recognizing that such views fail to provide a superior alternative to realism.

The first problem faced by anti-realist views is that they are inconsistent with a basic truth of classical logic (Wrenn, 2014). Consider the following proposition: <there is a teddy bear floating ten trillion light years from our galaxy or there is not a teddy bear floating ten trillion light years from our galaxy>. This proposition is of the form <*p* or not *p*>, and so it is something that everyone, even anti-realists, grant that we know to be true. Of course, any proposition that is constructed by combining two other propositions with "or" such as <*p* or not *p*> or <*p* or *q*> is true provided

[9]For various versions of the coherence theory of truth see Bradley (1914), Putnam (1981), and Young (1995).

[10]For more on the pragmatic theory of truth see Hookway (2013) and Misak (2007).

[11]For objections to the coherentist theory of truth see Russell (1907), Thagard (2007), and Walker (1989). For objections to both the coherentist theory of truth and the pragmatic theory of truth see Russell (1912/2001) and Wrenn (2014).

that at least one of the propositions connected with "or" is true. However, given anti-realism <there is a teddy bear floating ten trillion light years from our galaxy> is neither true nor false because we cannot know whether it is true. Similarly, <there is not a teddy bear floating ten trillion light years from our galaxy> is neither true nor false for the same reason according to anti-realism. This commits the anti-realist to claiming that propositions of the form <p or q> can be true even though neither p nor q is true. This is inconsistent with basic truths of classical logic. Thus, to accept anti-realism is to deny basic logical truths.

The second problem for anti-realism comes from the "Knowability Paradox".[12] The thrust of the problem the Knowability Paradox raises for anti-realism is that it shows anti-realism not only commits one to claiming that all truths are *knowable*, but it commits one to claiming that in fact, every truth is *known* by someone. To illustrate this let us assume there is a truth that no one knows (let us call this unknown truth "p"). So, p is true and not known to be true by anyone. This means that <p is true and not known> is true. According to anti-realism, since <p is true and not known> is true, <p is true and not known> is knowable. In other words, anti-realism is committed to it being possible for someone to both know that p is true *and* to know that p is not known. But, it follows from this that it is possible someone knows that p is true, and at the same time she knows that no one knows that p is true.[13] Of course, if someone knows that no one knows that p is true, then no one knows that p is true. After all, we saw in chapter two that knowledge entails truth—if S knows that p, then p is true. Thus, anti-realism entails that it is possible that at the same time someone knows that p is true and no one knows that p is true! This is a clear contradiction, so something must be wrong. But, what could it be?

The only assumptions made in the above argument are: (1) a very plausible principle (single premise closure), (2) the claim that knowledge entails truth, (3) there are truths that are not known, and (4) anti-realism about truth. At least one of these must be abandoned in order to avoid the above contradiction. (1) and (2) are both exceedingly plausible, so they should not be given up. The best bet for the anti-realist is to deny (3). Yet, this commits the anti-realist to claiming not only that all truths are knowable, but also that all truths are in fact known. This is a highly implausible claim. Of course, the most viable option when faced with this argument seems to be to give up (4)—deny anti-realism about truth.[14] Again, it appears that realism is superior to its anti-realist rival.

We have seen that there are strong reasons for thinking that among the three primary approaches to the nature of truth, realism is the best. Unsurprisingly, the debate does not end here. There are numerous forms that realism can take. It is

[12]For more in-depth discussion of this paradox see Fitch (1963), Kvanvig (2006), and Salerno (2009).

[13]This follows by an exceedingly plausible principle known as "single premise closure". Roughly, single premise closure is the idea that when someone knows that <p and q>, then that person knows that p and she knows that q. For more on single premise closure see Hawthorne (2004).

[14]For further discussion see Wrenn (2014).

because of this that the primary rivalries in the contemporary debates about truth are between supporters of opposing realist theories of truth. Fortunately, we do not need to adjudicate between these various versions of realism.[15] Instead, our purposes will be served by recognizing that a broadly realist approach is the best way to understand the nature of truth. However, before moving on to a final issue related to truth we will first briefly explore an approach to the nature of truth that attempts to cross the divides between each of the three approaches we have considered thus far.

4.3 Pluralist Theories of Truth

Although we have seen that realism is plausibly superior to its major rivals (relativism and anti-realism), recently some have argued that there is an even better approach to the nature of truth. These pluralist theories hold that there is not a single property that is had by all truths. Instead, pluralism holds that truth in different domains is yielded by different properties.

In general, pluralism is motivated by the claim that it avoids purported problems for various versions of the other approaches to truth. The primary problem that pluralism is constructed to avoid is the "scope problem" (Wrenn 2014). In order to appreciate this purported problem consider the following claims:

(a) The sun is more massive than Earth.
(b) The cars missed one another, but they could have collided.
(c) Ted's joke is funnier than Tom's joke.
(d) Causing needless suffering is wrong.

Essentially, the challenge of the scope problem is to account for the shared property that makes all of (a–d) true. Theories of truth are thought to have a scope problem when they work well in explaining the relevant property in some cases, but not others. Pluralists maintain that various forms of realism are well suited

[15]The three primary versions of realism on offer are correspondence theories, truth-maker theories, and deflationism. Roughly, they have the following to say about the nature of truth. Correspondence theories claim that truth consists of a proposition's "fitting" with reality. This is often put in terms of *p*'s corresponding to a particular fact about the world or to a particular state of affairs. The central idea behind truthmaker theories, which are sometimes considered to be a kind of correspondence theory, is that for any truth there is a truthmaker—something in the world that makes it true. Finally, deflationism is inspired by developments in modern logic. Deflationism is the view that all there is to understanding truth is to understand the logic and grammar associated with the predicate "is true". Of course, there is much more to be said about each of these varieties of realism. Nonetheless, as noted in the main text, for our purposes we do not need to go into the gritty details. For more on all three of these versions of realism see Burgess and Burgess (2011) and Wrenn (2014). For more on correspondence theories see Alston (1996), David (2009), Fumerton (2002), and Wittgenstein (1922/1990). For more on truthmaker theories see Armstrong (2004), Beebee and Dodd (2005), and Rodriguez-Pereyra (2006). For more on deflationism see Field (2001), Horwich (1998), Quine (1992), and Ramsey and Moore (1927).

to handle claims like (a), but not, say, (c). Similarly, various rivals to realism are well suited to account for claims like (c), but not those like (a). Pluralism has been proposed because it is thought that the best explanation as to why some truths are better explained by the various theories of truth than others is simply that there are different properties that make those disparate truths true.

There are two primary ways of cashing out the general commitments of pluralism. The first, *simple pluralism*, is just the view that there is not a single nature of truth.[16] Instead, according to simple pluralism, the nature of truth is different for various subject matters. A few philosophers have suggested this view, but most prominently it has been developed by Crispin Wright (1992, 2001).[17] Although Wright's view is interesting and worth serious consideration, we will not explore its finer points and the particular objections that it faces.[18] The reason for this is quite simple. While Wright does advocate pluralism about the nature of truth, he claims that when it comes to truths in science the nature of those truths is as realism describes. Since our primary focus is scientific knowledge, it is sufficient for our purposes that Wright accepts realism (or at least a particular version of it) when it comes to truths in science.

The second primary form of pluralism about truth is Michael Lynch's (2004, 2009) *alethic functionalism*. Lynch holds that truths in various discourses do not need to be true in the same way. However, he maintains that even though there are different ways of being true, the properties that make various claims true are all ways of manifesting a single property—the property of being true. An analogy will help make alethic functionalism clearer. Consider a famous role, the role of Santa Claus, say. Many actors can play, and have played, Santa Claus in many different film and theatre productions. Additionally, many people have dressed up as Santa Claus around the holidays. Although it is different people dressing up as Santa Claus in these cases, they are all dressing up as the same character—they are all playing a single role. Similarly, truth, according to alethic functionalism, is a role that can be played by various properties in different discourses.[19] Again, while alethic functionalism is an interesting theory of the nature of truth that warrants serious consideration, we do not need to explore it in detail. The reason for this is that Lynch, like Wright, maintains that truths in science are manifested by the sort of property that realists claim. So, even if pluralism about truth is correct as a general theory of the nature of truth, it seems that pluralists agree that truths in the domain of science are best understood in the way that realists understand all truths. Thus, for our purpose of understanding knowledge as it pertains to science it is safe to adopt a broadly realist conception of the nature of truth.

[16]This discussion follows Wrenn (2014) in referring to this view as "simple pluralism".

[17]Quine (1981) suggests that something like this view might be correct even though he primarily defends a deflationist view of truth.

[18]For more detailed critical discussion of Wright's view see Pedersen and Wright (2013) and Wrenn (2014).

[19]See Wrenn (2014) for further analogies that help explicate alethic functionalism.

4.4 Verisimilitude

At this point we have seen that the most promising way to understand truth, particularly as it pertains to science, is in broadly realist terms. This is very helpful in understanding the second component of the traditional account of knowledge. Before moving on to the final component of the traditional account it is worth pausing to briefly consider one further important facet of truth: verisimilitude ("truthlikeness").

As we already know, truth is a major component of knowledge—you cannot know that *p* unless *p* is true. In fact, truth is what we aim at in any legitimate inquiry. We inquire as to the truth of matters. Truth is what we aim at, but we do not always hit our target. Interestingly, we do not always miss our mark by the same amount. For example, if Jill says "$5 + 7 = 13$" and Jack says "$5 + 7 = 879$", both are clearly mistaken. Still, it seems that even though both Jill and Jack have missed the truth, Jill is much closer to it than Jack is. It seems that Jill's claim "$5 + 7 = 13$" has more verisimilitude, or truthlikeness, than Jack's.

It is not very surprising that some false claims such as Jill's are closer to the truth than others like Jack's. More surprising is the fact that some true claims are closer to the *whole* truth about an issue than others. The ultimate goal of our inquiry into any topic is not merely truth, but rather, the whole truth of the matter. Some truths are closer to the whole truth than others. For example, suppose that Billy's hitting a baseball through the window is why it broke. If we are inquiring into the cause of the window's breaking, it is true that <either a rock broke the window or a rock did not break the window>. However, <either a rock broke the window or a rock did not break the window>, while true, is farther from the whole truth than <a baseball broke the window> is. Notice though, both of these propositions are in fact true. This shows that some truths are closer to the full story than others—they have more verisimilitude to the whole truth of the matter.

More surprising still, it seems that sometimes a false claim can be closer to the whole truth about an issue than a true claim. Graham Oddie (2014) asserts that it may be that a truth such as <either electrons are fundamental particles or they are not> is farther from the whole truth than the falsehood that <electrons, protons, and neutrons are fundamental components of atoms>. The reason that this latter claim seems closer to the whole truth is that it is providing more information about the topic under consideration even though some of that information is not correct. While the total information provided is not exactly correct, it does seem to be getting us closer to the whole truth than the vacuously true claim that <either electrons are fundamental particles or they are not>. If this is correct, then it appears that sometimes falsehoods can be more verisimilitudinous than truths.

At this point we have seen a variety of things about verisimilitude. Some falsehoods are closer to the whole truth than other falsehoods. Some truths are closer to the whole truth than other truths. Some falsehoods seem closer to the whole truth than some truths. These facts about verisimilitude relate to our purpose of better understanding the truth component of knowledge in at least two ways.

First, recognizing that various truths (and falsehoods) can be closer to the whole truth than others helps to deepen our understanding of the nature of truth.[20] Second, appreciation of verisimilitude and recognition of the fact that knowledge requires truth, not truthlikeness but truth full stop, may give us reason to think that the proper focus when understanding NOS (and other issues of particular importance to science education) is not really knowledge in the strict sense at all. It seems that recognizing these facts about knowledge and verisimilitude gives us some reason to think that what really matters for scientific inquiry is the *evidence* we have in support of particular claims and theories rather than knowledge.[21] Fortunately, for our purposes we can continue to speak in terms of knowledge, but we are well served by keeping in mind that perhaps what we are really talking about is evidence for claims and whether we are justified in believing those claims.

4.5 Conclusion

We have seen that the best way to understand the nature of truth, at least with respect to our purposes and perhaps in general, is as realists claim. Truth depends on correctly describing an objective, mind-independent reality. It is this reality that we study in the sciences, and scientific knowledge is knowledge of the features of this objective, mind-independent reality. Now that we have a handle on the nature of truth it is time to turn our attention to the final component of the traditional account of knowledge: *justification*.

References

Alston, W. P. (1996). *A realist conception of truth*. Ithaca: Cornell University Press.

Aristotle. (1941). *The basic works of Aristotle* (trans: McKeon, R.). New York: Random House.

Armstrong, D. M. (2004). *Truth and truthmakers*. Cambridge, UK: Cambridge University Press.

Beebee, H., & Dodd, J. (Eds.). (2005). *Truthmakers: The contemporary debate*. Oxford: Clarendon Press.

Bird, A. (2007). *Nature's metaphysics: Laws and properties*. Oxford: Oxford University Press.

[20]There are very deep and important issues concerning how best to determine the degree of verisimilitude that a particular claim enjoys. The various measures of verisimilitude that have been offered are complex, and the debate concerning which is correct is quite intricate. Luckily, we do not need to enter into the details of these measures or the debate concerning their veracity here. A general appreciation of verisimilitude is all that is required for our purposes. For more on these measures and the surrounding debate see Newton-Smith (1981), Niiniluoto (1987), Oddie (1986, 2014), Popper (1963), Tichý (1978), and Zwart (2001).

[21]We will see in chapter eight that the fact that there is a serious problem with the traditional account of knowledge also gives us reason to think that our focus should be on evidence and justification rather than knowledge when we are trying to increase understanding of NOS.

Bradley, F. (1914). *Essays on truth and reality*. Oxford: Clarendon Press.
Burgess, A. G., & Burgess, J. P. (2011). *Truth*. Princeton: Princeton University Press.
David, M. (2009). Truth-making and correspondence. In E. J. Lowe & A. Rami (Eds.), *Truth and truth-making* (pp. 137–157). Montreal: McGill-Queen's University Press.
Fales, E. (1990). *Causation and universals*. London: Routledge.
Field, H. (2001). *Truth and the absence of fact*. Oxford: Oxford University Press.
Fitch, F. (1963). A logical analysis of some value concepts. *The Journal of Symbolic Logic, 28*, 135–142.
Fumerton, R. (2002). *Realism and the correspondence theory of truth*. Lanham: Rowman & Littlefield.
Hawthorne, J. (2004). *Knowledge and lotteries*. Oxford: Oxford University Press.
Hookway, C. (2013). Pragmatism. In E. N. Zalta (Ed.), *The Stanford encyclopedia of philosophy* (Winter 2013 Edition). http://plato.stanford.edu/archives/win2013/entries/pragmatism/
Horwich, P. (1998). *Truth*. Oxford: Oxford University Press.
James, W. (1907/2000). Pragmatism's conception of truth. In G. Gunn (Ed.), *Pragmatism and other writings* (pp. 87–104). New York: Penguin.
Kuhn, T. S. (1962). *The structure of scientific revolutions*. Chicago: University of Chicago Press.
Kvanvig, J. (2006). *The knowability paradox*. Oxford: Oxford University Press.
Lynch, M. P. (2004). *True to life: Why truth matters*. Cambridge, MA: MIT Press.
Lynch, M. P. (2009). *Truth as one and many*. Oxford, UK: Oxford University Press.
Misak, C. J. (Ed.). (2007). *New pragmatists*. Oxford: Oxford University Press.
Newton-Smith, W. H. (1981). *The rationality of science*. Boston: Routledge & Kegan Paul.
Niiniluoto, I. (1987). *Truthlikeness*. Dordrecht: Reidel.
Oddie, G. (1986). *Likeness to truth*. Dordrecht: Reidel.
Oddie, G. (2014). Truthlikeness. In E. N. Zalta (Ed.), *The Stanford encyclopedia of philosophy* (Summer 2014 Edition). http://plato.stanford.edu/archives/sum2014/entries/truthlikeness/
Pedersen, N. J. L. L. & Wright, C. (2013). Pluralist theories of truth. In E. N. Zalta (Ed.), *The Stanford encyclopedia of philosophy* (Spring 2013 Edition). http://plato.stanford.edu/archives/spr2013/entries/truth-pluralist/
Peirce, C. S. (1878/1982). How to make our ideas clear. In H. S. Thayer (Ed.), *Pragmatism: The classical writings* (pp. 79–100). Indianapolis: Hackett.
Popper, K. R. (1963). *Conjectures and refutations*. London: Routledge.
Putnam, H. (1981). *Reason, truth and history*. Cambridge, UK: Cambridge University Press.
Quine, W. V. O. (1981). *Theories and things*. Cambridge, MA: Harvard University Press.
Quine, W. V. O. (1992). *Pursuit of truth, revised edition*. Cambridge, MA: Harvard University Press.
Ramsey, F. P., & Moore, G. E. (1927). Symposium: Facts and propositions. *Proceedings of the Aristotelian Society, Supplementary Volumes, 7*, 153–206.
Rodriguez-Pereyra, G. (2006). Truthmakers. *Philosophy Compass, 1*, 186–200.
Russell, B. (1907). On the nature of truth. *Proceedings of the Aristotelian Society, 7*, 228–249.
Salerno, J. (Ed.). (2009). *New essays on the knowability paradox*. Oxford: Oxford University Press.
Shoemaker, S. (1998). Causal and metaphysical necessity. *Pacific Philosophical Quarterly, 79*, 59–77.
Soames, S. (1999). *Understanding truth*. New York: Oxford University Press.
Swoyer, C. (2014). Relativism. In E. N. Zalta (Ed.), *The Stanford encyclopedia of philosophy* (Summer 2014 Edition). http://plato.stanford.edu/archives/sum2014/entries/relativism/
Thagard, P. (2007). Coherence, truth and the development of scientific knowledge. *Philosophy of Science, 74*, 26–47.
Tichý, P. (1978). Verisimilitude revisited. *Synthese, 38*, 175–196.
Walker, R. C. S. (1989). *The coherence theory of truth: Realism, anti-realism, idealism*. New York: Routledge.
Wittgenstein, L. (1922/1990). *Tractatus logico-philosophicus*. London: Routledge.
Wrenn, C. B. (2014). *Truth*. Cambridge, UK: Polity Press.
Wright, C. (1992). *Truth and objectivity*. Cambridge, MA: Harvard University Press.

Wright, C. (2001). Minimalism, deflationism, pragmatism, pluralism. In M. P. Lynch (Ed.), *The nature of truth: Classic and contemporary perspectives* (pp. 751–787). Cambridge, MA: MIT Press.

Young, J. O. (1995). *Global anti-realism*. Aldershot: Avebury.

Zwart, S. D. (2001). *Refined verisimilitude*. Dordrecht: Kluwer.

Chapter 5
Justification

Abstract This chapter focuses on the final component of the traditional account of knowledge: justification. Traditionally, justification has been understood as having good reasons for believing that a particular claim is true. The nature of these good reasons is examined in this chapter. In particular, practical reasons for accepting a particular claim are distinguished from epistemic reasons for accepting a particular claim. It is the epistemic reasons that are necessary for knowledge. Additionally, a major contemporary debate in epistemology concerning whether one always needs good reasons in order to be justified is explored in this chapter. Internalists say "yes", but externalists say "no". Some of the major moves in this debate are explained in this chapter. However, it becomes clear by the chapter's end that whether internalists or externalists are correct in general, the sort of justification required for scientific knowledge does require good reasons, which are best understood as evidence.

We are now ready to explore the third, and final, component of the traditional account of knowledge: *justification*. As we have seen, in order to know that *p* you must believe that *p* and *p* must be true. Nonetheless, as we noted in chapter two, true belief is not sufficient for knowledge. After all, you might believe that there is an even number of stars in the Milky Way galaxy because of a guess, and you might be lucky and have guessed correctly. In such a case you would have a true belief, but clearly you would not *know* that there is an even number of stars in the Milky Way galaxy. What is missing? What more would you need in order to have knowledge? According to the traditional account of knowledge, you do not know that there is an even number of stars in the Milky Way galaxy because you do not have good reasons for your true belief. Since you believe what you do simply as a result of guessing, your belief lacks *justification*.

Intuitively, we recognize that some things provide good reasons to believe that *p* and others do not. So, we have at least some grasp on what it means to have justification for a belief. For example, your visual experience as of a tree in the yard in normal lighting conditions seems like a good reason to believe that there is a tree in the yard. Similarly, reading in a very reputable journal that a particular experiment yielded a certain result is a good reason to believe that the experiment had that result. Alternatively, your love for the Bears as your favorite football team is not a good reason, on its own, to think they will win the game. Your fear of

K. McCain, *The Nature of Scientific Knowledge*, Springer Undergraduate Texts in Philosophy, DOI 10.1007/978-3-319-33405-9_5

heights is not itself a good reason to believe that a tall bridge is unsafe. What is the difference between these two pairs of cases? Plausibly, the difference is that in the first two cases you have justification for the belief in question, but not in the second two cases. As we have already mentioned, traditionally justification is understood in terms of having good reasons. So, you have justification in the first two cases because you have good reasons for thinking that the propositions believed are true, but you lack such reasons in the second pair of cases. More generally, the traditional idea is that having justification for believing that *p* requires having good reasons for thinking that *p* is true.[1]

But, what sorts of things are these good reasons? This seemingly straightforward question is surprisingly difficult to answer. In part the difficulty arises from the fact that we may mean different things when asking this question. One thing we might mean, which we will set aside until chapter ten, is what does it take for some bit of evidence to provide a good reason to believe something? Another thing we might mean, which will be our focus in this chapter, is what is justification in general?

As will become clear in this chapter, there is still a lot of controversy concerning various aspects of justification despite the attention that it has garnered for hundreds of years in the field of epistemology. Although we are not likely to definitively settle any of the major debates concerning justification, this chapter will provide us with a sufficient grasp of the relevant issues to not only appreciate the traditional account of knowledge, but also, to begin to better understand scientific knowledge.

5.1 The Nature of Justification

Before beginning our investigation of three of the most well entrenched debates concerning the nature of justification it will be helpful to first briefly discuss several distinctions. While these distinctions will not make it obvious which side in the debates that follow is correct, they are very helpful for properly understanding those debates. Additionally, these distinctions are in many cases key to properly understanding justification.

[1]It is worth noting that this is different than the idea that justification is what is required to support our claims to know. Toulmin (2003) explains that when someone claims to know that *p* she is implicating that she can provide a justification for her knowledge claim. That is to say, she is committing herself to being able to provide adequate grounds for satisfactorily answering the question "how do you know?" While it may be true that one is expected to be able to justify claims to knowledge, this sort of dialectical justification is different than the justification required for knowledge. In order to know that *p* one simply needs to have sufficiently strong reasons for believing that *p*; one does not have to be able to fully and persuasively articulate those reasons to others.

5.1.1 Epistemic Justification Versus Pragmatic Justification

The first important distinction to draw is between *epistemic justification* and *pragmatic justification*. Both kinds of justification can be understood in terms of having good reasons. Nevertheless, they are very different. Epistemic justification consists of good reasons that are indicative of the truth of a proposition (McCain 2014). For example, your visual experience as of a tree in the yard provides you with epistemic justification for believing that there is a tree in the yard. This visual experience gives you epistemic justification because it is a good reason for thinking <there is a tree in the yard> is true. In other words, your visual experience in this case is indicative of the truth of the proposition <there is a tree in the yard>.[2]

Not all good reasons provide epistemic justification though. Seth might have a good reason to think that he will recover from an illness that has an extremely low survival rate because believing that he will recover helps him to better deal with his suffering. Now, if all of Seth's doctors tell him that it is unlikely that he will recover, all the medical studies Seth reads show that in cases like his the likelihood of recovery is extremely low, etc., then it does not seem that he has epistemic justification for believing that he will recover. Despite Seth's belief not being epistemically justified, it may still be pragmatically justified. After all, Seth has good pragmatic reasons for believing that he will recover—it is in his best interest to believe that he will recover because doing so eases his suffering. Unfortunately, it so happens that Seth's good pragmatic reasons are not indicative of the truth of the proposition <Seth will recover>. Consequently, while Seth might be pragmatically justified in believing that he will recover, he is epistemically unjustified in believing this.

It is epistemic justification that is necessary for knowledge. And, so it is epistemic justification that will be our focus throughout this chapter and the remainder of the book.[3]

5.1.2 Justification Versus Justifying

Another important distinction is that between *having justification* for a belief and *justifying* a belief. These are very different. Having justification for believing something, or one's belief being justified, is a particular state of a person. Ally's having a justified belief that the Bears won the game is a state of Ally—it involves her having a mental state with the appropriate representational content to be a belief, her having various other mental states that provide good reasons to believe <the Bears won the game>, and so on. Ally's justifying her belief that the Bears won

[2]We will explore what it takes for a particular mental state to be indicative of the truth of a particular proposition in chapter ten when we focus on the nature of evidential/epistemic support.

[3]From this point on we will typically drop the qualifier "epistemic" and simply speak of justification. However, it should be understood that unless specified "justification", "justified", and so on are being used in the epistemic senses of these terms.

the game is an action. When we justify a belief we explain to someone (possibly ourselves) the evidence or reasons we have in support of the truth of that belief. So, justification is a matter of having certain mental states/reasons, but justifying is the act of explaining to someone your reasons for what you believe.[4]

Although justification and justifying are very different, it is easy to see why we sometimes confuse them.[5] Often we might think that we are justified in believing something, but as we try to justify the belief we realize that we are not justified in believing as we do. For example, Eric may think that he is justified in believing that a particular celebrity is very ill. Yet, when Eric tries to justify his belief to his friend, Erin, he may come to realize that he does not have very good reasons for believing this at all because his only evidence in support of this belief is that he read in a tabloid that the celebrity was ill. When we attempt to justify our beliefs, even if we are only justifying them to ourselves, we sometimes realize that our reasons for believing as we do are not very good. Of course, there are also times when the process of justifying our beliefs helps us to realize that we have very good reasons for believing what we do. Jane may think that she is justified in believing that hydrogen is the lightest chemical element. As she begins to justify this belief to her friend, Neil, she realizes that she remembers reading this in her chemistry book, she has notes from her latest lecture where the professor said that hydrogen was the lightest chemical element, and so on. In this case, Jane comes to realize that she has very good reasons for her belief.

Given that the act of justifying may sometimes help us to recognize that we have good (or bad) reasons for believing as we do, it is not surprising that we sometimes confuse justifying with being justified. That is to say, we sometimes think that one's belief that *p* is not justified unless she can justify that belief. This is a mistake for at least two reasons, though. First, someone can be justified in believing that *p* without being able to justify that belief. Think of small children. Presumably, they have a number of justified beliefs. However, it is plausible that for many of their beliefs small children simply lack the conceptual capacities to adequately justify those beliefs to others. Yet, they seem to be justified all the same. Second, someone might come to be justified in believing something she was not originally justified in believing through the very act of trying to justify the belief. It is possible that Carrie believes that *p* without having good reasons to do so, but while explaining to Carl why she believes that *p* she comes up with what are in fact very good reasons to believe that *p*. In such a case, it seems that the act of justifying led Carrie to recognize reasons that make her justified in believing that *p*, but she was not justified in so believing before she began justifying her belief. In light of these reasons, we should be careful to clearly distinguish between justification and the act of justifying. It is only the former that is required for knowledge, and so, only justification will be our focus from this point on.

[4]We will discuss the nature of the mental states that are required for justification much more fully in chapter six when we discuss the nature of evidence.

[5]The act of justifying seems to be what Toulmin (2003) is interested in when he describes justification in terms of having an answer to the question "how do you know?"

5.1.3 *Justified in Believing Versus Justifiedly Believing*

Another distinction that is worth briefly mentioning here is the distinction between being justified in believing that *p* (propositional justification) and justifiedly believing that *p* (doxastic justification).[6] The idea behind this distinction is straightforward. As we have already noted, in order to be justified in believing that *p* one needs to have good reasons for thinking that *p* is true. However, merely having good reasons to think that p is true is not sufficient for one's belief that *p* to be justified.

To see the distinction between being justified in believing that *p* and justifiedly believing that *p* consider the following sort of situation:

> J.D. is playing a game of chess. It is J.D.'s move, and he has excellent reasons for thinking that he should not sacrifice his queen at this point in the game. He is aware that by sacrificing his queen his opponent will have much stronger pieces on the board, his own position will be inferior to his opponent's, and so on.

In light of these reasons, J.D. is justified in believing that he should not sacrifice his queen at this point in the game. In spite of this, it could be that J.D. believes that he should not sacrifice his queen at this point in the game for some silly reason—perhaps, it is simply his favorite piece to look at, and so he never wants to give up his queen. If J.D. believes that he should not sacrifice his queen at this point in the game because of this silly reason, then his belief is not justified. In this case, believing that he should not sacrifice his queen at this point in the game is justified for J.D., but his belief that he should not sacrifice his queen at this point in the game is not a justified belief. The problem is that while J.D. believes the correct thing, he believes it for the wrong reason. In order for one's belief that *p* to be justified one must not only have good reasons for believing that *p*, she must also hold the belief on the basis of those good reasons. This distinction will play a role in our discussion of internalism and externalism in this chapter, and it will be very important in later discussions.[7]

5.1.4 *Justification and Defeat*

A further point that should be clarified before moving on is that in order to be justified in believing that *p* it must be that the total information that one has on balance supports thinking that *p* is true. It is not enough for justification that merely a part of one's evidence supports thinking that *p* is true. For instance, part of the evidence that Marsha has might give her very good reasons for believing that

---

[6]This distinction can also helpfully be put in terms of having justification for believing that *p* versus having what Feldman and Conee (1985) term a "well-founded" belief that *p*.

[7]We will explore this distinction as well as what is required for a belief to be held on the basis of good reasons much more fully in chapter seven.

there is an elephant in front of her. Marsha could have a visual experience as of an elephant in clear lighting conditions. Along with her background knowledge of what elephants look like, that her vision typically works well, etc., this visual experience gives Marsha good reason to think that there is an elephant in front of her. Nevertheless, it may be that Marsha has additional evidence which defeats the justification her visual experience would normally provide. If in this situation Marsha also has good reason to think that there are no elephants in this area because she is sitting in the living room of her twentieth floor apartment, and she has good reason to think that she has recently taken medication which is known to cause elephant hallucinations as a side effect, then it seems that she is not justified in believing that there is an elephant in front of her. The justification that Marsha's visual experience would normally provide is defeated by the additional evidence she has about her location and the medication she has recently taken.

There are two lessons to be learned from cases like Marsha's. First, justification that is usually provided by particular reasons can be defeated by other information. Second, when determining whether a particular belief is justified we have to take into consideration the total amount of information possessed by the believer.[8]

An additional aspect of justification, which is related to the idea that we have to consider one's total information when assessing justification, is that justification comes in degrees. This simply means that it is possible for you to be more or less justified in believing various things. For example, say your friend, the expert mathematician, has just completed what is for her a simple proof with clearly true assumptions in support of *p*. Meanwhile, your other friend, the tabloid reader, has read in the latest issue of his favorite tabloid, which he knows to be fairly unreliable, that *p* is true. Clearly, your mathematician friend has very good reasons for believing that *p*. Perhaps your tabloid-reading friend has some reason to believe that *p* too (it depends on what he knows about how unreliable the tabloid is). Yet, it is clear that your mathematician friend is more justified in believing that *p* than your tabloid-reading friend. Similarly, you can be more or less justified in believing various things—you are likely more justified in believing that you know your own name than you are in believing that you know how long it takes to fly from New York to London. So, it is important to keep in mind that justification comes in degrees because you can have better or worse reasons for believing things.

Of course, it is natural at this point to wonder what degree of justification is required for knowledge. That is, you might be wondering how much justification do you need in order to satisfy the justification condition of the traditional account of knowledge. This is a very difficult question to answer. Perhaps the best way to go is with the "criminal standard" view of evidence/good reasons (Conee and Feldman 2004). According to this standard, in order to know that *p* you must have good reasons that make the truth of *p* beyond a reasonable doubt. Hence, knowledge does not require good reasons that make *p* absolutely certain for you, but it does seem that

[8]This is enough on this point for our current purposes. However, we will return to this issue and explore it more fully in chapter ten.

knowledge requires more than merely some reason to believe. Although "beyond a reasonable doubt" is somewhat vague, it will be sufficient for a working guide for the degree of justification required for knowledge.[9]

5.1.5 *Justification and Truth*

A final point to clarify concerns the relationship between justification and truth. Although we have already noted that justification for *p* requires having good reasons for thinking that *p* is true, a bit more should be said about justification and truth. In order to be justified in believing that *p* you have to have good reasons for thinking that *p* is true, however, it is possible to have such reasons even when *p* is false. In other words, being justified in believing that *p* does not entail that *p* is true. It is possible to have justified false beliefs. For example, you might have excellent reasons to believe that a particular mathematical theorem is true—a math professor, who you know to be trustworthy, told you that the theorem is true, you read in a reputable journal that the theorem is true, and you have just worked through what strikes you as a sound proof of the theorem. Intuitively, your belief that the theorem is true is justified in this case. After all, you have excellent evidence in support of thinking this theorem is true. Yet, it is still possible that the theorem is false. It could be that you have all of this excellent evidence despite the falsity of the theorem because the math professor was mistaken about which theorem you were discussing, the journal made an error, and you have made a subtle mistake in your proof. This is not to say that these are all likely to have occurred; they are probably very unlikely. Nevertheless, it is possible that all of these things did occur. In such a case you still have excellent reasons to believe as you do, but your belief is false. Justification requires that you have a good indication of the truth of *p*, but it does not require an indication that entails the truth of *p*. So, just as there are unjustified true beliefs, there are justified false beliefs.

5.2 Justification and Normativity

With the preliminaries of the previous section out of the way we are now ready to begin our examination of the first of three major debates about justification. This first debate concerns the relationship between justification and normativity. It seems pretty clear that justification is a normative concept. After all, when we say that a belief is justified/unjustified we seem to be saying that the belief is appropriate/inappropriate or favorable/unfavorable with respect to our ends of

[9]We will touch upon this issue again in chapter ten when we discuss how a body of evidence supports believing a particular proposition.

believing truths and disbelieving falsehoods, and so on. Although most philosophers agree that justification is a normative concept, there is disagreement about exactly how we should understand this normative aspect of justification.[10]

The debate over the normative aspect of justification typically comes down to whether one accepts a deontological view or a non-deontological view of the normativity of justification. Deontological views focus on things like duty, obligation, permissibility, and blame. Non-deontological views focus on goodness or appropriateness without reference to deontological concepts. Very roughly, a deontological view understands justification for believing that *p* in terms of it being obligatory for one to believe that *p* or it being permissible for one to believe that *p* or one's being blameless for believing that *p*. Non-deontological views understand justification in terms of a belief being good or appropriate without appealing to obligation or permissibility or blame. So, just as we might evaluate a knife as being appropriate for the purpose of cutting a pineapple, a belief might be judged appropriate (justified) for the purpose of believing truths while avoiding falsehoods. The non-deontologist will claim that duty, obligation, permissibility, blame, and so on do not enter into the picture.

Let us start by looking more closely at the deontological approach. This view of the nature of justification has a long history stemming from the works of philosophers like René Descartes (1641/1988) and John Locke (1690/1975). Many philosophers are convinced that "the whole notion of epistemic justification has its origin and home in this deontological territory of duty and permission . . . at bottom, epistemic justification *is* deontological justification" (Plantinga 1993b, p. 14).[11] The best way to understand the deontological conception of epistemic justification is in terms of two key ideas (Steup 1996). The first idea is that we are committed to the end of having true beliefs in virtue of the sort of intellectual/rational beings that we are. The second idea is that our having this end imposes upon us a duty to believe in accordance with the evidence that we have (Feldman 1988; Steup 1996). Simply put, according to the deontological conception of justification, a belief is justified when having that belief is what is dictated by one's epistemic duties.[12] Intuitively, this seems correct. When we say that Mary's belief is justified it does seem like we are claiming that she is believing as she *ought* to believe; she is following her epistemic obligations and doing her duty as a rational thinker. Likewise, when we say that Marty's belief is unjustified it seems like we are saying that he is not believing as he ought; he is violating his obligations as a rational thinker.

[10]Of course, not all philosophers agree that justification is normative. Maffie (1990) and Moser (1989) both suggest that justification can be separated from the normative connotations that usually accompany it.

[11]Philosophers with views as varied as BonJour (1985), Chisholm (1956, 1977), Feldman (1988), Ginet (1975), Goldman (1986)—though he appears to have changed his mind in his (1999)—Lycan (1988), Pollock (1986), and Steup (1996) accept a deontological view of justification.

[12]We will consider what counts as violating one's epistemic duties in this sense more in chapter ten when we explore the nature of evidential support for beliefs.

Despite its intuitive plausibility and its impressive philosophical pedigree, the deontological view of justification is controversial. In fact, the deontological view of justification faces at least two serious objections. The first of these is what we might call the "isolated community" objection. According to this objection, it is possible that one grew up in an isolated community where the traditions of the community are taken to be authoritative. As a result, if it is the consensus of the community that *p*, then everyone in the community believes that *p*. Suppose a child, Iso, grows up in this community and like everyone else does not question the consensus of the community. Iso comes to believe that *p* because it is the consensus of the community. However, the community's reasons for *p*, which Iso accepts, are really bad reasons for believing that *p*. According to William Alston (1985), it could be that Iso, and other people in such a situation, has done the best that she can with respect to *p* so that Iso is blameless for believing that *p*. Despite being blameless for believing that *p*, it seems that Iso is still "in a very poor epistemic position" with respect to *p* (Alston 1985, p. 34). In essence, Alston's objection is that merely being blameless in what one believes does not seem to be enough to make one justified in so believing.

Although Alston's objection seems worrisome and it does illustrate that perhaps justification should not be understood in terms of praise and blame, defenders of deontological approaches can reasonably respond that the primary notion of the deontological approach is not blame. Instead, they can claim that what matters is duty, obligation, and permissibility. Hence, deontologists may respond to Alston's objection by maintaining that his objection rests on a conflation of "conditions of epistemic *permissibility*, or *rightness*, with conditions of epistemic *blamelessness*" (Moser 1989, p. 39). In other words, deontologists about justification can allow that Iso's belief, and the beliefs of others in similar situations, are blameless, while continuing to hold that her belief is not permissible. Essentially, deontologists can plausibly maintain that it is possible that one's belief can be impermissible even if she is blameless for having that belief. This is similar to the ethical truth that someone might do the wrong thing, but not be blameworthy for doing so. For instance, say that it is very bad to push a particular button—doing so will cause a lot of bunnies to suffer intense pain. Jason has no reason to think that pushing the button will have this effect, plus someone he trusts has told him that pushing the button will cause a lot of bunnies to feel very happy. Jason pushes the button. Did he do something wrong? It is plausible that the answer to this question is "yes". Jason caused unnecessary suffering, which is wrong. Should we blame Jason for what he did? It seems equally plausible that the answer to this question is "no". After all, Jason was doing the best he could, and he did have good reasons for thinking that what he was doing was right, or at the very least not something that was wrong to do. The deontologist can plausibly claim that something similar happens with belief at times. There are times when someone believes wrongly, she violates her epistemic duties, but she is blameless for having the belief. Since the person believes wrongly, it seems that her belief is unjustified even though it is blameless.

Alston (1985, 1988) offers a second objection to the deontological conception of justification, which we might term the "involuntariness" objection. The gist of this objection is that in order to have a duty or obligation to do something it seems that we must be able to do that thing. As it is sometimes put in ethical discourses, *ought implies can*—you are obligated to do *X* only if you *can* do *X*. For instance, if you would need superhuman strength in order to save your neighbor's cat from a burning building, it seems that you are not obligated to save the cat because you cannot do it. Similarly, Alston claims that in order to have an obligation to believe or refrain from believing that *p* it has to be the case that you can so believe or refrain from believing. However, Alston points out that it seems we simply do not have this sort of voluntary control over our beliefs. Beliefs come to us involuntarily—we cannot simply choose by force of will what we will believe and what we will not believe. Since our beliefs are involuntary, Alston claims that we cannot be obligated to believe or refrain from believing things. But, if we lack such obligations, then the deontological view must be mistaken because it holds that justification is a matter of meeting or failing to meet one's epistemic obligations. Thus, Alston concludes that the deontological view of justification is mistaken because we lack voluntary control over what we believe.

There are at least two promising tactics that deontologists can make use of in responding to this second objection from Alston. The first is to argue, as Richard Feldman (1988) does, that one can have an obligation to believe that *p* (or refrain from believing that *p*) even if she lacks voluntary control over what she believes. Feldman points out that many obligations are such that they do not adhere to the "ought implies can" principle. As Feldman notes, when you take out a mortgage on a house you have a legal obligation to make your payments. This obligation does not go away if you become unable to make the payments. Many legal obligations are such that they hold even if you cannot meet them. Similarly, it seems that receiving credit for a particular course might require passing an exam with a score above a certain percentile (Steup 1996). The fact that Jeb cannot make a score above that percentile on the test does not mean that he will satisfy the course requirements because he is not obligated to do what he cannot. The obligation to score higher than the particular percentile remains; Jeb simply fails to meet his obligation in this case. Consequently, the deontologist may respond to Alston's objection by maintaining that one does not have to have voluntary control over one's beliefs in order to have obligations with respect to what one is to believe. Thus, one might maintain that the involuntariness objection is not a problem for the deontological view.

The second approach to Alston's objection involves following Matthias Steup (1996) in arguing that we do have sufficient voluntary control over our beliefs to have obligations regarding what we are to believe. According to Steup, while it is true that we cannot simply choose what to believe through force of will, we do have some voluntary control over our beliefs. In many cases it is within our voluntary control to deliberate more or less carefully over the evidence that we have. Making the decision to focus carefully on the evidence may lead to one having different beliefs even if once one has appreciated the evidence for *p* she cannot help but believe that *p*. So, one might respond to the involuntariness objection by arguing that we do have sufficient voluntary control in order to have the sort of obligations that the deontological view claims.

As we have seen the deontological view is an intuitively plausible view of the nature of justification with an impressive pedigree. Although there are serious objections to the deontological view, there are also promising responses to those objections. In light of these facts, the deontological view seems to remain very much a live option. Of course, the debate over the proper way to understand the normative aspect of justification is still raging.[13] Fortunately, our goal is to simply better understand justification and its role in knowledge—a goal for which it is enough to appreciate the basic features of this debate and some of the things that can be said about what seems to be the historically dominant view.

5.3 The Structure of Justification

The next debate concerns the structure of justification. This is the oldest of the three debates that we will consider. In fact, it is likely one of the oldest debates about justification, period, as it originates in ancient times in the work of Aristotle and Sextus Empiricus. The question of the structure of justification arises from consideration of the regress of reasons. Although there are numerous ways to express this regress, perhaps the simplest is to think of it in terms of the sort of questioning that one might receive from a child.[14] From a very young age children begin asking "why?" The regress of reasons can be understood as a continual series of why-questions of the sort that a child might ask you. Assume that you justifiedly believe that the Bears won yesterday's game. Now, someone might ask you, "why do you believe that the Bears won yesterday's game?"—that is, someone might ask you what your reason(s) for believing this is. You might answer that you believe this because you read about the score of yesterday's game in the *New York Times*. Hence, you have given your reason for believing that the Bears won; you read that they won in the *New York Times*. But, now your interlocutor might ask "why do you believe what the *New York Times* says?" Plausibly, you will respond that the *New York Times* is a reliable source for this sort of information. But, of course, your interlocutor can ask, "why do you think the *New York Times* is a reliable source of this sort of information?" By now you can see that this process can continue

[13]One very prominent alternative to the deontological conception of justification that has arisen in recent years is the virtue approach. Virtue epistemologists claim that justification is normative, but not in the sense of having to do with duty or obligation. Instead, virtue epistemologists claim that justification is normative in the sense of being connected to particular excellences or cognitive virtues. Exploration of this rich area of contemporary epistemology would take us too far afield from our present purposes. The interested reader is encouraged to consult some of the recent literature on virtue epistemology, such as: Baehr (2011), Greco (2000, 2010), Kvanvig (1992), Montmarquet (1993), Pritchard (2005, 2010), Sosa (1991, 2007, 2009, 2011), and Zagzebski (1996).

[14]For a sampling of the many alternative presentations of the regress of reasons see BonJour (2010), DePaul (2014), Feldman (2003), Klein (2014a, b), Olsson (2014a, b), and Steup (1996).

indefinitely. You keep providing reasons, and your interlocutor keeps asking you for reasons in support of those reasons. It seems that each of your reasons in this chain must itself be reasonable for you to hold or else nothing in the chain is reasonable, including your initial belief that the Bears won yesterday's game. This is the regress of reasons.

The debate over the structure of justification concerns how we should respond to the regress of reasons. In particular, the challenge is to explain how this regress might play out in such a way that our original assumption, that you justifiedly believe that the Bears won yesterday's game, is vindicated. Of course, one option is that there is no way to do this, and so, your belief is not justified. This form of skepticism is highly implausible because it would imply that none of our beliefs whatsoever are justified because this regress can apply to each of them. So, this is to be our last resort—only to be accepted if no other option is at all plausible. What are our other options for responding to the regress? We can sort the classic responses to the regress by asking a pair of questions: Do the reasons form a circle? Namely, is it the case that something like the following happens: your reason for p is q, your reason for q is r, and your reason for r is p? If the reasons do not form a circle, do they eventually end? (Turri 2014). This gives us three possibilities: *coherentism*—the reasons form a circle (as we will see below this is a somewhat misleading characterization of coherentism), *foundationalism*—the reasons do not form a circle, and they eventually end, and *infinitism*—the reasons do not form a circle, and they do not end.

Coherentism and foundationalism have long been the primary contenders with foundationalism being the dominant choice for much of the history of thought on the regress of reasons (BonJour 2010 and Kvanvig 2014). In fact, it seems that the consensus with respect to the regress is that coherentism and foundationalism are the only real contenders. D.M. Armstrong (1973, p. 155) claims that infinitism, like the skeptical response, is "a *desperate* solution, to be considered only if all others are clearly seen to be unsatisfactory." Laurence BonJour (2010, p. 179) goes so far as to question whether infinitism is really even logically possible. There is a strong tendency both historically and in contemporary discussions to regard infinitism as not really an option.[15] In light of this, we will limit our focus to the two primary approaches, coherentism and foundationalism, as well as hybrid approaches that seek to combine elements from these two leading responses. We will begin with foundationalism because as the dominant response to the regress historically it is easier to understand other responses as reactions to it.

[15] Fumerton and Hasan (2010) suggest that Peter Klein seems to be the only supporter of infinitism. Whether Klein is the only supporter or not, it is clear that he is its most prominent supporter and that he has attempted to defend it on numerous occasions (1999, 2005, 2007, 2014a, b). In spite of Klein's efforts, the consensus remains that infinitism faces devastating objections which render it an option that is only worth taking seriously if both foundationalism and coherentism are first *definitively* shown to be flawed, which they have not been. For a sampling of the many objections to infinitism see Audi (1988, 1993, 2011), BonJour (2010), Dancy (1985), Ginet (2005a, b), Post (1980), and Steup (1996).

5.3.1 Foundationalism

It is an understatement to say that foundationalism is simply the more dominant of two primary contenders. As Richard Fumerton and Ali Hasan (2010) say, "it is surely fair to suggest that for literally thousands of years the foundationalist's thesis was taken to be almost trivially true." As we saw above, this general foundationalist thesis is that the regress of reasons stops. A helpful way to get the general picture is to think of the structure of justification as analogous to the structure of a building. Just as a building has many parts that rest on a foundation, foundationalists hold that our beliefs and the reasons for those beliefs ultimately all rest on a foundation.[16] The foundation of our justification consists of *basic beliefs*. Basic beliefs are beliefs that are justified but do not receive their justification from any other beliefs. These basic beliefs are thought to receive their justification from something that provides justification without itself needing to be justified, e.g. perceptual experience (on most versions of foundationalism). Since basic beliefs receive their justification from something other than a belief, and this provider of justification does not itself admit of justification, basic beliefs are held to be regress stoppers. Once you get to a basic belief, such as, perhaps, your belief that you are having a particular sensory experience, the question of "why do you believe that you are having that particular sensory experience?" is simply answered with "because I am having it". That is to say, there is no further reason for you to point out aside from the experience itself. Foundationalists hold that it does not make sense to ask what reason you have for the experience because experiences are not the sort of thing for which one can have reasons. Despite the fact that one cannot have reasons for experiences, foundationalists hold that experiences can themselves provide reasons for basic beliefs. Consequently, the regress of reasons ends with basic beliefs. It is upon this foundation of basic beliefs that all of our other justified beliefs ultimately rest. Thus, the structure of justification is one of a foundation of basic beliefs with other justified beliefs being justified by these basic beliefs and by other beliefs which are themselves justified by basic beliefs.

Although all foundationalist views accept this general view of the structure of justification, there are various ways to spell out the foundationalist picture. Depending on what one allows as basic beliefs and how one allows justification to transmit from those basic beliefs to other beliefs, very different pictures of justification can arise. We will take a brief look at two of the major categories of foundationalist views.

A very extreme form of foundationalism, the sort that Descartes defended, limits basic beliefs to those about which we are infallible and requires that other beliefs be entailed by these basic beliefs in order to be justified. According to this classical form of foundationalism, the only basic beliefs are those that we cannot be wrong about. How many of our beliefs are like that though? It seems that there are few, if

[16]See DePaul (2014), Smithies (2014), Sosa (1980), and Steup (1996) for more on the building metaphor for foundationalism.

any. Consider your belief that you are seeing this page. Is it impossible that you are wrong about this? No. After all, you could be merely dreaming that you are seeing this page, or you could be hallucinating that there is a page. It seems that the only beliefs about which we might be infallible are beliefs about the contents of our own current mental states. For example, perhaps you are infallible when you believe that it seems to you that you are seeing this page. Even if we are infallible about such beliefs, and so they can be basic beliefs on this account, we still seem to be faced with two problems.[17]

First of all, it does not seem that we form beliefs about the contents of our own mental states very often (Feldman 2003). When you form the belief that there is a tree in the yard you typically just have a visual experience and form the belief that there is tree in the yard. You do not tend to first form the belief that it visually appears to you that there is a tree in the yard and then deduce from this belief that there is a tree in the yard. So, it seems that we simply do not form many of the basic beliefs that classical foundationalism requires for the beliefs we actually have to be justified.

Second, it is very hard to see how the beliefs that we normally form can be justified on the classical foundationalist picture. Assuming that we do form beliefs about the contents of our mental states, it is far from clear that such beliefs logically entail the many other beliefs that we commonsensically take to be justified. For instance, it does not seem that your belief that it seems to you that you are seeing this page entails that you are seeing this page. As we noted above, it could be that it seems to you that you are seeing this page while there is no page for you to see at all. This is the primary problem for classical foundationalism—it seems to yield the result that most all of our beliefs about the world outside of our own mental states are unjustified. This thoroughgoing skepticism is unpalatable. Thus, the majority of foundationalists opt for a less demanding form of foundationalism.[18]

These more moderate forms of foundationalism loosen the restrictions that classical foundationalism places on justification. First, moderate foundationalists allow that basic beliefs can transmit justification to other beliefs through non-deductive connections to those beliefs. Second, the foundation of basic beliefs is much broader in moderate views than in classical foundationalism. Moderate foundationalism allows that some basic beliefs can be about the external world,

[17]It is not clear that we really are infallible about such beliefs. Feldman (2003) and Williamson (2000) both offer arguments to the effect that we can be mistaken about the content of our own current mental states.

[18]This is not to say that there are no classical foundationalists. BonJour (2000, 2010), Fumerton (1995, 2000), and McGrew (1995) all defend versions of classical foundationalism. It is worth noting that each of their versions of foundationalism tends to weaken Descartes' requirement that justified beliefs be entailed by basic beliefs. Instead, they require the non-basic beliefs to be supported by basic beliefs in either a deductive or non-deductive fashion. Although this mitigates the threat of skepticism somewhat, each of these classical foundationalists acknowledges that their views face skeptical challenges that may be very difficult to overcome.

not just our own minds. An example of this is that many forms of moderate foundationalism will allow that your belief that you see this page (or at least that you are seeing something with a particular shape and color) is a basic belief. By allowing a broader foundation of basic beliefs (often perceptual beliefs, memorial beliefs, and introspective beliefs are included), moderate foundationalism seems to avoid both of the problems listed above for classical foundationalism. We actually form the beliefs that moderate foundationalism claims are basic. Further, it seems that moderate foundationalism has a much better chance of adequately responding to the threat of skepticism than classical foundationalism.

Although moderate foundationalism seems superior to classical foundationalism, it is not without its challenges. The primary challenge to moderate foundationalism is to account for how it is that basic beliefs are themselves justified. What is it about a visual experience, of this page say, that justifies a particular belief? Often this sort of problem is put in the form of a dilemma.[19] Either the experiences that justify basic beliefs have the same sort of content as a belief or they do not. If they do have the same sort of content, then it seems that they themselves admit of being justified or unjustified; and so, they seem to require justification. This, of course, would mean that they are not fitting stopping places for the regress. If these experiences do not have the same sort of content as belief, then while it seems that they cannot be justified or unjustified themselves, it is mysterious how they can justify beliefs.

Moderate foundationalists give various responses to this challenge. Some simply argue that if the dilemma is a problem, then it is a problem for all non-skeptical responses to the regress (Steup 1996). Others argue that this challenge to foundationalism makes a levels confusion. More precisely, the challenge conflates what is required for justified beliefs about the world with what is required for justified beliefs about which of our beliefs about the world are justified (Pryor 2001). We do not need to settle the issue here. It is enough for our purposes to recognize that there is a challenge and that moderate foundationalists have ways that they may attempt to respond.[20]

5.3.2 *Coherentism*

Foundationalism is the most widely held view of the structure of justification both historically and currently. It is not with out rivals, however. Chief among foundationalism's rivals is coherentism. Coherentism arises out of dissatisfaction with foundationalism (Kvanvig 2014). The sort of dilemma that we mentioned above leads some to believe that the basic beliefs posited by foundationalism are arbitrary (Kvanvig 2014). Others reject the idea that anything other than a belief

---

[19] BonJour (1978, 1985), Sellars (1956), and Klein (1999, 2005, 2007) all put forward versions of this sort of dilemma.

[20] For additional responses to this sort of worry see Smithies (2014).

can justify a belief—they simply deny that it makes sense to think that there can be basic beliefs (Davidson 1986). Finally, there is the fact that it just seems intuitive that when your beliefs cohere with one another they are more justified than when they do not.

As we mentioned above, coherentism offers a unique response to the regress. The coherentist response does not require the positing of basic beliefs or infinite regresses of reasons. Instead, it seems that coherentism responds to the regress of reasons by claiming that our reasons form a circle (or more accurately, that our reasons are not a matter of linear progression). On its face, this response is highly implausible. After all, how can circular reasoning lead to justified beliefs? This is one of the reasons that many find coherentism untenable.

Coherentists are quick to point out, however, that they are not suggesting that circular reasoning yields justification. In fact, they claim that it is only because of a mistaken assumption that one might construe coherentism as responding to the regress in this fashion. The mistaken assumption is that justification is a linear affair where a single belief, *p*, justifies another belief, *q*, and *q* justifies *r*, and so on. It is this picture of justification that gives rise to the regress in the first place. Coherentists maintain that instead of this linear picture we should understand justification holistically. Beliefs are justified because of how they fit with other beliefs—our beliefs form a web, rather than a line or chain, and it is a belief's placement in this web and the number of connecting strands within this web that yields its justification.[21] Once we recognize that beliefs are justified in sets or systems rather than individually coherentists claim the regress does not arise.[22]

Coherentism avoids the regress by denying the assumption that justification proceeds in a linear fashion from one belief to the next. It avoids the challenge facing moderate foundationalism by denying that there are basic beliefs, and instead holding that beliefs are justified only by other beliefs through a relation of coherence. So far things are looking good. However, coherentism faces challenges of its own. The first major obstacle for coherentism is to spell out exactly what it means for beliefs to cohere with one another. One early attempt at making this coherence relation precise holds that coherence is a matter of two things: consistency and logical entailment (Ewing 1934). The idea is that in order for a set of beliefs to be coherent, and so for the individual beliefs that make up the set to be justified, the set must be consistent (it must be possible for all of the beliefs to be true at the same time) and each member of the set must be logically entailed by the conjunction of the other members. Schematically, the set $\{p, q, r\}$ is coherent just in case it is possible that *p*, *q*, and *r* are all true and *p* & *q* entail *r*, *p* & *r* entail *q*, and *q* & *r* entail *p*.

[21]The web metaphor comes from Quine and Ullian (1970). Another common metaphor used to describe the coherentist approach is that of a raft (Neurath 1932; Sosa 1980).

[22]This way of responding to the regress seems to have been first suggested by Bosanquet (1920).

Although this early account of coherence is admirably clear, it is problematic. It seems that there can be sets of beliefs that are intuitively coherent (and justified), but are such that the individual beliefs in the set are not entailed by the conjunction of the other members. For example, consider the following set of beliefs[23]:

(A) "Rob was at the location of the robbery"
(B) "Rob owns the same kind of gun as that used by the robber"
(C) "Rob deposited a large amount of money in the bank the day after the robbery"

It seems that {A,B,C} is clearly coherent, however, this set fails to satisfy the requirement that each member be entailed by the other members. After all, A does not logically follow from B & C, B does not logically follow from A & C, and C does not logically follow from A & B.

The account of coherence developed by C.I. Lewis (1946) improves upon the one just described. Lewis' account requires consistency, but instead of logical entailment it only requires that each belief be probabilistically supported by the other beliefs in the set. So, on this account of coherence A does not have to be entailed by B & C, instead it simply must be the case that A is more probable given the truth of B & C (similar considerations apply to B and C). This seems to avoid the problem just described for the entailment view of coherence. In spite of this, there is an additional problem that remains for both accounts of coherence.

It seems that requiring consistency is too strong. It seems that your beliefs can be coherent, and justified, despite the fact that there is some inconsistency lurking among the members of the set. This seems particularly true in cases where you cannot see the inconsistency. An excellent example of this was present in the actual thinking of the great philosopher and mathematician, Gottlob Frege. After much careful study and reflection, Frege posed basic laws of arithmetic. He was justified in believing each of these proposed laws and presumably his beliefs were coherent. However, he did not realize that the laws were inconsistent until Bertrand Russell wrote him a letter explaining the inconsistency.[24] It is quite plausible that among the vast number of beliefs that we have some members of the set are inconsistent. Does this mean that all of our beliefs fail to be coherent? And, so on coherentism they fail to be justified? It seems not. Thus, it seems that consistency is too strong of a requirement for coherence.[25]

A final attempt to spell out the coherence relation that we will consider is in terms of explanatory relations. The idea here is that beliefs cohere with one another by explaining or being explained by one another.[26] This approach seems to avoid the problems of the previous two, and it does seem initially plausible. But, it faces its own difficulty. If coherence, and hence justification, is increased by the quality

[23]This example is borrowed from Olsson (2014a).

[24]For more on this example see Kornblith (1989) and Steup (1996).

[25]For more on this see Kvanvig (2012).

[26]This sort of view has been suggested as at least a component of coherence by BonJour (1985), Harman (1973), Lehrer (2000), Lycan (1988), Poston (2014), and Thagard (2000).

of the explanatory picture, this approach needs to be supplemented with an account of what makes one explanation better than another (Steup 1996). While this is not a trivial task by any means, it is one that can be accomplished. In fact, it is one that we will take steps toward accomplishing in chapters nine and ten.

It seems that coherentism has some hope of overcoming its first major obstacle by spelling out what coherence is. It faces additional problems though. Primary among these are what are known as the "alternative systems" and "isolation" objections. Although these objections are different in important ways, they expose the same general problem for coherentism—it seems that coherentism does not require enough of a connection to the world around you for justification. Let us take a closer look at these objections beginning with the alternative systems objections.

The alternative systems objection challenges whether coherence is related to truth. Essentially, the idea is that for any set of beliefs that someone has she could have a completely different set of beliefs that exhibits just as much coherence as the first set. It seems that there are indefinitely many sets like this. What is especially worrisome is that the coherence of these sets of beliefs has nothing to do with how the world around the person is. This objection makes it seem that coherence is not connected to the truth. After all, an indefinite number of vastly different sets of beliefs can each be equally coherent, and so given coherentism, equally justified. However, it is implausible that all of these vastly different sets of beliefs are equally likely to be true especially since their coherence has nothing to do with the world around the believer. Justification is supposed to help us to achieve our aim of believing the truth, but it is not clear that coherence of this sort is connected with truth in a way that helps with this aim. This objection challenges whether coherence can really be the correct way to understand justification.

The isolation objection again presses the issue of coherence not being sufficient for justification. According to this objection, it is possible that you have a very coherent, and so given coherentism well justified, set of beliefs even though your beliefs do not take account of the information your experiences provide about the world around you. Richard Feldman (2003, p. 68) offers an excellent example to illustrate this objection:

> Professor Feldman is a rather short philosophy professor with a keen interest in basketball. Magic Johnson (MJ) was an outstanding professional basketball player. While playing a game, we may suppose, MJ had a fully coherent system of beliefs. Magic Feldman (MF) is a possible, though unusual, character, who is a combination of the professor and the basketball player. MF has a remarkable imagination, so remarkable that while actually teaching a philosophy class, he thinks he is playing basketball. Indeed, he has *exactly* the beliefs MJ has. Because MJ's belief system was coherent, MF's belief system is also coherent.

The problem is that according to coherentism, it seems that Magic Feldman's beliefs are justified because they are coherent, but intuitively this is mistaken. The problem is that Magic Feldman's beliefs are isolated from the world around him—they are not at all responsive to the experiences that he is having while teaching class. In light of this, the isolation objection is a serious challenge to the idea that coherence is the correct way to understand justification.

Coherentists often try to resolve these problems by assigning a special role to beliefs that are close to experiences. The idea is that beliefs caused directly by experiences are special in some way or, alternatively, some beliefs have some initial credibility all on their own.[27] While this may avoid these problems by giving experience a more central role to play in coherentism, it becomes unclear whether these theories are really coherentist (Olsson 2014b). After all, this seems to be granting at least some beliefs something very similar to the status of the foundationalists' basic beliefs. For now, we will simply note that there seems to be serious problems for coherentism that can be avoided, but only at the price of giving up some of the "purity" of the coherence theory. As we will soon see, this is not necessarily a bad thing though.[28]

The final major obstacle to coherentism comes from recent formal work in probabilistic measures of coherence. The problem for coherentism is that two distinct impossibility theorems have been proven, and they seem to imply that no measure of coherence is truth conducive (Bovens and Hartmann 2003; Olsson 2005). Basically, these theorems state that it does not matter how we try to probabilistically measure coherence it will not be the case that more coherence amounts to a higher probability of truth. This seems to pose a very serious threat to the hope of coherence being sufficiently connected with the truth to provide an account of justification.

Thus far, coherentists have taken one of two approaches when responding to the proofs for these impossibility theorems (Olsson 2014b). The first approach is to argue that the proofs for these theorems are not successful because they rely on dubious premises or even that the entire project of expressing coherence in a probabilistic framework is a mistake.[29] The second approach is to accept that these impossibility results have been proven, but argue that this does not show that coherence is not a valuable epistemic property. Philosophers who take this approach argue that even if coherence is not truth conducive, it is still linked indirectly to confirmation (Dietrich and Moretti 2005), or it can be useful in inference to the best explanation by helping us to make comparative evaluations of explanations (Glass 2007), or it may still be that higher coherence entails a higher probability that the sources of information providing the elements in the coherent set are reliable (Olsson and Schubert 2007), or coherence is connected to the truth in a weaker, but still important sense (Angere 2007, 2008). Fortunately, we do not need to settle these issues; it is enough to recognize that coherentism faces a challenge from these impossibility theorems and that there are live options for responding to this challenge.

---

[27] BonJour (1985), Lewis (1946), Lycan (1988, 2012), Poston (2014), Rescher (1973), and Thagard (2000) take this sort of approach.

[28] Olsson (2014b) and Steup (1996) have both noted that various paradigm coherentist theories such as BonJour's (1985) and Lehrer's (1974, 2000) may be compatible with foundationalism.

[29] See Douven and Meijs (2007), Huemer (2007, 2011), and Schupbach (2008) for responses claiming that the theorems rely on dubious premises. See Thagard (2000, 2005) for arguments for thinking that expressing coherence in a probabilistic framework is simply a mistake.

5.3.3 Hybrid Responses

At this point we have seen that both foundationalism and coherentism have some intuitive plausibility. Nonetheless, we have also seen that both responses to the regress face challenges. Interestingly, we have seen that what has become the standard coherentist response to two of the classic objections to the theory, the alternative systems objection and the isolation objection, is to make the theory more similar to foundationalism (so similar, in fact, that some claim it makes the view really a kind of moderate foundationalism). It is worth noting that it is common for foundationalists to embrace coherentist elements in their theories too. Many foundationalists accept that coherence can increase the level of justification had by one's beliefs.[30] Some foundationalists even allow that coherence can increase the justification of basic beliefs (DePaul 2014). Perhaps by embracing some coherentist elements foundationalists can offer an explanation for how experiences provide justification for basic beliefs.

In light of these considerations it seems clear that "foundationalists and coherentists can mutually benefit from the respective virtues their theories have to offer" (Steup 1996, pp. 158). In fact, many philosophers have realized that not only can they benefit from this sort of cross-pollination; the best theory of justification is a combination of coherentism and foundationalism . At first this may seem to be impossible because foundationalism requires basic beliefs, and coherentism denies there are such a thing. Yet, if we understand the core claims of these views as Earl Conee (1988) suggests, the two are compatible. According to Conee, there are three core foundationalist claims:

1. Epistemic justification must be grounded in experience.
2. Sensory experiences can act as a constraint on justification without themselves having to be justified by beliefs.
3. Coherence among beliefs is insufficient for justification in the face of conflicting experiences.

Conee's take on foundationalism is slightly different than the standard presentation, however, it seems to capture the heart of moderate foundationalism. Experiences form a foundation that allows for the regress of reasons to be stopped.

Conee construes coherentism as consisting of two primary claims:

1. A sufficient level of explanatory coherence yields justification.
2. A system of beliefs can be justified by their coherence alone without the need for any of the beliefs in the system to be justified independently of its role in the system.

Again, it seems that Conee captures the heart of coherentism. Furthermore, there is no conflict between the core claims of foundationalism and coherentism.

---

[30] Chisholm (1966, 1977, 1989) in the various incarnations of his *Theory of Knowledge* accepts that coherence enhances justification; Audi (1988), Russell (1912a, b), and Steup (1996) agree.

These considerations suggest that perhaps the best theory of justification incorporates the core claims of both views. Such a view would hold that the regress of reasons is stopped by justifying experiences, but that in order for a belief to be justified it must be part of a coherent set of both beliefs and experiences. There are various ways that such a view could be spelled out.[31] As we will see in chapter ten, one of the most promising ways to spell out such a theory of justification is in terms of explanatory considerations that take both beliefs and experiences to be part of the explanatory set.

5.4 Internalism Versus Externalism

The final debate that we will explore in this chapter is the internalism versus externalism debate.[32] This debate concerns what sort of information can provide one with the good reasons required for justification. Very roughly, internalists hold that justification depends upon facts that can be known from a first-person perspective via reflection alone. Externalists deny this. Before delving into the details of this debate it is worth briefly mentioning how it relates to the previous two major debates we have discussed. Concerning the normativity debate, some philosophers claim that the deontological conception of justification provides strong support for internalism, but others deny this connection.[33] Concerning the debate over the structure of justification, there is a less controversial connection. All major coherentist theories are internalist while foundationalist theories come in both internalist and externalist varieties (the same can be said of hybrid theories). So, as is clear, there are some connections between this debate and the previous two, but for simplicity we can treat it as a fully autonomous debate here.

The first step in understanding this debate is to get clear about the range of internalist and externalist views. Before doing so, it is very important to recall the distinction drawn above between propositional justification and well-founded belief/doxastic justification. Internalist theories tend to be theories of propositional justification—most internalists agree that something beyond internal facts are required in order for a belief to be appropriately based on one's reasons.

---

[31]In addition to Conee (1988), see Haack (1993), McCain (2014), and Sosa (1980, 1991).

[32]As we noted in chapter four it is often an unfortunate fact that different philosophical debates are given the same name. There are at least two major internalism/externalism debates. The first, which we will be focusing on, concerns epistemic justification. The second concerns the nature of the content of our mental states. Roughly, this second debate centers on the question of whether one's external environment determines the content of her internal mental states. For more on the internalism/externalism debate about mental content see Burge (1986), Gertler (2012), Goldberg (2007), Kripke (1972), Lau and Deutsch (2014), and Putnam (1975).

[33]For discussion of this purported connection see Bergmann (2006), BonJour (1985), Ginet (1975), Goldman (1999), Kvanvig (2014), Poston (2008), and Vahid (2014).

Essentially, internalists think that propositional justification is an internal matter (we will see more about this very soon) and that propositional justification is only one necessary component of doxastic justification. Mistakenly taking internalism to be about well-founded belief/doxastic justification can lead to confusion, which at times leads to internalists and externalists talking past one another.[34]

Following James Pryor (2001) we can sort internalist and externalist views by first distinguishing between various forms of internalism. All internalist theories accept:

> Mentalism: Whether one is justified in believing *p* supervenes on facts which one is in a position to know about by reflection alone. (Pryor 2001, p. 104)[35]

To say that some set of facts, X, supervenes on some other set of facts, Y, is just to say that no two things can be different with respect to the X facts without being different with respect to the Y facts. So, Mentalism is the view that no two subjects can differ with respect to what is justified for them without differing with respect to what facts they are in a position to know by reflection alone. Alternatively, this may be expressed as "If any two possible individuals are exactly alike mentally, then they are alike justificationally, e.g., the same beliefs are justified for them to the same extent" (Feldman and Conee 2001, p.2).[36]

Some internalists add further requirements in addition to Mentalism. For example, some internalists are "access internalists" because they accept the following[37]:

> Access Internalism: One always has 'special access' to one's justificatory status. (Pryor 2001, p. 105)

It is important to be clear that Mentalism does not entail the stronger Access Internalism . Mentalism says that accessible facts, our mental states, form a supervenience basis for our justification. Access Internalism goes further by claiming that whether or not we are justified in believing a particular proposition is itself a fact that is accessible to us via reflection alone.

One additional requirement that is sometimes added to Mentalism is a weaker form of Access Internalism. This weaker, *Inferential Internalism*, only concerns inferential justification (justification that one belief receives via inference from other beliefs). According to Inferential Internalism, in order to be justified in believing that *p* on the basis of your belief that *q* you must have special access to the fact that *q*

[34] This mistake apparently occurs in Greco (2005).

[35] Pryor refers to this internalist thesis as "Simple Internalism", but we will follow Feldman and Conee (2001) in referring to this as "Mentalism" for two reasons. First, this makes it a bit easier to keep the thesis clear from other internalist theses. Second, "Mentalism" has become the much more common name in the philosophical literature.

[36] This is made somewhat more complicated by views that hold that knowledge is itself a mental state (Williamson 2000). For our purposes here we can set this complication aside.

[37] BonJour (1985) and Chisholm (1977) both seem to accept this stronger form of internalism.

is a good reason to believe that p.[38] Notice this is weaker than Access Internalism on the assumption that you can have justification for believing that p without inferring it from other beliefs, i.e. on the assumption of foundationalism.

With these various internalist theses in hand we can see that there is a spectrum of internalist views:

	Internalist views
Most extreme	1. Mentalism + access internalism
	2. Mentalism + inferential internalism
Least extreme	3. Mentalism[a]

[a]Of course, there are more fine-grained distinctions to be made here. For example, different versions of Inferential Internalism will require different kinds of access. One might require something very strong such as direct acquaintance with the fact that q makes p probable in order for q to justify believing that p (Fumerton1995), or one might require something much weaker such as a disposition for it to seem to oneself that p is part of the best explanation of one's evidence (McCain2014). We will set aside these finer grained distinctions for now

For the sake of simplicity, we can understand externalism as simply the denial of Mentalism and focus our discussion accordingly.

In what follows we will focus on a particular version of Mentalism, *Evidentialism* , and a particular version of externalism, *Reliabilism*, in order to make our discussion more concrete. The points we raise about these particular versions of internalism and externalism are generalizable to other versions.[39] We will begin with Evidentialism.

[38] Fumerton (1995) seems to accept a sophisticated version of this view.

[39]The externalist objections to Evidentialism may be leveled at any form of Mentalism. Since the other forms of internalism arise by adding additional constraints onto Mentalism, objections to Mentalism will generally apply to these other forms of internalism as well. The internalist objections to Reliabilism are generalizable to other forms of externalism. In particular this is because most of the prominent externalist theories that are not explicitly forms of Reliabilism contain reliabilist commitments. For example, proper functionalist theories like Plantinga (1993a) and Bergmann (2006), virtue epistemologies like Sosa (1991, 2007, 2009, 2011) and Greco (2000, 2010), and theories that include a safety condition (S's belief that p would not have easily been false) such as Sosa's view and Williamson (2000) or a sensitivity condition (if p were not true, s would not believe that p) such as Nozick (1981) and Roush (2005) all incorporate reliabilist elements. Objections to reliabilism constitute objections to the reliabilist components of these externalist theories.

5.4.1 Evidentialism

Although Evidentialism is more specific than Mentalism, it is in reality a family of mentalist views. Evidentialist theories share a commitment to Mentalism and to the idea that justification is a matter of the evidence that one has. More precisely, we may understand evidentialist theories as those which accept that:

> S is justified in believing that *p* at a particular time, *t* if and only if believing that *p* fits the evidence that S has at *t*[40]

On its face, Evidentialism seems exceedingly plausible. After all, it says that you are justified in believing things that fit the evidence you have. If your evidence supports believing that *p*, then you are justified in believing that *p*. If your evidence does not support believing that *p*, then you are not justified in believing that *p*. Evidentialism seems to be well suited to the traditional project of epistemology—determining for oneself what she ought to believe.[41] In light of its intuitive plausibility, it is not surprising that internalist views like Evidentialism were until fairly recently "assumed without question by virtually all philosophers who paid any serious attention to epistemological issues" (BonJour 2010, p. 203).[42] Rather than focus on the details of the positive case that can be made for Evidentialism we will explore some of the major objections that have been raised against this sort of view and what may be said in response.[43]

The first objection that is commonly raised against Evidentialism is that it over-intellectualizes justification. It seems clear that various unsophisticated thinkers, such as children, non-human animals, and generally unreflective adults, have a large number of justified beliefs. However, externalists claim that it does not seem that these unsophisticated thinkers have the sort of evidence that is required by Evidentialism. As a result, externalists maintain that Evidentialism cannot appropriately account for the justification had by unsophisticated thinkers.

This objection does seem to be a serious problem for views that accept Access Internalism. After all, it is very likely that unsophisticated thinkers will not have a concept of justification, and so they will fail to meet Access Internalism's requirement that they be in a position to know through reflection alone the justificatory status of their beliefs. Nonetheless, this objection does not seem to pose much of a problem for Evidentialism. The key to seeing this is to get clear on what Evidentialists mean by "evidence". While we will put off our in-depth discussion of this topic until the next chapter, it is worth noting now that typically evidence is

---

[40]This formulation of the basic evidentialist commitment is a slight modification of Feldman and Conee's (1985). Many of the variations among evidentialist theories come down to how they spell out this schema—there are different views on what evidence is, what it takes to have evidence, and what it means for a belief to fit one's evidence.

[41]See Poston (2008).

[42]Externalist theories of justification did not gain much prominence until roughly the 1970s.

[43]For positive support for Evidentialism see in particular Feldman and Conee (1985, 2001).

understood in terms of mental states, or the contents of mental states, that a subject has. So, saying that your pet Fido the dog has evidence that you are home consists of saying that Fido has mental states that give him good reason to think that you are home. What sort of mental states? Presumably, things like his mental state of having a visual experience of you, his olfactory experience of your particular scent, and so on. Insofar as it is reasonable to think that Fido is justified in believing that you are home, it is reasonable to think that he has this sort of evidence. Hence, it does not seem that Evidentialism faces much of a problem here.

The second common objection to Evidentialism is what we may call the "problem of forgotten evidence". Here is a recent expression of this objection:

> Years ago Ichabod formed a belief in proposition *q* by acquiring it in an entirely justified fashion. He had excellent evidence for believing it at that time (whether it was inferential or non-inferential evidence). After ten years pass, however, Ichabod has forgotten all of this evidence and not acquired any new evidence, either favorable or unfavorable. However, he continues to believe *q* strongly. Whenever he thinks about *q*, he (mentally) affirms its truth without hesitation. At noon today Ichabod's belief in *q* is still present, stored in his mind, although he is not actively thinking about it. I stipulate that none of his other beliefs confers adequate evidence for either believing *q* or for disbelieving it. (Goldman 2011b, p. 260)[44]

So, at noon today Ichabod believes that *q*, but he no longer has any evidence for (or against) it. Goldman (2011b, p. 260) adds, "Since Ichabod remembers *q*'s being the case, and since he originally had excellent evidence for *q*, which was never subsequently undermined, Ichabod's belief in *q* at noon today is justified".

At first glance this case might seem to pose a serious problem for Evidentialism. In fact, one might be tempted to agree with Goldman that if we deny Ichabod's belief is justified in this case, "there will be serious skeptical ramifications: people will fail to know a great many things that common sense credits them with knowing" (2011b, p. 260). The reason for this is that one might think that we cannot now recall our original evidence for a large number of our beliefs, and yet it seems that we are justified in believing those things. For instance, you may not be able to recall your original evidence for thinking that Columbus sailed in 1492, but it still seems that you are justified in believing that this is the year he sailed. Thus, one might be tempted to think that the problem of forgotten evidence is a serious threat to Evidentialism.

There are, however, plausible responses that Evidentialists can make to this objection. First, Earl Conee and Richard Feldman (2011) have recently responded to this objection by appealing to a particular kind of disposition. They argue that in the purported cases of forgotten evidence where it is clear that the person's belief that *p* remains justified it is plausible that she has a "disposition to recollect" that *p*. That is to say, in these cases the person is disposed to recall *p* as something she knows to be true. They argue that this disposition is evidence she has in support of *p*, so these cases pose no problem for evidentialist theories.[45]

---

[44] Also see Goldman (1999, 2009), Greco (2005, 2010, 2011), and Moon (2012a) for versions of this objection.

[45] For similar responses see Feldman (2005) as well as Feldman and Conee (2001). Also, see McCain (2015c) for further elaboration and defense of this sort of response.

A second way to respond to such cases is to question whether the subject really is justified. The idea here is that if we grant that these subjects have no evidence concerning the proposition in question, then it is far from clear that their beliefs are really like our stored beliefs. Consider a typical memory belief for which you have forgotten your original evidence, your belief that Columbus sailed in 1492 for instance. Quite plausibly you currently have a lot of evidence in support of this belief. You likely have some evidence that things you recall in the way you do the claim that Columbus sailed in 1492 are things you actually learned in the past, not things that you merely dreamed up. Additionally, it is likely that you have evidence which supports thinking that you are often right when it comes to basic historical facts. Finally, it is likely that you have meta-memories concerning your belief that Columbus sailed in 1492. That is to say, if someone asks you whether you know when Columbus sailed, it is likely that you would immediately answer affirmatively—even before you actually recall the particular date from memory. This sort of meta-memory is something that we commonly have for beliefs we have stored in our memory, and studies show that we tend to be reliable when we make judgments based on meta-memory.[46] So, even in cases where you cannot recall your original evidence for a particular belief you often have a large amount of evidence in support of that belief. In Goldman's case (and other cases used to press the problem of forgotten evidence), however, Ichabod lacks all of this evidence by stipulation. He has no evidence about how his memory works, no evidence about his track record when it comes to things like *q*, and no meta-memories concerning *q*. Ichabod's belief is clearly very different from the typical case where you have a belief for which you have forgotten your original evidence. It is so different that it seems reasonable to doubt the belief is justified at all. Thus, Evidentialists have at least two good ways of responding to this objection.[47]

The final objection to Evidentialism that we will consider here is that it cannot adequately respond to the threat of external world skepticism. Many externalists object to Evidentialism because they think that the sort of requirements Evidentialism places on justification makes it so that skeptical arguments, which claim we lack justification for a vast number of our beliefs, cannot be rebutted.[48] Fully responding to this objection would require sketching an Evidentialist response to skepticism, which would take us too far afield here. So, for present purposes we will simply note two important points. First, it is quite plausible that Evidentialism has the resources to rebut these skeptical arguments. In fact, we will explore in some detail how Evidentialism can be used to respond to both external world skepticism and skepticism about our inductive reasoning in chapters eleven and twelve, respectively. Second, it is worth pointing out that internalists often reply to this objection that responding to skepticism should not be overly easy, so

[46]See Cohen (1996) for further discussion of meta-memory and studies showing its reliability.

[47]For further arguments in support of this response to the problem of forgotten evidence and detailed discussion of the problem in general see McCain (2015b).

[48]This objection has been forcefully made by Greco (2000, 2005, 2010).

Evidentialism treats skepticism as it should be treated, as a legitimate philosophical issue. They often go on to charge externalists with taking the threat of skepticism too lightly. That is to say, internalists often respond to this sort of objection by arguing that externalist views do not treat skepticism as a legitimate philosophical problem, and thus, externalist views make justification and knowledge implausibly easy to get.[49]

We have seen that Evidentialism is intuitively plausible, and it is not adversely affected by the major objections that have been leveled against it. While our exploration is far from exhaustive, it is sufficient to establish that Evidentialism is a serious contender for the correct theory of justification and should be taken seriously. Also, by extension it shows that internalism is very much a live option when it comes to theories of justification. Now, let us take a look at the most prominent externalist theory on offer, Reliabilism.[50]

5.4.2 Reliabilism

Like Evidentialism, Reliabilism is more a family of theories than a particular theory. This is because there are a number of ways of spelling out the basic tenets of Reliabilism. However, rather than explore these particular incarnations individually we will look at the common core of Reliabilism and the problems it faces. The simplest way of understanding Reliabilism is in terms of what has been named "process reliabilism" (we will be referring to this as "Reliabilism" unless otherwise noted):

> S's belief that *p* is justified if and only if that belief is produced by a reliable cognitive process.[51]

Two points are worth mentioning immediately. First, Reliabilism, like many versions of externalism, is expressed in terms of justified belief (doxastic justification) instead of what is justified for one (propositional justification). This is different than Evidentialism, which is expressed in terms of propositional justification. But, this will not hinder our efforts here because the objections we will consider cannot be avoided by formulating Reliabilism in terms of propositional justification. Second, we should get clear about what is meant by a "reliable cognitive process". A cognitive process is a mental faculty of the believer. Basically, cognitive processes are information processing mechanisms within one's brain that take various mental

---

[49]For more on this charge see BonJour (2010), Cohen (2002), Fumerton (1995), Poston (2008), Vahid (2014), and Vogel (2000).

[50] Kvanvig (2014) even goes so far as claiming that Reliabilism is not only the most prominent externalist theory, but that it is currently the most popular of all theories of justification.

[51]This formulation is simplifying a great deal, but it is sufficient for our purposes. For more intricate versions of reliabilism see in particular Goldman (1979, 1986, 1988, 1992, 2008), Henderson and Horgan (2011), and Lyons (2009).

states of the believer—experiences, beliefs, desires, and so on—and maps those states to beliefs. For example, your visual experience of a red patch might be mapped via a cognitive process in your brain to the belief <there is something red in front of me>. So, you have the visual experience of a red patch, and this cognitive process results in your believing that there is something red in front of you. Although what it takes to be reliable is a bit vague, Reliabilists make it clear that while it does not require perfect accuracy, reliability does require more than mere chance. For our purposes we can understand a cognitive process to be reliable when it produces true beliefs the majority of the time.

Before considering a few of the major objections to Reliabilism it is worth noting its motivations. Reliabilism is motivated in large part by three considerations. The first is simply dissatisfaction with internalism because of some of the objections we explored above. The second is the desire to make the connection between justification and truth a close one.[52] The third is to provide a solution to the infamous Gettier problem for the traditional account of knowledge (a problem that will be the focus of our discussion in chapter eight). It is largely because of these motivations that Reliabilism has become a popular view in epistemology. Despite its popularity it is susceptible to some serious objections.

The first major objection to Reliabilism stems from the fact that it does not seem that reliability is necessary for justification. That is, it seems there can be cases in which your belief is justified even though it was not produced by a reliable cognitive process. To see this consider the fact that, there is a possible world in which "you have exactly the same experiences, apparent memories, and intuitions . . . go through exactly the same processes of reasoning, and form exactly the same beliefs" as you do in the actual world, but in that other possible world, your experiences are caused by the manipulations of a demon who is deceiving you (Wedgwood 2002, p. 349). This possibility is problematic for Reliabilism because intuitively you and your demon-world counterpart are equally justified in the beliefs you form; yet, only your beliefs are formed by reliable cognitive processes. Your beliefs are formed in the normal way, and so plausibly they are reliable. Your counterpart's beliefs, though, are formed by the tampering of a demon who is trying to deceive her—clearly an unreliable cognitive process. This "New Evil Demon Problem" demonstrates that, contra Reliabilism, reliability is not necessary for justification.[53]

Reliabilists have offered a number of responses to this objection. Some simply "bite the bullet" and claim that your counterpart's beliefs in the demon world are not justified, but yours are (Brewer 1997; Engel 1992). More plausible approaches which Reliabilists have taken seek to modify Reliabilism so that it yields the

[52]In fact, sometimes externalists object to internalism because it does not make the connection between justification and truth close enough. For discussion of this issue as well as internalist responses to the challenge see Cohen (1984), Lehrer and Cohen (1983), and Poston (2008).

[53]This problem is called the New Evil Demon problem in order to differentiate it from the older problem of external world skepticism, which is often put forward by appealing to the activity of a deceptive demon. See Cohen (1984) and Lehrer and Cohen (1983) for canonical discussions of this problem for Reliabilism.

intuitively correct result that both your beliefs and your counterpart's are justified. These various approaches seek to modify what is meant by "reliable" in various ways. For example, one early approach sought to relativize reliability to "normal worlds" (Goldman 1986). On this approach, whether a cognitive process counts as reliable depends on whether it is a process that is reliable in normal worlds. Normal worlds on this approach are worlds in which our general beliefs about the actual world are true. For example, normal worlds are worlds in which our perceptual experiences are good things to use when forming beliefs about the world around us. This approach was quickly abandoned, and many others were proposed.[54] These various approaches share the idea that reliability needs to be relativized in some sense in order to account for the New Evil Demon problem. It remains to be seen if any of these approaches can solve the problem.[55] It is enough for our purposes to note that Reliabilism faces a genuine problem here and that Reliabilists are attempting to find a solution.[56]

The second major objection to Reliabilism comes in the form of cases that intuitively show reliability is not sufficient for justification. In other words, these cases seem to show one's belief that *p* can be the result of a reliable cognitive process, and yet fail to be justified. Although several of these cases have been put forward, the most famous is the case of Norman the clairvoyant.[57] Here is Norman's story:

> Norman, under certain conditions that usually obtain, is a completely reliable clairvoyant with respect to certain kinds of subject matter. He possesses no evidence or reasons of any kind for or against the general possibility of such a cognitive power, or for or against the thesis that he possesses it. One day Norman comes to believe that the President is in New York City, though he has no evidence either for or against his belief. In fact the belief is true and results from his clairvoyant power, under circumstances in which it is completely reliable. (BonJour 1980, p. 62)

Intuitively, Norman's belief about the President's whereabouts is unjustified. This is a problem for Reliabilism because it is stipulated in the example that the cognitive process that produces this belief in Norman is very reliable. Thus, it seems that reliability is not sufficient for justification.

Like the previous objection there are a number of ways that Reliabilists can respond to cases like Norman the clairvoyant. Of course, one way to respond is to simply "bite the bullet" and maintain that Norman's belief about the President really is justified, but this is not very plausible. A better way of responding is to introduce

---

[54] Goldman (1988) himself argued that the normal worlds approach is flawed.

[55] For nice overviews of some of the various approaches to the New Evil Demon Problem and their merits see Goldman (2011a) and Littlejohn (2009).

[56] It is worth noting that Moon (2012b) attempts to show that internalist theories face their own version of the New Evil Demon problem. See McCain (2015a) for decisive reasons to think that Moon's attempt fails.

[57] Other influential versions of this sort of case can be found in Lehrer (2000) and Plantinga (1993b).

the sort of relativizing move that we mentioned above.[58] Perhaps the most promising response is to supplement Reliabilism with a "no-defeater" condition. This sort of condition would need to be something to the effect that one's justification is defeated when she has sufficiently strong evidence to the contrary of her belief or when she has an absence of supporting evidence altogether (Steup 1996). Adding such a "no-defeater" condition seems to allow Reliabilists to avoid the problematic consequences of cases like Norman the clairvoyant. Nevertheless, some internalists argue that this way of handling the problem really just sneaks in Evidentialism (Steup 1996). Once some such Evidentialist requirements are allowed it is not clear why the rest should not be adopted (BonJour 2006). This might be exactly what some Reliabilists want though. Recently, there has been a move by some Reliabilists to combine Evidentialism and Reliabilism into a single theory.[59] Perhaps such a synthesis is the best bet for resolving these sorts of problems for Reliabilism.

The final major objection to Reliabilism that we will consider is perhaps the most devastating, the Generality Problem. This problem threatens to undermine Reliabilism's claim to being a genuine theory of epistemic justification.[60] In order to understand this problem it is important to first draw a general distinction between *types* and *tokens*. Types are general kinds of things, and tokens are particular instances of those kinds. For example, your copy of this book is a token of a particular type. Your copy is an individual thing—it only exists right where you have the book. Alternatively, the type of this book exists wherever there are token copies of the book. Hence, all of the printed copies of this book are tokens of a single type. A similar thing is true of cognitive processes. There are token cognitive processes and types of cognitive processes. Recognizing this is the first step to appreciating the Generality Problem.

Recall from above that Reliabilism tells us that a belief is justified if and only if it is produced by a reliable cognitive process. The Generality Problem for Reliabilism is the problem of specifying what the relevant cognitive process is. Take Matt's belief that there is a tree in his yard, which he forms after taking a look outside. This belief is the product of a cognitive process that is a token of indefinitely many types. For example, the cognitive process that produces Matt's belief is a token of all of the following types: vision, vision in sunlight, forming beliefs about trees, forming beliefs about plants, forming beliefs on such and such day of the week, forming beliefs quickly, and so on. Some of these types are reliable, and some are not. In order for Reliabilism to give a genuine answer to the question of whether Matt's belief is justified it must have a principled way of identifying the relevant cognitive

[58] Bergmann (2006), Goldman (1992), and Sosa (1991) offer this sort of response.

[59] See Comesaña (2010) and Goldman (2011b).

[60] The Generality Problem for Reliabilism has been noted by several philosophers, including Alvin I. Goldman (1979) when he first formulated the version of Reliabilism that we have been discussing. The Generality Problem has been most carefully and forcefully put forward by Feldman (1985) and Conee and Feldman (1998). Other philosophers who recognized this problem early in the development of Reliabilism include Chisholm (1982) and Pollock (1984).

process type. It is only after the relevant cognitive process type has been identified that it can be determined whether that cognitive process is reliable or not. So, before Reliabilism can tell us anything about whether a particular belief is justified it must first provide a principled way of determining the relevant cognitive process type for evaluation. According to Laurence BonJour (2010, p. 215), "without some way of answering this question [what the relevant process type for evaluation is] in a specific and nonarbitrary way, the reliabilist has not succeeded in offering a definite position at all." Answering this question, providing a principled way of selecting the relevant cognitive process type, is the Generality Problem, and it is an extremely difficult problem.[61]

In fact, the Generality Problem is so difficult that "though reliabilists have struggled with this problem, no solution has yet been found that even a majority of reliabilists find acceptable" (BonJour 2010, p. 215). Even the most prominent defender of Reliabilism, Alvin I. Goldman, admits that the Generality Problem remains unsolved (2011a). The difficulty presented by this problem has led some Reliabilists to attempt to establish that this problem is shared by all theories of justification.[62] Establishing that every theory faces the Generality Problem would not solve the problem, but it would make it so that the Generality Problem does not provide a reason to abandon Reliabilism for some other theory of justification. Unfortunately for Reliabilists, it does not seem that this is a problem shared by Evidentialism.[63] As a result, it seems that Reliabilists may be sole proprietors in the business of needing to find a solution to the Generality Problem. Currently, the problem seems to be without a solution.

5.5 Conclusion

In this chapter we have explored the nature of justification. As we have seen there are a number of important distinctions to keep in mind when thinking about this component of knowledge. Additionally, we have seen that there are a number of live debates concerning various aspects of justification—fortunately, we do not need to settle these debates for our purposes. This being said, it will be helpful to adopt a particular epistemic framework for our discussions in the rest of this book. Given the results of our discussion of Evidentialism and Reliabilism, particularly Reliabilism's major unsolved problem, the most promising approach to justification seems to be Evidentialism. So, we will adopt this sort of approach throughout

[61] See Conee and Feldman (1998) for discussion of some of the more promising responses to this problem that have been given and why each is inadequate.

[62] See Comesaña (2006) and Bishop (2010).

[63] See Conee (2013) and Matheson (2015) for responses to Reliabilist arguments that all theories face the Generality Problem. Both Conee and Matheson convincingly argue that Evidentialism does not face this problem.

the remainder of our discussions. Our adoption of an Evidentialist framework is supported by three very good reasons. First, it will simply be helpful to work with a particular epistemic theory as a guide in our later discussions. Doing so will help us to orient on the relevant points of the discussions more easily. Second, as already noted it seems that Evidentialism is better suited to respond to the objections pressed against it than Reliabilism is for its objections. Third, even if one remains unconvinced that Evidentialism is the superior epistemic theory, most everyone agrees (Reliabilists included!) that when we are discussing the acceptance of scientific theories and our scientific knowledge we are concerned with the evidence that we have. We are not concerned so much with the reliability of the cognitive processes of the theorists who come up with a scientific theory. Instead, we are concerned with the evidence that the theorists have and the evidence that can be provided in support of their theories. So, even if one does not approve of Evidentialism as a general picture of the nature of justification, the sort of picture of justification that Evidentialism provides does seem to clearly be what we are concerned with when discussing scientific knowledge. In the next chapter we turn toward discussing two key components of Evidentialism that are especially relevant to our discussion of scientific knowledge: the nature of evidence and what it takes to be in possession of evidence.

References

Alston, W. P. (1985). Concepts of epistemic justification. *The Monist, 68*, 57–89.
Alston, W. P. (1988). The deontological conception of epistemic justification. *Philosophical Perspectives, 2*, 257–299.
Angere, S. (2007). The defeasible nature of coherentist justification. *Synthese, 157*, 321–335.
Angere, S. (2008). Coherence as a heuristic. *Mind, 117*, 1–26.
Armstrong, D. M. (1973). *Belief, truth and knowledge*. London: Cambridge University Press.
Audi, R. (1988). *Belief, justification, and knowledge*. Belmont: Wadsworth.
Audi, R. (1993). Contemporary foundationalism. In L. Pojman (Ed.), *The theory of knowledge: Classic and contemporary readings* (pp. 206–213). Belmont: Wadsworth.
Audi, R. (2011). *Epistemology: A contemporary introduction to the theory of knowledge* (3rd ed.). New York: Routledge.
Baehr, J. (2011). *The inquiring mind: On intellectual virtues and virtue epistemology*. Oxford: Oxford University Press.
Bergmann, M. (2006). *Justification without awareness: A defense of epistemic externalism*. New York: Oxford University Press.
Bishop, M. (2010). Why the generality problem is everybody's problem. *Philosophical Studies, 151*, 285–298.
BonJour, L. (1978). Can empirical knowledge have a foundation? *American Philosophical Quarterly, 15*, 1–13.
BonJour, L. (1980). Externalist theories of empirical knowledge. *Midwest Studies in Philosophy, 5*, 53–73.
BonJour, L. (1985). *The structure of empirical knowledge*. Cambridge, MA: Harvard University Press.
BonJour, L. (2000). Toward a defense of empirical foundationalism. In M. DePaul (Ed.), *Resurrecting old-fashioned foundationalism* (pp. 21–40). Lanham: Rowman & Littlefield.

BonJour, L. (2006). Replies. *Philosophical Studies, 131*, 743–759.
BonJour, L. (2010). *Epistemology: Classic problems and contemporary responses* (2nd ed.). Lanham: Rowman & Littlefield.
Bosanquet, B. (1920). *Implication and linear inference*. London: Macmillan.
Bovens, L., & Hartmann, S. (2003). *Bayesian epistemology*. Oxford: Clarendon Press.
Brewer, B. (1997). Foundations of perceptual knowledge. *American Philosophical Quarterly, 34*, 41–55.
Burge, T. (1986). Individualism and psychology. *Philosophical Review, 95*, 3–45.
Chisholm, R. M. (1956). Epistemic statements and the ethics of belief. *Philosophy and Phenomenological Research, 16*, 447–460.
Chisholm, R. M. (1966). *Theory of knowledge*. Englewood Cliffs: Prentice Hall.
Chisholm, R. M. (1977). *Theory of knowledge* (2nd ed.). Englewood Cliffs: Prentice Hall.
Chisholm, R. M. (1982). *The foundations of knowing*. Minneapolis: University of Minnesota Press.
Chisholm, R. M. (1989). *Theory of knowledge* (3rd ed.). Englewood Cliffs: Prentice Hall.
Cohen, G. (1996). *Memory in the real world* (2nd ed.). East Sussex: Psychology Press.
Cohen, S. (1984). Justification and truth. *Philosophical Studies, 46*, 279–295.
Cohen, S. (2002). Basic knowledge and the problem of easy knowledge. *Philosophy and Phenomenological Research, 65*, 309–329.
Comesaña, J. (2006). A well-founded solution to the generality problem. *Philosophical Studies, 129*, 27–47.
Comesaña, J. (2010). Evidentialist reliabilism. *Nous, 44*, 571–600.
Conee, E. (1988). The basic nature of epistemic justification. *Monist, 71*, 389–404.
Conee, E. (2013). The specificity of the generality problem. *Philosophical Studies, 163*, 751–762.
Conee, E., & Feldman, R. (1998). The generality problem for reliabilism. *Philosophical Studies, 89*, 1–29.
Conee, E., & Feldman, R. (2004). Making sense of skepticism. In E. Conee & R. Feldman (Eds.), *Evidentialism* (pp. 277–306). New York: Oxford University Press.
Conee, E., & Feldman, R. (2011). Replies. In T. Dougherty (Ed.), *Evidentialism and its discontents* (pp. 428–501). New York: Oxford University Press.
Dancy, J. (1985). *Introduction to contemporary epistemology*. Oxford: Blackwell.
Davidson, D. (1986). A coherence theory of knowledge and truth. In E. Lepore (Ed.), *Truth and interpretation* (pp. 307–319). Oxford: Blackwell.
DePaul, M. (2014). Foundationalism. In S. Bernecker & D. Pritchard (Eds.), *The Routledge companion to epistemology* (pp. 235–244). New York: Routledge.
Descartes, R. (1641/1988). *Meditations on first philosophy*. In J. Cottingham, R. Stoothoff, & D. Murdoch (Trans.), *Descartes: Selected philosophical writings* (pp. 73–123). Cambridge, UK: Cambridge University Press.
Dietrich, F., & Moretti, L. (2005). On coherent sets and the transmission of confirmation. *Philosophy of Science, 72*, 403–424.
Douven, I., & Meijs, W. (2007). Measuring coherence. *Synthese, 156*, 405–425.
Engel, M. (1992). Personal and doxastic justification. *Philosophical Studies, 67*, 133–151.
Ewing, A. C. (1934). *Idealism: A critical survey*. London: Methuen.
Feldman, R. (1985). Reliability and justification. *Monist, 68*, 159–174.
Feldman, R. (1988). Epistemic obligations. *Philosophical Perspectives, 2*, 235–256.
Feldman, R. (2003). *Epistemology*. Upper Saddle River: Prentice Hall.
Feldman, R. (2005). Justification is internal. In M. Steup & E. Sosa (Eds.), *Contemporary debates in epistemology* (pp. 270–284). Malden: Blackwell.
Feldman, R., & Conee, E. (1985). Evidentialism. *Philosophical Studies, 45*, 15–34.
Feldman, R., & Conee, E. (2001). Internalism defended. *American Philosophical Quarterly, 38*, 1–18.
Fumerton, R. (1995). *Metaepistemology and skepticism*. Lanham: Rowman & Littlefield.
Fumerton, R. (2000). Classical foundationalism. In M. DePaul (Ed.), *Resurrecting old-fashioned foundationalism* (pp. 3–20). Lanham: Rowman & Littlefield.

Fumerton, R. & Hasan, A. (2010). Foundationalist theories of epistemic justification. In E. N. Zalta (Ed.), *The Stanford encyclopedia of philosophy* (Summer 2010 Edition). http://plato.stanford.edu/archives/sum2010/entries/justep-foundational/

Gertler, B. (2012). Understanding the internalism-externalism debate: What is the boundary of the thinker? *Philosophical Perspectives, 26*, 51–75.

Ginet, C. (1975). *Knowledge, perception, and memory*. Dordrecht: Reidel.

Ginet, C. (2005a). Infinitism is not the solution to the regress problem. In M. Steup & E. Sosa (Eds.), *Contemporary debates in epistemology* (pp. 140–148). Malden: Blackwell.

Ginet, C. (2005b). Reply to Klein. In M. Steup & E. Sosa (Eds.), *Contemporary debates in epistemology* (pp. 153–155). Malden: Blackwell.

Glass, D. H. (2007). Coherence measures and inference to the best explanation. *Synthese, 157*, 257–296.

Goldberg, S. (Ed.). (2007). *Internalism and externalism in semantics and epistemology*. Oxford: Oxford University Press.

Goldman, A. I. (1979). What is justified belief? In G. Pappas (Ed.), *Justification and knowledge* (pp. 1–23). Dordrecht: Reidel.

Goldman, A. I. (1986). *Epistemology and cognition*. Cambridge, MA: Harvard University Press.

Goldman, A. I. (1988). Strong and weak justification. *Philosophical Perspectives, 2*, 51–69.

Goldman, A. I. (1992). Epistemic folkways and scientific epistemology. In A. I. Goldman (Ed.), *Liaisons: Philosophy meets the cognitive and social sciences* (pp. 155–175). Cambridge, MA: MIT Press.

Goldman, A. I. (1999). Internalism exposed. *Journal of Philosophy, 96*, 271–293.

Goldman, A. I. (2008). Immediate justification and process reliabilism. In Q. Smith (Ed.), *Epistemology: New essays* (pp. 63–82). New York: Oxford University Press.

Goldman, A. I. (2009). Internalism, externalism, and the architecture of justification. *Journal of Philosophy, 106*, 309–338.

Goldman, A. I. (2011a). Reliabilism. In E. N. Zalta (Ed.), *The Stanford encyclopedia of philosophy* (Spring 2011 Edition). http://plato.stanford.edu/archives/spr2011/entries/reliabilism/

Goldman, A. I. (2011b). Toward a synthesis of reliabilism and evidentialism? or: Evidentialism's troubles, reliabilism's rescue package. In T. Dougherty (Ed.), *Evidentialism and its discontents* (pp. 393–426). New York: Oxford University Press.

Greco, J. (2000). *Putting skeptics in their place: The nature of skeptical arguments and their role in philosophical inquiry*. Cambridge, UK: Cambridge University Press.

Greco, J. (2005). Justification is not internal. In M. Steup & E. Sosa (Eds.), *Contemporary debates in epistemology* (pp. 257–270). Malden: Blackwell.

Greco, J. (2010). *Achieving knowledge*. Cambridge, UK: Cambridge University Press.

Greco, J. (2011). Evidentialism about knowledge. In T. Dougherty (Ed.), *Evidentialism and its discontents* (pp. 167–178). New York: Oxford University Press.

Haack, S. (1993). *Evidence and inquiry*. Oxford: Blackwell.

Harman, G. (1973). *Thought*. Princeton: Princeton University Press.

Henderson, D. K., & Horgan, T. (2011). *The epistemological spectrum: At the interface of cognitive science and conceptual analysis*. Oxford: Oxford University Press.

Huemer, M. (2007). Weak Bayesian coherentism. *Synthese, 157*, 337–346.

Huemer, M. (2011). Does probability theory refute coherentism? *Journal of Philosophy, 108*, 35–54.

Klein, P. (1999). Human knowledge and the infinite regress of reasons. *Philosophical Perspectives, 13*, 297–325.

Klein, P. (2005). Infinitism is the solution to the epistemic regress problem. In M. Steup & E. Sosa (Eds.), *Contemporary debates in epistemology* (pp. 131–140). Malden: Blackwell.

Klein, P. (2007). Human knowledge and the infinite progress of reasoning. *Philosophical Studies, 134*, 1–17.

Klein, P. (2014a). Infinitism. In S. Bernecker & D. Pritchard (Eds.), *The Routledge companion to epistemology* (pp. 245–256). New York: Routledge.

Klein, P. (2014b). No end in sight. In R. Neta (Ed.), *Current controversies in epistemology* (pp. 95–115). New York: Routledge.
Kornblith, H. (1989). The unattainability of coherence. In J. W. Bender (Ed.), *The current state of the coherence theory: Critical essays on the epistemic theories of Keith Lehrer and Laurence BonJour, with replies* (pp. 207–214). Dordrecht: Kluwer.
Kripke, S. (1972). *Naming and necessity*. Oxford: Blackwell.
Kvanvig, J. (1992). *The intellectual virtues and the life of the mind: On the place of the virtues in contemporary epistemology*. Savage: Rowman & Littlefield.
Kvanvig, J. (2012). Coherentism and justified inconsistent beliefs: A solution. *Southern Journal of Philosophy, 50*, 21–41.
Kvanvig, J. (2014). Epistemic justification. In S. Bernecker & D. Pritchard (Eds.), *The Routledge companion to epistemology* (pp. 25–36). New York: Routledge.
Lau, J. & Deutsch, M. (2014). Externalism about mental content. In E. N. Zalta (Ed.), *The Stanford encyclopedia of philosophy* (Summer 2014 Edition). http://plato.stanford.edu/archives/sum2014/entries/content-externalism/
Lehrer, K. (1974). *Knowledge*. Oxford: Clarendon Press.
Lehrer, K. (2000). *Theory of knowledge* (2nd ed.). Boulder: Westview Press.
Lehrer, K., & Cohen, S. (1983). Justification, truth, and coherence. *Synthese, 55*, 191–207.
Lewis, C. I. (1946). *An analysis of knowledge and valuation*. LaSalle: Open Court.
Littlejohn, C. (2009). The new evil demon problem. *Internet Encyclopedia of Philosophy.*http://www.iep.utm.edu/evil-new/#SH2a
Locke, J. (1690/1975). *An essay concerning human understanding*. Oxford: Clarendon Press.
Lycan, W. G. (1988). *Judgement and justification*. Cambridge, UK: Cambridge University Press.
Lycan, W. G. (2012). Explanationist rebuttals (coherentism defended again). *Southern Journal of Philosophy, 50*, 5–20.
Lyons, J. (2009). *Perception and basic beliefs: Zombies, modules, and the problem of the external world*. New York: Oxford University Press.
Maffie, J. (1990). Recent work on naturalizing epistemology. *American Philosophical Quarterly, 27*, 281–293.
Matheson, J. (2015). Is there a well-founded solution to the generality problem? *Philosophical Studies, 172*, 459–468.
McCain, K. (2014). *Evidentialism and epistemic justification*. New York: Routledge.
McCain, K. (2015a). A new evil demon? No problem for moderate internalists. *Acta Analytica, 30*, 97–105.
McCain, K. (2015b). Is forgotten evidence a problem for evidentialism? *Southern Journal of Philosophy, 53*, 471–480.
McCain, K. (2015c). No knowledge without evidence. *Journal of Philosophical Research, 40*, 369–376.
McGrew, T. (1995). *The foundations of knowledge*. Lanham: Littlefield Adams Books.
Montmarquet, J. (1993). *Epistemic virtue and doxastic responsibility*. Lanham: Rowman & Littlefield.
Moon, A. (2012a). Knowing without evidence. *Mind, 121*, 309–331.
Moon, A. (2012b). Three forms of internalism and the new evil demon problem. *Episteme, 9*, 345–360.
Moser, P. K. (1989). *Knowledge and evidence*. Cambridge, UK: Cambridge University Press.
Neurath, O. (1932). Protokollsätze. *Erkenntnis, 3*, 204–214.
Nozick, R. (1981). *Philosophical explanations*. Cambridge, MA: Harvard University Press.
Olsson, E. (2005). *Against coherence: Truth, probability, and justification*. Oxford: Clarendon Press.
Olsson, E. (2014a). Coherentism. In S. Bernecker & D. Pritchard (Eds.), *The Routledge companion to epistemology* (pp. 257–267). New York: Routledge.
Olsson, E. (2014b). Coherentist theories of epistemic justification. In E. N. Zalta (Ed.), *The Stanford encyclopedia of philosophy* (Spring 2014 Edition). http://plato.stanford.edu/archives/spr2014/entries/justep-coherence/

Olsson, E., & Schubert, S. (2007). Reliability conducive measures of coherence. *Synthese, 157*, 297–308.
Plantinga, A. (1993a). *Warrant and proper function*. New York: Oxford University Press.
Plantinga, A. (1993b). *Warrant: The current debate*. New York: Oxford University Press.
Pollock, J. (1984). Reliability and justified belief. *Canadian Journal of Philosophy, 14*, 103–114.
Pollock, J. (1986). *Contemporary theories of knowledge*. Totowa: Rowman & Littlefield.
Post, J. (1980). Infinite regress of justification and of explanation. *Philosophical Studies, 38*, 32–37.
Poston, T. (2008). Internalism and externalism in epistemology. *Internet Encyclopedia of Philosophy*.http://www.iep.utm.edu/int-ext/
Poston, T. (2014). *Reason & explanation: A defense of explanatory coherentism*. New York: Palgrave-MacMillan.
Pritchard, D. (2005). *Epistemic luck*. Oxford: Oxford University Press.
Pritchard, D. (2010). Knowledge and understanding. In D. Pritchard, A. Millar, & A. Haddock (Eds.), *The nature and value of knowledge: Three investigations* (pp. 5–88). Oxford: Oxford University Press.
Pryor, J. (2001). Highlights of recent epistemology. *British Journal for the Philosophy of Science, 52*, 95–124.
Putnam, H. (1975). *Philosophical papers, vol. 2: Mind, language, and reality*. Cambridge, UK: Cambridge University Press.
Quine, W. V. O., & Ullian, J. S. (1970). *The web of belief*. New York: Random House.
Rescher, N. (1973). *The coherence theory of truth*. Oxford: Oxford University Press.
Roush, S. (2005). *Tracking truth: Knowledge, evidence, and science*. Oxford: Oxford University Press.
Russell, B. (1912a). *The problems of philosophy*. London: Williams and Norgate.
Russell, B. (1912b/2001). Truth and falsehood. In M. P. Lynch (Ed.), *The nature of truth: Classic and contemporary perspectives* (pp. 17–24). Cambridge, MA: MIT Press.
Schupbach, J. N. (2008). On the alleged impossibility of Bayesian coherentism. *Philosophical Studies, 141*, 323–331.
Sellars, W. (1956). Empiricism and the philosophy of mind. *Minnesota Studies in the Philosophy of Science, 1*, 239–252.
Smithies, D. (2014). Can foundationalism solve the regress problem? In R. Neta (Ed.), *Current controversies in epistemology* (pp. 73–94). New York: Routledge.
Sosa, E. (1980). The raft and the pyramid. *Midwest Studies in Philosophy, 5*, 3–25.
Sosa, E. (1991). *Knowledge in perspective*. Cambridge, UK: Cambridge University Press.
Sosa, E. (2007). *A virtue epistemology: Apt belief and reflective knowledge* (Vol. I). Oxford: Oxford University Press.
Sosa, E. (2009). *Reflective knowledge: Apt belief and reflective knowledge* (Vol. I). Oxford: Oxford University Press.
Sosa, E. (2011). *Knowing full well*. Princeton: Princeton University Press.
Steup, M. (1996). *An introduction to contemporary epistemology*. Upper Saddle River: Prentice Hall.
Thagard, P. (2000). *Coherence in thought and action*. Cambridge, MA: MIT Press.
Thagard, P. (2005). Testimony, credibility, and explanatory coherence. *Erkenntnis, 63*, 295–316.
Toulmin, S. E. (2003). *The uses of argument*. Cambridge, UK: Cambridge University Press.
Turri, J. (2014). *Epistemology: A guide*. Malden: Wiley Blackwell.
Vahid, H. (2014). Externalism/internalism. In S. Bernecker & D. Pritchard (Eds.), *The Routledge companion to epistemology* (pp. 144–155). New York: Routledge.
Vogel, J. (2000). Reliabilism leveled. *Journal of Philosophy, 97*, 602–623.
Wedgwood, R. (2002). Internalism explained. *Philosophy and Phenomenological Research, 65*, 349–369.
Williamson, T. (2000). *Knowledge and its limits*. Oxford: Oxford University Press.
Zagzebski, L. (1996). *Virtues of the mind: An inquiry into the nature of virtue and the ethical foundations of knowledge*. Cambridge, UK: Cambridge University Press.

Chapter 6
Evidence

Abstract This chapter explores two central issues when it comes to the concept of evidence. The first issue concerns the nature of evidence itself as it pertains to justification and knowledge. There are two primary theories of the nature of evidence. The first claims that evidence consists of non-factive mental states, and the second claims that evidence consists of propositions. This chapter explains both of these theories and considers some of the major challenges facing each. The second issue this chapter explores is that of what it takes for someone to have an item of information as evidence. After all, one cannot have knowledge on the basis of evidence that one does not possess. The extreme views of evidence possession each have serious problems, however, moderate views face challenges too. After elucidating some of the challenges facing the various views, this chapter demonstrates that there are some promising ways of providing a moderate account of what it takes to have evidence.

In the previous chapter we examined the issue of the nature of justification. Despite the various disagreements about justification, even the disagreement about whether the justification of all beliefs requires one to have evidence, all sides of these disagreements accept that scientific knowledge is grounded in evidence. So, in order for a scientific claim to be justified for us we have to have evidence in support of that claim. This prompts us to ask two very important questions. What exactly *is* evidence? When do we *have* something as evidence? We will explore answers to both of these questions in this chapter.

6.1 What Is Evidence?

At times we might cite things like United Nations reports as evidence that a particular country possesses chemical weapons. We might say that Dorothy's recent study habits are evidence that she has gotten serious about her education. We might say of a weapon with the defendant's fingerprints on it presented at a trial that it is evidence of his guilt. We might say that the darkening sky and the increased wind are evidence that it will rain soon. We might claim that the perihelion precession of Mercury is evidence that supports General Relativity over Newtonian mechanics.

K. McCain, *The Nature of Scientific Knowledge*, Springer Undergraduate Texts in Philosophy, DOI 10.1007/978-3-319-33405-9_6

Although each of the senses of "evidence" is important, none of them are evidence that pertains directly to the justification of our beliefs. The reason for this is that all of these things might exist, but you may fail to have them as evidence. In fact, some of them might exist without anyone having them as evidence. For example, it could be that Dorothy's study habits are a sign that she is serious about her education, but if you are not aware that she has those habits, *you* do not have evidence that she is serious about her education. In order for any of these items to be evidence in the sense relevant for justification you have to be aware of them in some way. The mere fact that Mercury has a perihelion precession does not provide you with evidence in support of General Relativity over Newtonian mechanics. If you are completely unaware of this fact about Mercury or of how this fact supports General Relativity and not Newtonian mechanics, then you do not have this evidence in support of General Relativity. In order to be evidence for you to accept some claim as true the item in question must be something that you can use in forming your doxastic attitudes (your beliefs, disbeliefs, suspensions of judgment, and perhaps degrees of belief). The sorts of things mentioned above are not themselves something that you can use in forming doxastic attitudes.

But, what is it that you can use when forming your doxastic attitudes? *Reasons*—the very thing we noted in the previous chapter as required for your beliefs to be justified. These reasons can be good or bad. For example, consider two situations in which you believe that there is a bush in your yard. In the first situation, you have a visual experience as of a bush in your yard in normal viewing conditions. However, in the second situation you have no such visual experience. In this second situation you believe there is a bush in your yard simply because you really want one to be there. You have *a reason* for your belief that there is a bush in your yard in both situations, but you only have a *good reason* in the first situation. It is only in the first situation that you have a reason for your belief that is indicative of the truth of the claim that there is a bush in your yard. In terms of evidence, we would say that you only have evidence in the sense relevant for justification in the first situation. So, we now have an approximate answer to our question for this section: *evidence is good reasons that are indicative of the truth concerning the proposition that is the object of your doxastic attitude.*[1] In other words, you have evidence for believing that *p* when you have good reasons that indicate that *p* is true. This is some progress, but it is not enough because it is not yet clear what good reasons are.

---

[1]See Conee (2004), Conee and Feldman (2008), and Kim (1988) for expressions of this view of evidence. As we noted in chapter five, the reason we need to restrict evidence to good reasons that are indicative of truth is that it is possible to have good reasons for adopting a particular doxastic attitude toward a proposition that are not related to the truth of the proposition. For example, in cases of dismal medical prognosis it might be that believing that one will recover can help ease one's suffering. In such a case it may be that the person has good reason to believe that she will recover—after all, doing so eases her suffering. However, the prognosis indicates that it is very unlikely that this person will recover. As a result, the good reason that she has for thinking that she will recover has nothing to do with the actual likelihood of recovery. Such reasons are not evidence. One's evidence in such a case consists of the prognosis, which gives good reason to think that recovery is unlikely.

There are two main categories of views of the nature of the kind of good reasons that constitute evidence: *psychologism* and *propositionalism*. Psychologism is the view that evidence consists solely of non-factive mental states or events.[2] Non-factive mental states are mental states that you can be in even if those states are misleading. For instance, seeing that there is a bush in your yard is a factive mental state because you can only be in the mental state of *seeing* that there is a bush in your yard when there actually is a bush in your yard. However, *seeming to see* that there is a bush in your yard is a non-factive mental state. You can seem to see that there is a bush in your yard even though you are mistaken. Even if you were only hallucinating and there were no bush, you could still be in the non-factive mental state of seeming to see a bush in your yard.

Of course, psychologism does not entail that every non-factive mental state you have is part of your evidence. Imagining a bush in your yard and desiring a bush in your yard are both non-factive mental states directed toward there being a bush in your yard, but they are not evidence that there is such a bush. The non-factive mental states that count as evidence are those that represent the world as actually being a certain way—things like beliefs, introspective experiences, perceptual experiences, memorial experiences, and perhaps intuitions and rational insights are evidence. One way to understand the distinction between non-factive mental states that are evidence and non-factive mental states that are not is that only those that are evidence have what James Pryor (2000, p. 547) calls "phenomenal force". That is, these mental states represent certain content to us in such a way that it "feels as if" the content is actually true. For instance, Diane's imagining that there is a pink elephant in the room is not the sort of non-factive mental state that is evidence because it lacks this kind of phenomenal force. Diane's visual experience of a tree is the sort of non-factive mental state that is evidence though.

Propositionalism is the view that evidence consists of propositions, not mental states.[3] Although propositionalism holds that mental states are not themselves evidence, they accept that the way we have evidence is by having certain mental states. Very roughly, according to propositionalism, propositions are evidence, and we have that evidence by having various mental states which have those propositions as part of their representational content. In terms of what it takes for us to have evidence, psychologism and propositionalism agree that we have to have particular mental states in order to have evidence. Hence, the disagreement between these two views comes down to a disagreement about what evidence is, not what it takes to have evidence. We will compare these views below.

[2]Littlejohn (2012) provides this name for the view. Turri (2009) refers to this as "statism", while Kelly (2008) and Williamson (2000) both refer to it as the "phenomenal conception of evidence". Brueckner (2009), Chisholm (1977), Cohen (1984), Conee and Feldman (1985, 2004, 2008, 2011), McCain (2014), Pollock (1974), and Turri (2009) all endorse psychologism.

[3]Propositionalists are not a unified group. Some, such as Dancy (2000), Hyman (1999), Kvanvig (2007), Littlejohn (2011, 2012, 2013), Schroeder (2008), and Williamson (2000), think that only true propositions can be evidence. Others, such as Arnold (2011), Comesaña and McGrath (2014), and Rizzieri (2011), argue that false propositions can be evidence too.

6.1.1 Psychologism Versus Propositionalism

It will be helpful for us to begin by considering why we might be inclined to accept psychologism. One reason is that the view seems intuitively plausible. It seems that our experiences provide us with evidence. Your experience of being tired is evidence for you that you are tired. Your experience of feeling sad is evidence for you that you are sad. Your experience of a bush's looking green is evidence for you that the bush is green.[4] And so on. Earl Conee and Richard Feldman (2008, p. 87) put the general point this way, "Experience is our point of interaction with the world—conscious awareness is how we gain whatever evidence we have". Now, this is not to say that psychologism entails that *only* experiences are evidence. Psychologism allows that other non-factive mental states such as beliefs are evidence too. Nonetheless, consideration of the role that experiences play in providing us with evidence provides some reason in support of psychologism.

Another reason that we might be inclined to accept psychologism is because of John Turri's (2009, p. 504) "master argument". According to the master argument, when we want to understand your evidence for believing that *p*, it is necessary to be aware of both the particular mental states that you have and how these mental states are related to one another. Turri claims that this information is also sufficient for understanding your evidence for believing that *p*. To clarify, Turri says, "Consider Barry. Barry has an ordinary visual experience as of a bear in his yard, which in conjunction with his habit of taking experience at face value causes him to believe that there's a bear out there. That description allows you to understand Barry's reasons" (2009, p. 504). Psychologism offers us a very good explanation of why being aware of your having certain mental states related in particular ways is necessary and sufficient for understanding your evidence. Your evidence just is your non-factive mental states. Turri maintains that propositionalism unduly complicates things. Propositionalism introduces something beyond what is necessary and sufficient for explaining your reasons for believing that *p*—namely, it introduces the propositions that are the content of your mental states. Turri claims that since all we need is information about your mental states, adding propositions to the story makes for an unnecessarily complicated explanation. Thus, psychologism provides the superior explanation of your reasons for believing that *p*.

Given these considerations in favor of psychologism, we might question why anyone would prefer propositionalism as a theory of the nature of evidence. There are two main reasons that might make one inclined to accept propositionalism over psychologism. The first of these reasons is that some worry psychologism will commit us to a thoroughgoing skepticism. According to Thomas Kelly (2008, p. 945), "if our evidence consists exclusively of non-factive mental states, it is far from clear that it provides us with an epistemic foothold in the world sufficient to

[4]See Conee and Feldman (2008) and Dougherty and Rysiew (2013) for additional examples of this sort.

underwrite the knowledge that we ordinarily take ourselves to have." Kelly goes on to point out that historically psychologism "has consistently aroused suspicions that it ultimately leads to skepticism" (2008, p. 945). The reason for these suspicions is that some question whether a connection strong enough for us to have justification for believing propositions about the external world can be established between our non-factive mental states and the state of the external world.

Although the threat of skepticism is a real issue for psychologism, supporters of this view of evidence are not without plausible responses. First, they may point out that skepticism about the external world is a significant philosophical problem. They may reasonably claim that it would in fact be a mark against a theory of evidence's plausibility if that theory simply did away with this problem by defining evidence in a particular manner. As we saw in chapter five, it is exactly this sort of dismissal of the skeptical problem that leads some internalists to argue that externalist theories of justification are unacceptable.

Second, it is far from clear that psychologism makes responding to external world skepticism impossible. For example, the response to external world skepticism that we will consider in chapter eleven, the *Explanationist Response*, is consistent with psychologism. As we will see in that chapter, this response to external world skepticism is very plausible. Furthermore, there are additional ways of responding to external world skepticism that are consistent with psychologism as well. One such way would be to adopt Reliabilism about justification and argue that our non-factive mental states are reliably connected to the external world. It is consistent with psychologism to claim that when you have the sort of visual experience you typically associate with seeing a bush in your yard, the belief that there is a bush in your yard will be true more often than not. Hence, one could plausibly accept psychologism and claim that the cognitive process that takes your visual experience of a bush as an input and yields your belief that there is a bush as an output is a reliable process. In light of these various responses, it does not seem that the worry that psychologism inevitably leads to external world skepticism is well founded.

The second reason that some might be inclined to accept propositionalism instead of psychologism is that they find Timothy Williamson's (2000) arguments concerning these views persuasive. Williamson offers three arguments in favor of propositionalism over psychologism. Although the arguments are distinct, they have the same basic form:

1. All evidence has property X.
2. Only propositions can have property X.
3. Therefore, all evidence is propositions.[5]

Williamson's first argument appeals to the nature of explanations. Specifically, Williamson says, "evidence is the kind of thing which hypotheses explain. But

[5]This presentation of the general structure of Williamson's arguments is similar to Dougherty's (2011).

the kind of thing which hypotheses explain is propositional" (2000, p. 195). So, "property X" in our skeleton of Williamson's argument above should be understood as "the property of being explained by hypotheses". In support of his argument Williamson maintains, "we may seek to explain why Kosovo rather than Bosnia was peaceful in 1995. The evidence in question would be the propositions that Kosovo was peaceful in 1995 and that Bosnia was not" (2000, p. 195). He goes on to claim that when we explain a particular event we "explain why it occurred, or had some distinctive feature", not the event itself (2000, p. 195).

In his second argument, Williamson appeals to probabilistic reasoning. He asserts, "what has probability is a proposition; the probability is the probability that" (2000, p. 196). Here "property X" should be understood as "the property of having probability". Williamson elaborates that "what gives probability must also receive it", so he claims that evidence is the sort of thing that can give and receive probability (2000, p. 196). And, of course, according to Williamson, what makes something probable and what is made probable by other things must be a proposition.

Williamson relies on the notion of logical consistency in his third argument. He correctly notes, "our evidence sometimes rules out some hypotheses by being inconsistent with them . . . But only propositions can be inconsistent in the relevant sense" (2000, p. 196). Thus, in this final argument "property X" should be understood as "the property of being inconsistent with some hypotheses".

Although each of these arguments might initially seem promising, there are plausible responses on behalf of psychologism. Concerning the first argument, Williamson is correct that typically when we seek to explain an event we are seeking answers as to why the event occurred or why it had the features that it did. He is also correct when he claims that the answers to these questions will be expressed with propositions. Yet, it does not follow from these facts that it is a proposition that is explained rather than the event itself. As Earl Conee and Richard Feldman (2011, p. 322) point out, when explaining WWI one does not explain a proposition, "but the occurrence of the war that the proposition asserts to have occurred". Although our explanatory reasoning is represented by propositions, we are not explaining propositions—we are explaining events. Consider a similar practice that we have, the practice of describing events. When Dana describes a touchdown that occurred in Saturday's game to Randy, she will do so by using sentences that express propositions. She will say things like "the offense ran a pass play", "the receiver was all alone in the end zone", and so on. The fact that Dana describes the touchdown using propositions does not mean that she is actually describing a proposition. Clearly, Dana is describing the event of a particular touchdown occurring, not a proposition. Similarly, simply because an explanation is put in terms of propositions does not entail that what is explained must be a proposition. Thus, the second premise in Williamson's first argument is dubious.

In order for Williamson's argument to be convincing he needs to provide strong reasons to think that we only explain propositions, but he has not done this. In fact, it seems that the most reasonable view is that we use propositions to explain events/phenomena, but we do not explain propositions. Thus, Williamson's argument does not seem to provide a convincing case against psychologism. In light of this, we do not seem to have good reason to think that evidence construed as mental states rather than propositions cannot be the sort of thing that hypotheses explain.

Now, if it is plausible that we explain events by way of propositions, one might argue that something similar holds in cases of probabilistic and deductive reasoning. So, supporters of psychologism can reasonably maintain that even if our evidence consists of only non-factive mental states, we can still make sense of evidence standing in various probabilistic and logical relations to propositions. Probability relations can be understood in terms of the probability that one's evidence occurs or that it has a particular feature. Thus, it seems that psychologism can allow for all of the same probabilistic relations that propositionalism can. Plausibly, similar considerations apply in the case of deductive reasoning. When our evidence is inconsistent with a particular hypothesis it is because the proposition that our evidence exists or that it has a particular feature is inconsistent with the hypothesis. Consequently, the fact that evidence can be inconsistent with hypotheses does not seem to entail that evidence cannot be non-factive mental states. Therefore, supporters of psychologism seem to have solid responses to Williamson's second and third arguments too.[6]

At this point we have seen some of the considerations for and against both psychologism and propositionalism. Obviously, we have not seen enough here to settle the issue between these two theories of the nature of evidence.[7] Nonetheless, we have gained an understanding of the debate over these theories and some of the important considerations on both sides. For our purposes it is enough that we understand that evidence should be understood in the way that one or the other of these theories maintain. This is enough for us to recognize that the sense of evidence that is relevant for knowledge must be something mental—either mental states or propositions that are the content of mental states. With this understanding of the nature of evidence in hand we can now turn to our second question.

6.2 When Do We Have Evidence?

It will be helpful to start our discussion of having evidence by introducing some terminology. We can distinguish between "total possible evidence" and "total evidence".[8] Your total possible evidence at any give time is all and only the information that you have stored in your mind at that time. Hence, any information, conscious or unconscious, retrievable or irretrievable, that is in your mind at a time is part of your total possible evidence at that time. Although this includes quite a lot

[6]Conee and Feldman (2008) make similar points in response to Williamson's arguments.

[7]For more extensive discussion of these theories see McCain (2014).

[8]This distinction comes from Feldman (2004).

of information, it does rule out some things from being part of your total possible evidence. For instance, things that you used to remember, but now have completely forgotten, are not part of your total possible evidence now. Of course, things that you have never been made aware of are not part of your total possible evidence either. Even if other people are aware of these things, if you have not been made aware of them, they are not part of *your*total possible evidence. After all, the mere fact that some evidence exists for *p*, in the sense that it is evidence that someone has, does not entail that it is evidence that *you* have.[9]

Your total evidence is simply the portion of your total possible evidence that you possess in the sense that is relevant for justification. It is possible that you have something as part of your total possible evidence without having it as part of your total evidence. It seems plausible that information which is so deeply stored in your memory that you could only access it after years of intense therapy is not part of your total evidence before you have undergone the necessary therapy.[10]

Now that we have the terminology of total possible evidence and total evidence in hand, it is worth mentioning some plausible constraints on evidence possession. These constraints will help us to determine whether a given theory offers a plausible account of when an item of your total possible evidence is part of your total evidence.[11] The first constraint concerns psychological accessibility. According to this constraint, in order for x to be part of your total evidence you must have the right sort of psychological access to x.[12] This psychological access might require that you are currently conscious of x or that you are able to recall x in a suitable way—we will see more about these options as we consider accounts of evidence possession. Accordingly, any account of evidence possession that does not require you to have some psychological access to x in order for it to be part of your total evidence is unacceptable.

The second constraint concerns epistemic acceptability. According to this constraint, an acceptable account of evidence possession should not yield the result that we lack sufficient evidence for justification when it is clear that we are justified. Similarly, an acceptable account of evidence possession should not yield the result that we have sufficient evidence for justification when it is clear that we are unjustified. In other words, an acceptable account of evidence possession should yield the intuitively correct results concerning the evidence possessed in clear cases of justified/unjustified belief.

We can now employ this terminology and these constraints to evaluate various accounts of evidence possession.

---

[9]You may later learn of this evidence in which case it would become part of your total possible evidence. However, the point here is that at this time, before you have learned any of this information, this information is not part of your total possible evidence.

[10]See Feldman (2004) for more on this point.

[11]Feldman (2004) proposes these constraints.

[12]We should understand x to be neutral between psychologism and propositionalism. So, x might be a non-factive mental state, or x may be the propositional content of a mental state.

6.2.1 Extreme Views of Having Evidence

When it comes to accounts of evidence possession there are two extreme views that have been discussed in the literature. The first is extremely permissive, and the second is extremely restrictive. We will look at both of these and the challenges they face.

The extremely permissive view simply equates total evidence with total possible evidence. According to this view, any information that is stored in your mind is evidence that you have. Of course, this is too permissive, and it is not surprising that this view has no serious defenders. In order to see the problem with this account of evidence possession let us consider the following sort of example:

> Michael is a normal adult. He has many memories of his childhood. Some of these memories Michael can easily recall, and some he can only recall with very special prompting. He has a particularly painful memory—a memory of being forgotten and left at a store when he was a small child. The painful nature of this memory has made it so that it is very deeply stored in Michael's memory. It is possible that Michael could bring this memory to consciousness, but he could only do so after undergoing years of intense training and psychological therapy. Currently, Michael has not undergone any of this training or psychological therapy.[13]

It seems that in this case Michael's deeply stored memory of being left at the store is not evidence that he currently has. That is to say, Michael's memory is part of his total possible evidence, but not part of his total evidence at this time. After all, it would be implausible to think that if Michael suddenly believed he was left at the store when he was a small child, his belief would be justified. Alternatively, it seems that if Michael suddenly believed that he was not left at a store when he was a small child, that belief would not be justified either. It seems that Michael simply does not at the current time have evidence about this issue—at least not the sort of evidence that is relevant for justification (total evidence). This, however, means that the very permissive account of having evidence is mistaken. We cannot simply equate total evidence with total possible evidence. Total evidence requires more stringent constraints on psychological accessibility than simply being stored somewhere in one's memory.

Of course, the problems with the extremely permissive account of having evidence can be avoided by accepting an extremely restrictive account. One such extremely restrictive account, which at least at one time had a serious defender, claims that the only time x is part of your total evidence is when you are currently thinking of x.[14] On this view, no stored information at all is part of your total evidence. There are various problems with this view as well. It will be instructive to consider both an objection to this view and how one might try to respond to that objection.

---

[13]Feldman (2004) and McCain (2014) both use similar examples to argue against the extremely permissive account of having evidence.

[14]Feldman (2004) defended this account. However, it seems that he has since changed his view or at least become less sure of it (see Conee and Feldman (2008)).

One particularly strong objection to the extremely restrictive account comes from an example that Richard Feldman (2004, p. 221) describes:

> Suppose my friend Jones tells me that the hike up to Precarious Peak is not terribly strenuous or dangerous, that it is the sort of thing that I can do without undue difficulty. Assume that Jones knows my abilities with respect to these sorts of things and that he seems to be an honest person. On the basis of his testimony, I believe that the hike is something I can do. It seems that it is rational for me to believe this proposition. But suppose I've failed to think about the time Jones told me that I could paddle my canoe down Rapid River, something he knew to be far beyond my abilities. He just gets a kick out of sending people off on grueling expeditions. If you were to say to me, "Remember when Jones lied about the canoe trip?" I'd say "Yes! How could I have failed to think about that?"

We can all agree that after Feldman is reminded of Jones' past lie he should not trust Jones' testimony. Consequently, at that point it is clear that his total evidence does not justify him in thinking that he can hike up Precarious Peak. Feldman claims that before he is currently thinking of the fact that Jones has lied to him about this sort of thing in the past, he should believe that he can hike up Precarious Peak without much difficulty though. In other words, Feldman maintains that when he is not currently thinking of Jones' previous deceit he does not have evidence for doubting Jones' testimony. Of course, this is consistent with the extremely restrictive account of evidence possession that Feldman defends, but should we think this is correct?

It does not seem that we should accept Feldman's assessment of the case. In order to help see this let us take a look at a modified version of Feldman's example:

> Suppose that at 1pm Feldman recalls the time that Jones lied to him about the difficulty of canoeing down Rapid River, something Jones knew to be far beyond Feldman's abilities. Feldman also remembers that Jones just gets a kick out of sending people off on grueling expeditions. So, Feldman remembers that he should not trust Jones' testimony. At 1:01pm Feldman reads in his guidebook that the hike up to Precarious Peak is an extremely difficult hike. Feldman easily recognizes that this hike is something that would be nearly impossible for him to do. At 1:02pm Feldman is eating lunch and no longer thinking about the times Jones lied to him or about what he just read concerning the hike to Precarious Peak. Of course, Feldman could very easily recall this information—all he has to do is think about whether he can make the hike. While Feldman is having lunch Jones tells him that the hike up to Precarious Peak is not terribly strenuous or dangerous, that it is the sort of thing that Feldman can do without undue difficulty. Feldman is aware that Jones knows Feldman's abilities with respect to these sorts of things and Jones is not acting particularly suspicious.[15]

We might reasonably maintain that Feldman has evidence in this case for thinking that he should not trust Jones' testimony about this hike. The reason we might think this is that it seems Feldman's total evidence does include things he is not currently thinking of. In particular, it seems that Feldman's total evidence includes the information that he just read in the guidebook a minute ago and his memory of Jones' past lies, which he recalled only two minutes ago. If this is

[15]This example is similar to one presented in McCain (2014).

correct, then the extremely restrictive view of evidence possession is mistaken. Thus, it seems that both extreme views of evidence possession face some serious challenges.[16]

6.2.2 *Moderate Views of Having Evidence*

As we have seen, both extreme accounts of evidence possession face some serious problems. In light of this, we might be well advised to seek out some more moderate view of what it takes to have evidence. Moderate views of having evidence hold that your total evidence is not simply all of your total possible evidence, but your total evidence is not restricted to only the things that you are currently thinking either. Instead, moderate views of evidence possession hold that both the things you are currently thinking of and some of the information that you have stored in memory constitute your total evidence at any given time.

Of course, there is a challenge here. Moderate views need some principled way of distinguishing between the memories that are accessible enough to count as part of your total evidence and those that are not. Feldman (2004, p. 232) challenges moderate views by pointing out "whether a person will think of some fact depends largely upon how the person is prompted or stimulated." According to Feldman, almost any memory can be recalled given suitable prompting. This would seem to make it very difficult to draw a principled distinction between memories that count as part of your total evidence now and those that do not.

At this point there has not been a lot of effort to give a precise moderate account of evidence possession. For the most part, epistemologists have seemed content to handle the issue of evidence possession on a case-by-case basis. Although this approach may be able to do the work that we need, it is not as intellectually satisfying as having a precise account of evidence possession. Fortunately, there is at least one promising moderate approach to settling when a memory is part of your total evidence and when it is not.[17]

This promising moderate account of evidence possession suggests that we should first rethink a basic assumption that is often taken for granted in discussions of what it takes to have evidence. Throughout our own discussion we have made this assumption by taking it for granted that having evidence is a two-place relation. We have simply been considering whether you have x as part of your total evidence. We might make important strides toward giving a successful moderate account of evidence possession by doing away with this assumption. Instead, we would understand your total evidence as relative to a topic of inquiry. This idea would

---

[16]For further discussion of the challenges facing these extreme views of having evidence see McCain (2014).

[17]For further elaboration and defense of this moderate account of evidence possession see McCain (2014).

make having evidence a three-place relation. In other words, when trying to figure out what total evidence you have at a particular time we would be asking whether *x* is part of your total evidence with respect to the truth of *p*.

This change in the way we approach the topic of evidence possession has some initial plausibility. It seems plausible that you might possess *x* as evidence for *p*, but not possess *x* as evidence for *q*. This is not simply to state the obvious—that *x* may be evidence in support of *p* while not supporting *q*. Instead, the idea here is that even if *x* supports *p*, and *x* also supports *q*, it could be that at a particular time you have *x* as evidence for *p*, but not for *q*. This might be the case because you have the appropriate sort of psychological access to *x* with respect to *p*, but not with respect to *q*. So, a first step toward providing an acceptable moderate account of evidence possession may be to change how we view evidence possession in general.

Now, simply understanding evidence possession as a three-place rather than a two-place relation is not enough to really get a moderate account of evidence possession off the ground. We still need some way of deciding when you have *x* available relative to a particular topic of inquiry. One plausible answer is that it comes down to what you can recall when reflecting on that topic. In other words, we might say that you have *x* as part of your total evidence with respect to *p* when you can recall *x* by merely thinking about the question of *p*'s truth.

We can see how such an account of evidence possession might work by considering another example of Feldman's. According to Feldman (2004, p. 232), "If I ask my childhood friend if he remembers the time we spray-painted my neighbor's dog, I may get an embarrassed 'Yes.' If I ask him if he remembers any of our childhood pranks, this one may fail to come to mind." Feldman goes on to claim that this case is a problem for moderate accounts of having evidence. He says, "Is the fact that we spray-painted the dog easily accessible? There seems to be no clear answer" (2004, p. 232). According to the moderate account of evidence possession we have been considering, Feldman's question is not an appropriate question to ask about evidence possession at all. We should instead ask things like: does Feldman's friend have his memory of spray-painting the dog available as evidence pertaining to the truth of whether he and Feldman spray-painted the dog? Does Feldman's friend have his memory of spray-painting the dog available as evidence pertaining to the truth of whether he and Feldman committed childhood pranks? In the first case, the answer seems to be "yes" because he does remember spray-painting the dog when reflecting on this topic. In the second instance the answer depends on whether or not this memory comes to mind when he reflects on the broader topic. If it does come to mind, then he has the memory as part of his total evidence with respect to whether he and Feldman committed childhood pranks. If it does not come to mind, then he does not have this memory as part of his total evidence with respect to that topic.

It is clear that more work is needed before this moderate account of evidence possession could be considered very precise. There are various questions that we still need answers to: Do you have to immediately think of *x* when you consider the topic, or is *x* part of your total evidence with respect to that topic even if it takes a while for you to recall *x*? How widely should we understand a topic of inquiry?

Is a topic of inquiry simply a proposition? Is it a general question? And so on. Importantly, for our purposes we do not need to develop an account with this level of precision. It is enough that we recognize that there is at least one promising way of developing a moderate account of what it takes to have evidence.

6.3 Conclusion

We have examined answers to two of the key questions about evidence—what it is and what it takes to have it. There are live debates concerning the correct answers to both of these questions. As we have seen, the two main approaches to the nature of evidence, psychologism and propositionalism, both face challenges. However, it is plausible that one or the other of these theories is correct. Also, we have seen that there are problems for extreme views of evidence possession, but moderate views face challenges of their own. Again, we have seen that there is hope in this debate. It seems that there is at least one initially promising way of developing a moderate account of evidence possession. Both of these questions are important for our understanding of the justification that evidence can provide for believing a proposition. Of course, this means that they are important for our understanding of scientific knowledge too. Fortunately, we do not need to settle the debates about the nature of evidence and its possession in order to understand how evidence can provide us with knowledge of scientific claims. It is enough that we understand the basics of these issues and some of the major moves in these debates.

References

Arnold, A. (2011). Some evidence is false. *Australasian Journal of Philosophy, 91*, 165–172.

Brueckner, A. (2009). E = K and perceptual knowledge. In P. Greenough & D. Pritchard (Eds.), *Williamson on knowledge* (pp. 5–11). Oxford: Oxford University Press.

Chisholm, R. M. (1977). *Theory of knowledge* (2nd ed.). Englewood Cliffs: Prentice Hall.

Cohen, S. (1984). Justification and truth. *Philosophical Studies, 46*, 279–295.

Comesaña, J., & McGrath, M. (2014). Having false reasons. In C. Littlejohn & J. Turri (Eds.), *Epistemic norms: New essays on action, belief, and assertion* (pp. 59–80). Oxford: Oxford University Press.

Conee, E. (2004). First things first. In E. Conee & R. Feldman (Eds.), *Evidentialism* (pp. 11–36). New York: Oxford University Press.

Conee, E., & Feldman, R. (2004). Making sense of skepticism. In E. Conee & R. Feldman (Eds.), *Evidentialism* (pp. 277–306). New York: Oxford University Press.

Conee, E., & Feldman, R. (2008). Evidence. In Q. Smith (Ed.), *Epistemology: New essays* (pp. 83–104). Oxford: Oxford University Press.

Conee, E., & Feldman, R. (2011). Replies. In T. Dougherty (Ed.), *Evidentialism and its discontents* (pp. 428–501). New York: Oxford University Press.

Dancy, J. (2000). *Practical reality*. Oxford: Oxford University Press.

Dougherty, T. (2011). In defense of propositionalism about evidence. In T. Dougherty (Ed.), *Evidentialism and its discontents* (pp. 226–235). New York: Oxford University Press.

Dougherty, T., & Rysiew, P. (2013). Experience first. In M. Steup, E. Sosa, & J. Turri (Eds.), *Contemporary debates in epistemology* (2nd ed., pp. 17–22). Malden: Wiley Blackwell.
Feldman, R. (2004). Having evidence. In E. Conee & R. Feldman (Eds.), *Evidentialism* (pp. 219–241). New York: Oxford University Press.
Feldman, R., & Conee, E. (1985). Evidentialism. *Philosophical Studies, 45*, 15–34.
Hyman, J. (1999). How knowledge works. *Philosophical Quarterly, 49*, 433–451.
Kelly, T. (2008). Evidence: Fundamental concepts and the phenomenal conception. *Philosophy Compass, 3*, 933–955.
Kim, J. (1988). What is naturalized epistemology? *Philosophical Perspectives, 2*, 381–405.
Kvanvig, J. (2007). Propositionalism and the metaphysics of experience. *Philosophical Issues, 17*, 165–178.
Littlejohn, C. (2011). Evidence and knowledge. *Erkenntnis, 74*, 241–262.
Littlejohn, C. (2012). *Justification and the truth-connection*. Cambridge, UK: Cambridge University Press.
Littlejohn, C. (2013). No evidence is false. *Acta Analytica, 28*, 145–159.
McCain, K. (2014). *Evidentialism and epistemic justification*. New York: Routledge.
Pollock, J. (1974). *Knowledge and justification*. Princeton: Princeton University Press.
Pryor, J. (2000). The skeptic and the dogmatist. *Nous, 34*, 517–549.
Rizzieri, A. (2011). Evidence does not equal knowledge. *Philosophical Studies, 153*, 235–242.
Schroeder, M. (2008). Having reasons. *Philosophical Studies, 139*, 57–71.
Turri, J. (2009). The ontology of epistemic reasons. *Nous, 43*, 490–512.
Williamson, T. (2000). *Knowledge and its limits*. Oxford: Oxford University Press.

Chapter 7
Basing a Belief on the Evidence

Abstract In this chapter the important distinction between having justification for believing a proposition (propositional justification) and justifiedly believing a proposition (doxastic justification) is drawn. Very roughly, this distinction tracks the idea that simply believing the right thing is not sufficient for one's belief to be justified. After all, one might believe the right thing for the wrong reasons. In order to have a justified belief one must believe the right thing for the right reasons. Since justified belief is a necessary condition of knowledge, it is extremely important to understand what is required to move from merely having propositional justification to having doxastic justification. This chapter explores the relation that one's belief has to bear to one's propositional justification in order to be doxastically justified—what epistemologists call the "basing relation". Accounts of the basing relation fall into three categories: causal accounts, doxastic accounts, and hybrid accounts. The general features of each of these kinds of accounts, as well as the challenges they face, are explored in this chapter.

In earlier chapters we discussed the fact that knowledge requires justified belief and in the previous chapter we noted that scientific claims require evidence to be justified. Although this is correct, we uncovered an extremely important distinction in Chap. 5. There we distinguished between being justified in believing a proposition (propositional justification) and having a well-founded belief/justifiedly believing that proposition (doxastic justification). This is very important because one might have propositional justification without having doxastic justification. For instance, it is possible that you possess really good evidence in support of *p*, and you believe that *p*, but instead of believing *p* because of your evidence you believe it simply because you really want *p* to be true. In such a case we would say that you believe the right thing, your evidence does support believing *p* after all, but you believe it for the wrong reasons. The problem is that your belief is not based on your evidence. Instead, you believe because of your wishful thinking. So, although *p* is justified for you (you have propositional justification), your belief that *p* is not justified (you lack doxastic justification). Since, as we have seen, knowledge requires justified true belief, you must have *doxastic justification* in order to have knowledge. Thus, in order to know scientific claims you must *base* your beliefs in those claims on evidence that on balance supports them.

K. McCain, *The Nature of Scientific Knowledge*, Springer Undergraduate Texts in Philosophy, DOI 10.1007/978-3-319-33405-9_7

Of course, we are now faced with an important question. What does it take for a belief to be based on evidence? There are three broad categories of answers to this question. Some claim that in order for you to believe on the basis of some evidence that evidence has to cause your belief in the appropriate way. Others claim that your believing on the basis of some evidence requires you to have additional beliefs about the relationship between your belief and your evidence. Still others claim that it is some combination of both of these views that accounts for your believing on the basis of your evidence. We will explore each of these kinds of accounts of the basing relation in this chapter.

7.1 Causal Accounts of the Basing Relation

By far the most prevalent view among epistemologists is that the basing relation is a causal relation.[1] That is to say, most epistemologists think that in order for your belief that *p* to be doxastically justified the evidence which propositionally justifies *p* for you has to cause your belief that *p*.[2] Some, such as Alvin Goldman (2011), go so far as to claim that "there is no hope of elucidating a suitable basing relation without giving it a causal interpretation". We can see why someone might be inclined to agree with Goldman. After all, when we claim of someone, Cindy say, that her belief that *p* is based on her evidence we will tend to say things like "Cindy believes *p* because she has evidence for it", "Cindy's belief that *p* is the result of her having good evidence", and so on. These sorts of statements seem to be expressing causal claims. It is not hard to see why one might think that when we claim that Cindy's belief is based on her evidence we mean that Cindy's evidence in support of *p*, rather than something like a desire that *p* be true, is the cause of her believing that *p*.

We might also be inclined to accept a casual account of the basing relation because of its close connection with explanation. As we noted, saying that Cindy's

[1] As we noted in Chap. 5, the debate between internalists and externalists is over the correct way to understand propositional justification. Causal relations are not themselves "internal" in the sense of internalism about epistemic justification—although mental states may bear causal relations to one another and various other things, facts about causation do not supervene upon one's mental states. This may lead some to erroneously think that various internalist theories of justification are in fact externalist because they accept a causal account of the basing relation. It is important to keep in mind that internalists are only committed to claiming that propositional justification, which is a necessary condition for doxastic justification, supervenes on internal/mental factors. Internalists can happily embrace a causal view of the basing relation while retaining their commitment to internalism about propositional justification. More generally, being an internalist or an externalist about epistemic justification does not commit one to a particular view of the basing relation.

[2] Korcz (1997, 2000) claims that understanding the basing relation as a causal relation is the "standard view" of the basing relation. Mittag (2002), Turri (2011), and Vahid (2009) all agree with Korcz that this is the standard view. McCain (2012, 2014), Moser (1989), Pollock and Cruz (1999), and Wedgwood (2006) all argue in support of causal accounts of the basing relation, but they do not take a stand on what view of the basing relation is standard.

belief that p is based on her evidence is in some sense claiming that she believes p because she possesses that evidence. However we understand the basing relation, it is exceedingly plausible that Cindy's belief is based on her evidence when her having that evidence explains why she has that belief. Many find it clear that the sense of explanation at work here is the same sort of causal explanation that we employ in our daily lives. Why did the glass break? Because it fell to the floor. Why did Bobby miss the party? Because he was at home studying all night. These events explain the glass's breaking and Bobby's absence from the party because they provide information about the causes for the glass breaking and Bobby's absence. In particular, the glass's falling and Bobby's staying home are *difference-makers*. These events make a difference to whether the glass breaks and to whether Bobby attends the party, respectively. We might think that something similar applies to the basing relation. Cindy's having some evidence explains why she has a particular belief when her having that evidence or failing to have that evidence makes a difference as to whether she has that belief. So, we might think that Cindy's having that evidence has to be a cause of her having the belief in order for her belief to be based on the evidence.[3]

Additionally, we might be inclined to accept a causal account of the basing relation because such accounts provide a simple explanation for why there is doxastic justification in some cases and not in others. When your evidence causes your belief (and that evidence propositionally justifies the belief), your belief is doxastically justified. When something else causes your belief, then your belief is not doxastically justified. This seems to track our intuitions pretty well. After all, why do we tend to think that in a case where an experimenter has strong evidence for a scientific claim, but her belief is caused by her strong desire that the claim is true rather than by her evidence, her belief is not justified? Causal accounts of the basing relation provide a straightforward answer: the experimenter's belief is not justified because her evidence does not cause her to have that belief.

In light of the simple explanation that causal accounts provide of the basing relation and the fact that such views are dominant in the literature, we might be inclined to wonder why there are other accounts of the basing relation at all. The reason that causal accounts are not universally accepted is primarily the result of two challenges that have been raised for such accounts of the basing relation. The first challenge is that causal accounts of the basing relation, like causal accounts of any phenomenon, must adequately deal with the possibility of deviant causal chains. The second challenge comes from what have come to be known as "Superstitious Lawyer" examples.[4] We will examine each of these challenges.

---

[3]Davidson (1980) and Harman (1973) argue that basing requires one's evidence/reasons to explain her belief/action. Turri (2011) offers an argument in support of causal accounts of the basing relation based on the notion of evidence being a difference-maker.

[4]These examples draw their name from Lehrer's (1971) original presentation of such an example, however, the name has been slightly modified in recent years to correct for the racist undertones of the original name.

The challenge that causal deviancy poses for causal accounts of the basing relation can be illustrated by considering an example put forward by Alvin Plantinga (1993, p. 69):

> Suddenly seeing Sylvia, I form the belief that I see her; as a result, I become rattled and drop my cup of tea, scalding my leg. I then form the belief that my leg hurts; but though the former belief is a (part) cause of the latter, it is not the case that I accept the latter on the evidential basis of the former.

The point of Plantinga's example is that although Plantinga's evidence from seeing Sylvia is a cause of his belief that his leg hurts, his belief is not based on that evidence. Intuitively, Plantinga's belief is based on the sensation he has of his leg scalding, his visual evidence is merely a link in a causal chain that produces his scalding sensation and, ultimately, his belief that his leg hurts. The causal chain that leads from Plantinga's experience of seeing Sylvia to his belief that his leg hurts is deviant in an important sense. That is, his seeing Sylvia does not cause his belief that his leg hurts in the right way for his belief to be based on that experience.

Cases of causal deviancy, like Plantinga's, demonstrate that causal accounts of the basing relation have to be able to rule out deviant causes from counting as the bases of one's beliefs. Of course, causal theorists have long recognized that cases like Plantinga's show that simplistic causal accounts of the basing relation cannot be correct. Some causal theorists have responded to the challenge of causal deviancy by simply admitting that it is a problem and adding a requirement to their theories that the evidence must non-deviantly cause one's belief in order for that belief to be based on the evidence.[5] While this might be a step in the right direction, it is not very helpful. Other causal theorists have tried to provide accounts of what yields the deviancy in cases like Plantinga's and to provide principled ways of ruling out such causal deviancy. We will not explore these various accounts here. Instead, we will note two important facts. First, at least some of these accounts are promising.[6] Second, most of our concepts that involve causation face a challenge from causal deviancy. Importantly, the fact that properly analyzing these concepts presents a challenge from causal deviancy does not show that the concepts are mistaken. As John Turri (2011, p. 390) explains:

> Doubtless a causal account of murder is correct. To murder someone you must cause his death. But that's not all. You must cause his death in the right way. You must intend to kill him, and your intention must appropriately figure into the causal explanation of his death. What does it mean for your intention to figure appropriately? The deviance problem strikes again... The same goes for a theory of perception. An object must cause you to have certain sensations for you to see it. But that's not all. It must cause your sensations in the right way.

Hence, the fact that causal accounts of the basing relation face a challenge with respect to causal deviancy does not necessarily show that such accounts are incorrect.

---

[5]See Moser (1989).

[6]See Korcz (2000), McCain (2012), and Turri (2011) for such responses to the challenge of causal deviancy.

The second challenge facing causal accounts of the basing relation comes from Superstitious Lawyer examples. Since Keith Lehrer (1971, pp. 311–12) provides the canonical version of this sort of example, it is worth considering his example in its entirety here:

> The example involves a lawyer who is defending a man accused of committing eight hideous murders. The murders are similar in character, in each case the victim is an Oxford philosophy student who has been choked to death with a copy of Philosophical Investigations. There is conclusive evidence that the lawyer's client is guilty of the first seven murders. Everyone, including the lawyer, is convinced that the man in question has committed all eight crimes, though the man himself says he is innocent of all.
>
> However, the lawyer is a gypsy with absolute faith in the cards. One evening he consults the cards about his case, and the cards tell him that his client is innocent of the eighth murder. He checks again, and the cards give the same answer. He becomes convinced that his client is innocent of one of the eight murders. As a result he studies the evidence with a different perspective as well as greater care, and he finds a very complicated though completely valid line of reasoning from the evidence to the conclusion that his client did not commit the eighth murder. (He could not have obtained an eighth copy of Philosophical Investigations.) This reasoning gives the lawyer knowledge. Though the reasoning does not increase his conviction—he was already completely convinced by the cards—it does give him knowledge. Moreover, he claims that it is this reasoning that gives him knowledge.[7]

Lehrer insists that in this case it is the lawyer's trust in the tarot cards that causally sustains the lawyer's belief that his client is innocent, not his newfound evidence. As he says, "in my example a man comes to believe something and continues to believe it because of groundless superstition" (1971, p. 311). Lehrer insists that the reasons that the lawyer discovers "do not potentially explain his belief, because he would not hold the belief for those reasons if he were to become doubtful of his superstitious reasons for belief" (1971, p. 311).

Lehrer's Superstitious Lawyer example is thought to be a counterexample to causal accounts of the basing relation because, as we have noted, in order for someone to know that *p* her belief that *p* must be a justified belief, and in order for her belief that *p* to be justified it must be based on the evidence which propositionally justifies *p* for her. Lehrer asserts that in his example the lawyer comes to know that his client is innocent, as we are well aware at this point, this means that his belief must be based on his evidence. Nevertheless, Lehrer stipulates in the example that the lawyer's evidence is not causally relevant to his having the belief that his client is innocent. Consequently, Lehrer's example is claimed to illustrate a case where someone has a belief that is based on her evidence, but her evidence does not cause her to have that belief. Thus, if Lehrer is correct, his example shows that one's believing on the basis of the evidence does not require the evidence to cause her belief.

[7]See Foley (1987), Lehrer (1974, 2000), Korcz (2000), and Kvanvig (2003) for variations of this example.

Many epistemologists have simply failed to find Lehrer's example convincing.[8] Causal theorists claim that it is not clear why we should think that the lawyer knows that his client is innocent in Lehrer's example. Admittedly, the lawyer has strong evidence for thinking that his client is innocent, however, causal theorists maintain it is far from clear that this is enough for the lawyer to know (or have a justified belief). Often, causal theorists note that it is perfectly consistent to think that the lawyer fails to adequately use his evidence in Lehrer's example; and so, fails to have knowledge (or justified belief). After all, we might think that the very intuition which motivates the necessity of a basing requirement for justified belief is that in cases like Lehrer's the person fails to have a justified belief. So, although it is possible that Superstitious Lawyer examples such as Lehrer's provide a challenge for causal theories, it is not clear that they are much of a challenge.

7.2 Doxastic Accounts of the Basing Relation

Even though causal accounts dominate the literature on the basing relation, and it is not clear that the challenges they face are insurmountable, some think that these challenges are sufficiently difficult that we should look elsewhere for an adequate account of the basing relation. Since doxastic accounts usually arise from dissatisfaction with the causal approach, it is not surprising that among supporters of such accounts are those who find Lehrer's assessment of the Superstitious Lawyer example (mostly) correct.[9]

Of course, it is not simply the Superstitious Lawyer example that provides a reason to take doxastic accounts of the basing relation seriously. An additional motivation for such views comes from the intuition that when someone bases her belief on the evidence this involves her taking account of the evidence in an appropriate way. Keith Allen Korcz (2000, p. 527) suggests, "I may properly take into account the epistemic import of a reason intentionally, by means of an appropriate meta-belief to the effect that the reason is a good reason to hold the belief." That is to say, one might think that it is possible to take account of the evidence in the appropriate way by satisfying the conditions of doxastic accounts of the basing relation. Doxastic accounts of the basing relation hold that what it means

[8] Goldman (1979), McCain (2012), Pollock and Cruz (1999), Swain (1981), Turri (2011), and Wedgwood (2006) all simply deny that the Superstitious Lawyer example shows what Lehrer claims it does. Audi (1983) and Mittag (2002) both provide extensive arguments against the effectiveness of Lehrer's example. McCain (2012), (2014) argues that while Kvanvig's (2003) variation of the Superstitious Lawyer example is superior to Lehrer's original, it too is ultimately ineffective.

[9] Supporters of doxastic views of the basing relation include Foley (1987), Lehrer (1971, 1974, 2000), Kvanvig (2003), and Tolliver (1982). Although Korcz (2000) defends the effectiveness of the Superstitious Lawyer example, he is not included among these supporters because he accepts a hybrid account of the basing relation rather than a pure doxastic account.

for you to base your belief that *p* on your evidence is for you to have an additional belief that your evidence provides you with good reason to believe *p*. This is exactly the sort of meta-belief that Korcz suggests can be the appropriate way of taking account of the evidence. Also, such an account can provide the result that Lehrer claims is correct in the Superstitious Lawyer example—it implies that the lawyer's belief is based on his evidence since he does have the meta-belief that his evidence provides good reason to think that his client is innocent. Accordingly, if we were convinced by Lehrer's assessment of the Superstitious Lawyer example and Korcz's suggestion of what it means to take appropriate account of the evidence, a doxastic account of the basing relation would seem to be a very good option for us.

Unfortunately, doxastic accounts of the basing relation face serious challenges. It is implausible that we form the meta-beliefs (beliefs about the quality of evidence and what it supports believing) required by doxastic accounts of the basing relation very often. Consider the many perceptual beliefs that you have right now or those that you form in an average day. You believe that the sky is blue today because it looks that way to you. You believe that the radio is on because you hear particular sounds. It seems unlikely that you tend to also form beliefs like "the visual experience that I am having is a good reason to think that the sky is blue today" or "the sounds I am hearing provide good reason to think that the radio is on". Nonetheless, it seems that you have very good evidence for believing the sky is blue today or that the radio is on—namely, the experiences that you have in these situations. Additionally, it seems clear that your beliefs are based on your evidence. After all, if you did not have the visual experience that you do, you would not believe that the sky is blue today, and if you did not hear the sounds you do, you would not believe that the radio is on. As a result, doxastic accounts of the basing relation face a major challenge. They require you to have the additional beliefs that your evidence supports believing the sky is blue today or the radio is on in order for your beliefs about the sky and the radio to be based on your evidence. However, you rarely (if ever) form these additional meta-beliefs. Yet, it seems fairly obvious that your beliefs in these sorts of cases are based on your evidence.[10]

Of course, this challenge becomes even more daunting when we consider individuals who lack your conceptual sophistication. Children and unsophisticated adults pose a special problem for doxastic accounts of the basing relation. These individuals at least sometimes form beliefs on the basis of their evidence. After all, children and unsophisticated adults have at least some justified beliefs. But, it is likely that these individuals lack the concepts necessary to even form the sort of meta-beliefs that doxastic accounts of the basing relation require for beliefs to be based on evidence. Thus, there is a serious problem for doxastic accounts of the basing relation.[11]

---

[10] See Korcz (1997) and Wedgwood (2006) for similar arguments concerning the problem that perceptual beliefs pose for doxastic accounts of the basing relation.

[11] See Korcz (1997) and Vahid (2009) for discussion of this issue.

Although we will explore additional problems for doxastic accounts of the basing relation in the next section, at this point it is worth mentioning that the problems we have considered so far for doxastic accounts of the basing relation only seem to be a problem because these accounts claim that meta-beliefs are *necessary* for basing a belief on one's evidence.[12] In other words, we have only seen that it is doubtful that the mere fact that someone lacks such a meta-belief guarantees that her belief is not based on her evidence. However, we might think that although such meta-beliefs are not necessary, they are *sufficient* for basing a belief on one's evidence. After all, allowing that meta-beliefs are sufficient for one's belief to be based on the evidence would allow us to accommodate Lehrer's claim about the Superstitious Lawyer example as well as Korcz's claim about taking appropriate account of the evidence. Unfortunately, only giving a sufficient condition for a belief to be based on the evidence would not provide a full account of the basing relation. Consequently, simply accepting that the sort of meta-belief which doxastic accounts require is sufficient for basing a belief on the evidence would leave us with an incomplete account of the basing relation. It is because of this that some opt for hybrid accounts of the basing relation—views that attempt to combine elements of both causal and doxastic accounts of the basing relation.

7.3 Hybrid Accounts of the Basing Relation

There are a variety of ways that one might try to combine elements of causal and doxastic accounts of the basing relation . Hence, there are a variety of hybrid accounts of the basing relation. Some hybrid accounts require a causal connection between one's evidence and her belief as well as the sort of meta-beliefs required by doxastic accounts of the basing relation.[13] Of course, this sort of hybrid account does not seem to offer any advantage when it comes to the challenges discussed above. After all, such accounts must deal with all of the challenges of both causal accounts and doxastic accounts.

Other hybrid accounts are more plausible. These accounts have causal requirements for basing when it comes to certain kinds of beliefs, such as perceptual beliefs, and doxastic requirements when it comes to other kinds of beliefs, such as beliefs that are inferred from other beliefs.[14] Perhaps the most plausible kind of hybrid account is the sort which claims that so long as either one's belief is appropriately

---

[12]We will explore these additional problems for doxastic accounts in the next section because these problems are equally challenging for pure doxastic views and the most promising hybrid accounts of the basing relation.

[13]This seems to be the sort of view of the basing relation that Leite (2004) suggests. See Vahid (2009) for convincing criticisms of Leite's account.

[14]See Audi (1993) for this sort of view.

caused by her evidence or one has the sort of meta-belief required by doxastic accounts her belief is based on the evidence.[15]

At first, hybrid accounts might seem to offer the best of both worlds. They accommodate the intuition that some have about the Superstitious Lawyer example by allowing that the lawyer's belief is based on his evidence. Yet, they do not require us to have meta-beliefs for basing so they do not seem to imply that children and adults, when they are unreflective, cannot have beliefs based on their evidence. This sounds great—everyone gets what she wants. Unfortunately, hybrid accounts also face challenges.[16]

The first challenge facing hybrid accounts comes in the form of a dilemma. Since hybrid accounts allow that having the appropriate meta-belief is sufficient for justification, it is natural for us to ask a question about this meta-belief. Does this meta-belief itself have to be based on the evidence? If "yes", then it seems that there must be some meta-meta-belief to the effect that the evidence provides a good reason to accept the meta-belief. This meta-meta-belief will also have to be based on the evidence, and we are stuck with an infinite regress of higher-order beliefs. If, on the other hand, the meta-belief does not have to be based on the evidence, then it is unclear why such a meta-belief would be sufficient for basing. To see this, consider two cases. In the first case S has evidence in support of *p*, and she has the meta-belief that the evidence provides good reason to believe *p*, but S has no evidence for this meta-belief—it is simply the result of wishful thinking. The second case is exactly like the first except that S does not form any meta-belief about her evidence providing a good reason to believe that *p*. If we accept the second option of this dilemma, we seem to be committed to claiming that S's belief that *p* is based on her supporting evidence in the first case, but not the second. This is very counterintuitive. It does not seem that mere wishful thinking can render a belief based on the evidence. Thus, either way supporters of hybrid accounts of the basing relation go, the status of the required meta-belief seems to pose a problem for their accounts.[17]

The second challenge hybrid accounts of the basing relation face is that they seem to make it impossible for someone to be aware of two good reasons for believing *p*, but only believe that *p* for one of those reasons. John Turri (2011, p. 386) nicely illustrates this challenge:

> Martin believes that Mars contains significant amounts of water buried just below its surface (Q). He judges that this is good evidence to believe that life exists elsewhere in the universe (P). Martin also is certain that the conditions for life are overwhelmingly abundant throughout the universe (S). He judges that this too is good evidence to believe that life exists elsewhere in the universe. But Martin is utterly exhausted and in despair from several grueling and fruitless months on the academic job market, which understandably

[15]See Korcz (2000) for the most developed version of this sort of account to date. See Mittag (2002) for criticisms of Korcz's account.

[16]It is worth keeping in mind that both of the challenges we will consider for hybrid accounts here are equally challenges for doxastic accounts of the basing relation.

[17]For further discussion see McCain (2014) and Wedgwood (2006).

and predictably impairs his cognitive functioning, especially at the present moment. He consequently neglects his evidential judgment about the relevance of subterranean Martian water, and bases his belief that life exists elsewhere solely on his belief that the conditions for life are abundant throughout the universe.

According to hybrid accounts of the basing relation, since Martin believes that Q and he believes that Q is good evidence for P, Martin's belief that P is based on Q. Yet, it seems clear that in Turri's example Martin's belief that P is not based on Q. So, if Turri's example depicts a possible scenario, there is a major problem for hybrid accounts of the basing relation. And, as Turri (2011, p. 386) says, "it certainly seems possible. The job market may be bad enough to make Martin slightly irrational. But it's not bad enough to make him impossible." Therefore, it seems that hybrid accounts of the basing relation face a second serious challenge.

7.4 Conclusion

In this chapter we have explored the key distinction between two kinds of justification: propositional and doxastic. We also noted that the basing relation is what is required to bridge the gap between merely having justification for believing that *p* (propositional justification) and having a justified belief that *p* (doxastic justification). There are various accounts of the basing relation, but as we have seen they fall into three primary categories: causal, doxastic, and hybrid. Unfortunately, each kind of account faces challenges. Yet, given the importance of the basing relation it is not surprising that epistemologists are still working to overcome those challenges. We do not need to overcome these challenges or settle the debate concerning which sort of account is correct for our purposes. It is enough that we are aware of a few facts. First, there are various kinds of accounts of the basing relation. Second, the basing relation is very important because doxastic justification is required for knowledge, and proper basing is required for doxastic justification. So, we should understand the justification component of the traditional account of knowledge in terms of doxastic justification. Third, each of the three kinds of accounts of the basing relation is consistent with Evidentialism (and any of the other accounts of justification that we discussed in the previous chapter).

References

Audi, R. (1983). The causal structure of indirect justification. *Journal of Philosophy, 80*, 398–415.
Audi, R. (1993). *The structure of justification*. Cambridge, UK: Cambridge University Press.
Davidson, D. (1980). *Essays on actions and events*. Oxford: Oxford University Press.
Foley, R. (1987). *The theory of epistemic rationality*. Cambridge, MA: Harvard University Press.
Goldman, A. I. (1979). What is justified belief? In G. Pappas (Ed.), *Justification and knowledge* (pp. 1–23). Dordrecht: Reidel.

Goldman, A. I. (2011). Reliabilism. In E. N. Zalta (Ed.), *The Stanford encyclopedia of philosophy* (Spring 2011 Edition). http://plato.stanford.edu/archives/spr2011/entries/reliabilism/
Harman, G. (1973). *Thought*. Princeton: Princeton University Press.
Korcz, K. A. (1997). Recent work on the basing relation. *American Philosophical Quarterly, 34*, 171–191.
Korcz, K. A. (2000). The causal-doxastic theory of the basing relation. *Canadian Journal of Philosophy, 30*, 525–550.
Kvanvig, J. (2003). Justification and proper basing. In E. Olsson (Ed.), *The epistemology of Keith Lehrer* (pp. 43–62). Dordrecht: Kluwer.
Lehrer, K. (1971). How reasons give us knowledge, or the case of the Gypsy lawyer. *Journal of Philosophy, 68*, 311–313.
Lehrer, K. (1974). *Knowledge*. Oxford: Clarendon Press.
Lehrer, K. (2000). *Theory of knowledge* (2nd ed.). Boulder: Westview Press.
Leite, A. (2004). On justifying and being justified. *Philosophical Issues, 14*, 219–253.
McCain, K. (2012). The interventionist account of causation and the basing relation. *Philosophical Studies, 159*, 357–382.
McCain, K. (2014). *Evidentialism and epistemic justification*. New York: Routledge.
Mittag, D. (2002). On the causal-doxastic theory of the basing relation. *Canadian Journal of Philosophy, 32*, 543–560.
Moser, P. K. (1989). *Knowledge and evidence*. Cambridge, UK: Cambridge University Press.
Plantinga, A. (1993). *Warrant and proper function*. New York: Oxford University Press.
Pollock, J., & Cruz, J. (1999). *Contemporary theories of knowledge* (2nd ed.). Lanham: Rowman & Littlefield.
Swain, M. (1981). *Reasons and knowledge*. Ithaca: Cornell University Press.
Tolliver, J. (1982). Basing beliefs on reasons. *Grazer Philosophische Studien, 15*, 149–161.
Turri, J. (2011). Believing for a reason. *Erkenntnis, 74*, 383–397.
Vahid, H. (2009). *The epistemology of belief*. New York: Palgrave-MacMillan.
Wedgwood, R. (2006). The normative force of reasoning. *Nous, 40*, 660–686.

Chapter 8
A Problem for the Traditional Account of Knowledge

Abstract This chapter begins by briefly recapping what has been discovered about the traditional account of knowledge throughout the previous chapters. The traditional account of knowledge holds that one has knowledge of some proposition just in case one justifiedly believes the proposition, and the proposition is true. As earlier chapters showed, believing a proposition requires having an appropriate mental representation of the proposition, being justified in believing the proposition requires believing on it on the basis of sufficiently strong evidence, and a proposition's being true consists of it accurately describing objective reality. After the traditional account of knowledge has been made clear, a decisive objection to that account of knowledge is explained: the Gettier Problem. In addition to explaining how the Gettier Problem shows that the traditional account of knowledge is incomplete this chapter explores some promising responses to the Gettier Problem. Finally, the chapter concludes by noting that even without an answer to the Gettier Problem we can use the traditional account of knowledge as a framework for understanding scientific knowledge.

Now that we have covered the various components of the traditional account of knowledge in some detail we are in a position to spell out the traditional account more fully than we did initially.

Traditional Account of Knowledge
S knows that *p* if and only if:
1) S believes that *p* (has a mental representation of *p* and is guided by that representation—takes *p* as true)
2) *p* is true (*p* correctly describes objective, mind-independent reality)
3) S's belief that *p* is justified (S has sufficiently strong evidence for thinking that *p* is true, and S bases her belief that *p* on that evidence)

Philosophers almost universally accepted this conception of the nature of knowledge for hundreds of years (BonJour 2010a). In fact, as we noted in chapter two, something like the traditional account of knowledge has formed the foundation for thinking about knowledge for more than 2000 years. Despite its long and impressive history almost all philosophers now agree that the traditional account of knowledge has been refuted (BonJour 2010a; Feldman 2003; Hetherington 2005; Jenkins-Ichikawa and Steup 2014; Steup 1996). The refutation of the traditional account of knowledge comes at the hand of Gettier cases, named after Edmund Gettier

K. McCain, *The Nature of Scientific Knowledge*, Springer Undergraduate Texts in Philosophy, DOI 10.1007/978-3-319-33405-9_8

who is credited with bringing this class of problems for the traditional account to the forefront of discussions of the nature of knowledge.[1] In this chapter we will explore the Gettier Problem and its implications for our understanding of the nature of knowledge.

8.1 Gettier's Cases

The best place to start exploring the Gettier Problem is with Edmund Gettier's original cases. Here is the first:

> COINS[2]
>
> Suppose that Smith and Jones have applied for a certain job. And suppose that Smith has strong evidence for the following conjunctive proposition:
>
> (d) Jones is the man who will get the job, and Jones has ten coins in his pocket.
>
> Smith's evidence for (d) might be that the president of the company assured him that Jones would in the end be selected, and that he, Smith, had counted the coins in Jones's pocket ten minutes ago. Proposition (d) entails:
>
> (e) The man who will get the job has ten coins in his pocket.
>
> Let us suppose that Smith sees the entailment from (d) to (e), and accepts (e) on the grounds of (d), for which he has strong evidence. In this case, Smith is clearly justified in believing that (e) is true. But imagine, further, that unknown to Smith, he himself, not Jones, will get the job. And, also, unknown to Smith, he himself has ten coins in his pocket. (Gettier 1963, p. 122)

Although Smith's belief that (e) is justified and true, it does not seem that Smith *knows* that (e), <the man who will get the job has ten coins in his pocket>.

Here is Gettier's other case:

> TRIP
>
> Let us suppose that Smith has strong evidence for the following proposition:
>
> (f) Jones owns a Ford.
>
> Smith's evidence might be that Jones has at all times in the past within Smith's memory owned a car, and always a Ford, and that Jones has just offered Smith a ride while driving a Ford. Let us imagine, now, that Smith has another friend, Brown, of whose whereabouts he is totally ignorant. Smith selects three place-names quite at random, and constructs the following three propositions:
>
> (g) Either Jones owns a Ford, or Brown is in Boston;
>
> (h) Either Jones owns a Ford, or Brown is in Barcelona;

[1] Interestingly, Gettier's famous paper describing these problems was published in 1963, but Bertrand Russell (1912) described what is essentially a Gettier case involving a stopped clock. Chisholm (1977) notes that Alexius Meinong discussed these sorts of cases even earlier in 1906. Although Meinong and Russell first discussed the sort of case that Gettier made famous, Gettier is credited with discovering them. Plausibly, the reason for this is that Gettier emphasized the impact that such cases have on the traditional account of knowledge, but Meinong and Russell did not. Whatever the reason, the subsequent literature has credited Gettier with raising the issue and even refers to fixing the traditional account of knowledge so that it yields the intuitive correct results in Gettier cases as the "Gettier Problem".

[2] Here I give names to both of Gettier's cases in order to make them easier to refer to in the text. In the original he simply refers to them as "*Case I*" and "*Case II*".

> (i) Either Jones owns a Ford, or Brown is in Brest-Litovsk.
>
> Each of these propositions is entailed by (f). Imagine that Smith realizes the entailment of each of these propositions he has constructed by (f), and proceeds to accept (g), (h), and (i) on the basis of (f). Smith has correctly inferred (g), (h), and (i) from a proposition for which he has strong evidence. Smith is therefore completely justified in believing each of these three propositions. Smith, of course, has no idea where Brown is. But imagine now that two further conditions hold. First, Jones does not own a Ford, but is at present driving a rented car. And secondly,
>
> by the sheerest coincidence, and entirely unknown to Smith, the place mentioned in proposition(h) happens really to be the place where Brown is. (Gettier 1963, pp. 122–23)

Again, Smith's belief that (h) is justified and true, but it does not seem that he knows (h).

COINS and TRIP both seem to clearly demonstrate that the traditional account of knowledge is flawed. In both cases Smith has a justified true belief that, intuitively, is not an instance of knowledge. Admittedly, both of these cases are a bit strange and a little far-fetched. However, the traditional account of knowledge is supposed to account for knowledge in *all possible* cases. COINS and TRIP, while strange, are not impossible. Since they are possible situations, the traditional account, if true, must hold in them. Yet, in both of these cases it seems that the traditional account of knowledge fails. In other words, COINS and TRIP seem to clearly demonstrate that it is possible to satisfy all three of the conditions of the traditional account of knowledge and still fail to have knowledge. Gettier's cases show that justified true belief is not sufficient for knowledge. Thus, Gettier's cases have refuted the traditional account of knowledge.

8.2 Initial Responses to the Gettier Problem

Despite the widespread agreement that Gettier's cases show that the traditional account of knowledge is mistaken, most philosophers accept that something *like* the traditional account is correct. In fact, although it is accepted that Gettier's cases show that the three conditions of the traditional account are not sufficient for knowledge, it is widely accepted that they are necessary for it (BonJour 2010a; Feldman 2003; Hetherington 2005; Steup 1996). That is to say, Gettier showed that having a justified true belief that *p* is not enough to know that *p*, but his cases do not show that you can know that *p* without having a justified true belief that *p*. It is still commonly held that justified true belief is necessary for knowledge.

Since most philosophers hold that justified true belief is necessary for knowledge, the majority of responses to the Gettier Problem involve adding a fourth condition to the three conditions of the traditional account of knowledge. The purpose of this fourth condition is to rule out beliefs like Smith's in Gettier's cases from counting as knowledge. Before examining some of the fourth conditions that have been proposed, it will be instructive to consider a couple responses that seek to fix the traditional account of knowledge without adding a further condition to the original three.

It is apparent that both COINS and TRIP rely on two important assumptions. First, in both cases it is assumed that it is possible for someone to have a justified false belief. In both cases Smith infers the true proposition from a justified false belief that he has—(d) in COINS and (f) in TRIP. Second, in both cases it is assumed that it is possible to become justified in believing a proposition by deducing it from things you justifiedly believe. That is to say, both cases involve Smith deducing other propositions from his justified false beliefs. Given that Gettier's cases rely on two assumptions, two initial responses to the Gettier Problem jump out at us. Deny one or the other of the assumptions at work in Gettier's cases. Unfortunately, neither of these responses is very plausible. Let us take a look at each.

One way to avoid the Gettier Problem is to simply deny that it is possible to have justified false beliefs.[3] The idea here would be to maintain that one must have infallible evidence in support of *p* in order to be justified in believing that *p*. Infallible evidence is evidence that is so strong that it is not possible to have that evidence for *p* when *p* is false. This would avoid the Getter Problem because in Gettier's cases Smith does not have infallible evidence in support of either (d) or (f). On this view of what is required for justification Smith would not be justified in believing (d) or (f). Without justified beliefs in these propositions Smith's inferences to the problematic propositions in these cases would not be justified. Thus, Smith would not be justified in believing (e) or (h). Therefore, there would be no justified true belief which fails to be knowledge in either of these cases because the problematic belief would not be justified at all. And so, there would be no problem for the traditional account of knowledge.

While it might be tempting to think that we can avoid the Gettier Problem by accepting that justification requires infallible evidence, this cannot be correct. At least, it cannot be correct if we have hardly any justified beliefs at all. Requiring infallible evidence for justification straightforwardly leads to a thoroughgoing skepticism. At times our perceptual experiences can fail to be accurate. That is to say, sometimes we are subject to illusions, hallucinations, or simply misleading perceptual information. A clear example of this occurs in the famous Müller-Lyer illusion. In this illusion line segments of the same length appear to be different lengths because of the placement of arrow "heads" or "tails" on the line segment. This illusion and many others demonstrate that our perceptual experiences do not provide us with infallible evidence. Moreover, it is exceedingly plausible that, for any given perceptual experience you have, it is possible that the experience is misleading. This creates a severe problem for the view that justification requires infallible evidence. If you must have evidence of this strength in order to be justified in believing something, you are not even justified in believing that you are currently reading this book! After all, it is possible that you have your current evidence (perceptual experiences), which supports believing that you are currently reading

[3]This is the sort of view of justification made famous by Descartes (1641/1988). The Gettier Problem prompted a couple other epistemologists to seriously consider this view of justification too. See Lehrer (1971) and Unger (1971) for endorsements of this view.

this book, and yet you are not reading a book at all. It could be that you are having a very vivid hallucination as of reading this book, or you are in fact dreaming that you are currently reading this book. Admittedly, neither of these possibilities is at all likely to be true, but they are *possibilities* nonetheless. Since they are possible, the evidence that you have for thinking you are reading this book is not infallible. If we accept that infallible evidence is required for justification, then you are not justified in believing that you are reading this book. This sort of problem arises for all of your other beliefs about the world around you. Further, the evidence you gain from memory does not satisfy this infallibility requirement either. So, you would not be justified in believing the things you seem to remember. Similar considerations apply to most all of your beliefs. Thus, if we attempt to avoid the Gettier Problem by requiring infallible evidence for justification, we are stuck with a very extreme form of skepticism. In this case we would be avoiding one problem by embracing a much worse one.

A second way to avoid the Gettier Problem that might seem attractive, at least initially, is to deny that it is possible to become justified in believing a proposition by deducing it from things you justifiedly believe. However, once we think about this more carefully it becomes clear that this is not a very plausible response either. Intuitively, it seems that if you justifiedly believe that p, and you justifiedly believe that <p entails q>, then you should be in a position to justifiedly believe that q on the basis of an inference from p and <p entails q>. This response to the Gettier problem would deny this intuition though. According to this response, when you justifiedly believe that you are reading this book, and you justifiedly believe that if you are reading this book, then you know how to read, you cannot justifiedly believe that you know how to read by inferring it from these two justified beliefs. Denying that you can become justified in believing a proposition by deducing it from things you justifiedly believe is highly implausible. Again, such a response seems to avoid one problem by embracing an even worse one. What is more, this sort of response cannot handle all Gettier style cases—it fails in both Chisholm's (1977) "sheep in the field" case and Goldman's (1976) "fake barn" case, both of which are described below. Not surprisingly, few philosophers have accepted either of these responses.

8.3 A Fourth Condition for Knowledge?

Since the initial "quick fixes" to the Gettier Problem are implausible, philosophers have tried other responses. As we noted above, the typical strategy is to add a fourth condition designed to block the problem that arises for the traditional account of knowledge in Gettier cases. An early attempt at such a fourth condition which has some initial plausibility is what is sometimes called the "no false lemmas" (Jenkins-Ichikawa and Steup 2014) or "no false grounds" (Feldman 2003) response. Michael Clark (1963) put this response forward shortly after the publication of Gettier's original presentation of the problem. According to this response, the

Gettier Problem can be solved by adding the following condition to the traditional account of knowledge:

> All of S's grounds for believing *p* are true (Feldman 2003, p. 31).

The idea encapsulated in this condition is that the reasons S has for believing that *p* must all be true. In other words, S cannot know that *p* if she bases her belief that *p* on something false.

At first glance the no false grounds response seems to fare well. In both of Gettier's cases Smith has false grounds. In COINS Smith bases his belief that (e), <the man who will get the job has ten coins in his pocket>, on (d), <Jones is the man who will get the job, and Jones has ten coins in his pocket>. But, (d) is false. Hence, according to the no false grounds response, Smith fails to know that (e), which is the intuitively correct result. Similarly, in TRIP Smith bases his belief on a false ground. In this case Smith bases his belief that (h), <either Jones owns a Ford, or Brown is in Barcelona> on his false belief that (f), <Jones owns a Ford>. Thus, again the no false grounds response yields the intuitive result that Smith does not know that (h) is true.

Despite its ability to handle Gettier's original cases, the no false grounds response fails for two reasons. First of all, there are other Gettier style cases for which it does not give the right results.[4] Here is one from Roderick Chisholm (1977, p. 105):

> A man takes there to be a sheep in the field and does so under conditions which are such that, when a man does thus take there to be a sheep in the field, then it is evident to him that there is a sheep in the field. The man, however, has mistaken a dog for a sheep and so what he sees is not a sheep at all. Nevertheless, unsuspected by the man, there is a sheep in another part of the field.

Chisholm's example seems to be a Gettier style case—the man has a justified true belief that there is a sheep in the field, but intuitively he lacks knowledge. The no false grounds response does not seem to rule out the man's knowing there is a sheep in the field in this case though. The reason for this is that the man does not seem to be basing his belief that there is a sheep in the field on a false ground. He is basing his belief directly on his visual experience and his background knowledge of what sheep look like. As a result, it seems that the no false grounds approach is not up to the task of saving the traditional account of knowledge from the general threat of Gettier style cases.

An additional problem for the no false grounds response is that it seems to lead to skepticism. It is plausible that there will often, if not always, be some false ground that plays a role in the beliefs we form in ordinary cases (Hetherington 2005). If this is true, then according to the no false grounds response most, if not all, of our justified beliefs will fall short of knowledge. This skeptical concern has led some epistemologists to modify the no false grounds approach to the "no *essential* dependence on a falsehood" response (Feldman 2003; Harman 1973; Lehrer 2000; Lycan 2006).

---

[4]For additional Gettier style cases where it does not seem that the person is relying on a false ground see Feldman (1974), Goldman (1976), and Lehrer (1979).

According to the no essential dependence on a falsehood response, having knowledge does not require that you have no false grounds for the known proposition. Instead, this response adds the following fourth condition to the traditional account of knowledge:

> S's justification for *p* does not essentially depend on any falsehood (Feldman 2003, p. 37).

The idea here is that even if S bases her belief on something false that does not automatically rule out her having knowledge. In order for S to fail to know it has to be that the bases of her belief that *p* which remain after we have taken away the falsehoods are not sufficient for justification. For instance, if Marisa believes that there is a tree in her yard because she has a visual experience of a tree and because she believes it is a logical impossibility for someone to have a yard and not have a tree in it, her belief could still count as knowledge. According to the no essential dependence on a falsehood response, Marisa's belief may still be an instance of knowledge because even without her false belief that it is a logical impossibility for someone to have a yard and not have a tree in it, Marisa's belief is still justified by her visual experience.

The no essential dependence upon a falsehood response improves upon the no false grounds response because it does not seem to have the same skeptical problems. Yet, one might worry that it still faces a problem with Gettier style cases like Chisholm's sheep in the field case. Richard Feldman (2003) claims that the no essential dependence on a falsehood response yields the intuitively correct result in Chisholm's case though. He maintains that the man's belief that there is a sheep in the field "depends essentially on the proposition that what he sees is a sheep", and that belief is false (Feldman 2003, p. 36). So, Feldman claims the no essential dependence on a falsehood response gives the intuitive result that the man does not know there is a sheep in the field.

Perhaps this is correct, though it is not clearly so (Steup 1996). After all, it is not clear that the man forms a belief that what he sees is a sheep. It is plausible that he simply has a visual experience as of a sheep and immediately forms the belief that there is a sheep in the field. Even if this response works for Chisholm's case, another Gettier style case seems to pose a problem for the no essential dependence on a falsehood response though:

> Henry is driving in the countryside with his son. For the boy's edification Henry identifies various objects on the landscape as they come into view. "That's a cow," says Henry, "That's a tractor," "That's a silo," "That's a barn," etc. Henry has no doubt about the identity of these objects; in particular, he has no doubt that the last-mentioned object is a barn, which indeed it is. Each of the identified objects has features characteristic of its type. Moreover, each object is fully in view, Henry has excellent eyesight, and he has enough time to look at them reasonably carefully, since there is little traffic to distract him . . . unknown to Henry, the district he has just entered is full of papier-mâché facsimiles of barns. These facsimiles look from the road exactly like barns, but are really just facades, without back walls or interiors, quite incapable of being used as barns. They are so cleverly constructed that travelers invariably mistake them for barns. Having just entered the district, Henry has not encountered any facsimiles; the object he sees is a genuine barn. But if the object on that site were a facsimile, Henry would mistake it for a barn. (Goldman 1976, pp. 772–73)[5]

[5]Although Goldman was the first to publish this example, he credits Carl Ginet with coming up with the fake barn case.

In this "fake barn" Gettier-style example it seems that Henry has a justified true belief that the object is a barn, but the belief does not amount to knowledge. It is far from clear that the no essential dependence on a falsehood response can yield the intuitively correct result in this case. No falsehood immediately stands out as being essential for Henry's justification for thinking that he sees a barn. Thus, it seems the no essential dependence on a falsehood response is problematic as well.[6]

While there are a number of responses that seek to provide the elusive fourth condition that will fix the traditional account of knowledge, none of them have been widely accepted. Rather than catalog these various responses and their problems, we will turn our attention to responses to the Gettier Problem that do not attempt to provide a fourth condition.[7]

8.4 Other Responses to the Problem

The difficulty of the Gettier Problem and the lack of consensus concerning proposed responses have led some to think that a solution of the sort we are after is simply not to be had. Linda Zagzebski (1994) argues that the only way to avoid the Gettier Problem is to either make justification so strict that it entails the truth of the relevant proposition (the infallibilist response from above) or to divorce justification from truth to the point that even in the Gettier cases the person has knowledge. We have already noted that the infallibilist response has implausible skeptical consequences. Although some have taken the second approach that Zagzebski mentions, it is far from popular.[8] There is good reason for this—the intuition that the subject fails to know is one of the most widely held intuitions in contemporary epistemology; almost everyone agrees that the subject lacks knowledge in these cases.[9] In light of these considerations, neither of these options seems to be all that promising.

---

[6]Most philosophers think that fake barn cases are Gettier cases, but see Gendler and Hawthorne (2005) for some reasons to think that fake barn cases are not genuine Gettier cases. Of course, if such cases are not genuine Gettier cases, then the no essential dependence on a falsehood response is in better shape than it appears at this point.

[7]See Lycan (2006), Shope (1983), and Turri (2012a) for overviews of the various responses to the Gettier Problem that have been put forward and the difficulties facing each of these responses.

[8]See Butchvarov (1970) for a view of knowledge that is amenable to this sort of response. See Hetherington (1998, 2001) and Turri (2012b) for recent attempts to make this approach more palatable.

[9]Almost all philosophers accept the intuition at play in the Gettier style cases (that the subject lacks knowledge of the proposition in question). Nonetheless, some have argued that survey research suggests that non-philosophers may not always share this intuition, see Weinberg et al. (2001). These results have been challenged on the ground that there is merely a verbal disagreement between what non-philosophers are claiming and the intuition held by philosophers (Sosa 2007, 2009). Others argue that the intuitions of non-philosophers are not important in this case because they lack the sort of training that makes one good at evaluating abstract cases like Gettier style cases (Ludwig 2007; Williamson 2007). Still others have conducted additional survey research with the result that non-philosophers share the same intuition as philosophers about this sort of case (Turri 2013). So, at present it appears safe to say that the intuition that the subject in a Gettier-style case lacks knowledge is on pretty firm ground.

Others have taken the history of the Gettier Problem and the failure of various attempts to solve the problem to show that it cannot be solved. This has helped give rise to the "knowledge first" approach to epistemology (Williamson 2000). Essentially, this approach to epistemology holds that knowledge is an unanalyzable concept—we can never give an acceptable full analysis of the necessary and sufficient conditions for knowledge. Despite its being unanalyzable, supporters of this approach to epistemology hold that knowledge is an important concept that can be used to analyze other important epistemic concepts such as justification. Importantly, those who accept this approach, such as its most prominent defender, Timothy Williamson (2000), allow that even though knowledge is not susceptible to complete analysis there are still necessary and sufficient conditions for knowledge that can be identified, and it can be quite informative to identify them. Not surprisingly, the break with traditional epistemology that the knowledge first approach advocates is very controversial.[10]

The rise of the knowledge first approach as well as a general dissatisfaction with responses to the Gettier Problem have led some to think that philosophers should not be focused on analyzing knowledge at all. Several philosophers have argued that it is not knowledge that we should be concerned about anyway, but rather, we should be concerned with things like justification and evidence.[11] We tentatively reached a similar conclusion in chapter four when discussing verisimilitude. Recognizing the importance of the concept of verisimilitude and appreciating the fact that knowledge requires truth, not truthlikeness but truth full stop, provides reason for thinking that it is not really knowledge in the strict sense that we should be concerned with at all. These facts about knowledge and verisimilitude give us some reason to think that what really matters for scientific inquiry, and inquiry in general, is the *evidence* we have in support of particular claims and theories rather than knowledge. Of course, this does not mean that we cannot use the term "knowledge" to signify a particularly high cognitive ideal, but it does mean that we may want to rethink our general focus.

8.5 The Move Away from Knowledge

When faced with the difficulty of solving the Gettier Problem many philosophers opt for picking an initially plausible response (one that seems to handle most of the central Gettier-style cases) and continuing to work with the traditional account (Feldman 2003). This sort of approach would suffice for our purposes. Nevertheless, it is worth briefly considering here what effect moving away from concentrating on knowledge to focusing on justification and evidence would have on our understanding of NOS.

[10]For criticisms of this approach see the essays collected in Greenough and Pritchard (2009).

[11]See BonJour (2010a, b), Goodman and Elgin (1988), Elgin (1996), Kaplan (1985), and Kvanvig (1998, 2003).

Some commonly held aspects of NOS seem to make more sense when construed in terms of evidence and justification rather than knowledge. For instance, it is much more plausible that a theory which we are justified in believing to be true is tentative than it is that a theory we *know* to be true is tentative. Plausibly, when we speak of "scientific knowledge" what we really mean is scientific claims for which we have sufficient evidence to think they are true (or approximately true to a specific degree). Often, we are not careful to distinguish between knowledge and justified belief—as we have seen, there is perhaps a good reason for this since it is so difficult to say exactly what knowledge is. It seems that our focus is really not on knowledge at all, but rather on the sort of evidence, and methods of gaining that evidence, that can justify us in believing particular scientific claims are true.

In spite of the potential theoretical benefit of moving away from a strict focus on knowledge and the fact that we seem to be most concerned with evidence and justification anyway, it seems that there are practical reasons for continuing to use the *term* "scientific knowledge". Shifting from talking about scientific knowledge to theories and claims that are justified or reasonable to believe in light of the evidence may lead to the mistaken thought that many of our best scientific theories and laws are "just theories". This is a common, misguided objection to evolutionary theory.[12] Additionally, it is useful to continue to talk about scientific knowledge because doing so may help to avoid the problem of failing to distinguish well-supported facts in science from things that one merely believes.[13] In light of these concerns, it may be best to continue to employ the term "scientific knowledge" even though when we do so we are not particularly concerned with whether the state we are describing satisfies whatever condition might be necessary to solve the Gettier Problem. It is evidence and justification that we really care about. Nevertheless, the term "scientific knowledge" may be employed to signify beliefs that satisfy the conditions of the traditional account of knowledge. Of course, as we have noted, even without a complete analysis of knowledge it is exceedingly plausible that the three conditions of the traditional account are required for knowledge. It is because of this that often our discussions of beliefs for which we satisfy the conditions of the traditional account are discussions of knowledge. This is so because these discussions are concerned with whether a particular belief satisfies various necessary components of knowledge.

8.6 Conclusion

In this chapter we have explored a major problem facing the traditional account of knowledge, the Gettier Problem. The Gettier Problem shows that the three conditions of the traditional account are not sufficient for knowledge. We have

[12] See McCain and Weslake (2013) for discussion of this and other misguided objections to evolution.

[13] See Kampourakis (2014) for discussion of this problem.

considered some of the many responses to this problem and seen that they are problematic. Unfortunately, this seems to be the case with all responses to the Gettier Problem. Despite the difficulty of responding to the Gettier Problem we can still continue on our path of exploring the philosophical foundation for understanding scientific knowledge. Although we have seen that it is possible to change our focus from scientific knowledge to evidence and justification for scientific claims, we will continue to use the term "scientific knowledge" and will focus on the key components of the traditional account of knowledge. In doing this we can work with the assumption that something along the lines of the traditional account of knowledge is correct—if one is unwilling to grant this assumption, we can still continue to focus on knowledge; we simply need to do so with the understanding that it is the justification component and its reliance upon evidence that is most central to our investigation.

We have now completed Part I of this book. Throughout the chapters of this part we have explored many issues related to the traditional account of knowledge. We have seen that there are a number of live philosophical debates concerning various important components of knowledge. Although we often had to leave these debates unresolved, merely appreciating the foci of these debates and some of the major positions in the debates has helped to provide us with a fairly broad basis from which to understand the general features of knowledge. The understanding of the general features of knowledge afforded by our examination of the facets of the traditional account of knowledge provides us with a solid framework for better understanding NOS and the debates surrounding this important concept. We will now begin to narrow our focus from knowledge in general to an exploration of aspects of scientific knowledge in particular.

References

BonJour, L. (2010a). *Epistemology: Classic problems and contemporary responses* (2nd ed.). Lanham: Rowman & Littlefield.

BonJour, L. (2010b). The myth of knowledge. *Philosophical Perspectives, 24*, 57–83.

Butchvarov, P. (1970). *The concept of knowledge*. Evanston: Northwestern University Press.

Chisholm, R. M. (1977). *Theory of knowledge* (2nd ed.). Englewood Cliffs: Prentice Hall.

Clark, M. (1963). Knowledge and grounds: A comment on Mr. Gettier's paper. *Analysis, 24*, 46–48.

Descartes, R. (1641/1988). *Meditations on first philosophy*. In J. Cottingham, R. Stoothoff, & D. Murdoch (Trans.), *Descartes: Selected philosophical writings* (pp. 73–123). Cambridge, UK: Cambridge University Press.

Elgin, C. Z. (1996). *Considered judgment*. Princeton: Princeton University Press.

Feldman, R. (1974). An alleged defect in Gettier counter-examples. *Australasian Journal of Philosophy, 52*, 68–69.

Feldman, R. (2003). *Epistemology*. Upper Saddle River: Prentice Hall.

Gendler, T. S., & Hawthorne, J. (2005). The real guide to fake barns: A catalogue of gifts for your epistemic enemies. *Philosophical Studies, 124*, 331–352.

Gettier, E. L. (1963). Is justified true belief knowledge? *Analysis, 23*, 121–123.

Goldman, A. I. (1976). Discrimination and perceptual knowledge. *Journal of Philosophy, 73*, 771–791.

Goodman, N., & Elgin, C. Z. (1988). *Reconceptions in philosophy & other arts & sciences*. Indianapolis: Hackett.

Greenough, P., & Pritchard, D. (2009). *Williamson on knowledge*. Oxford: Oxford University Press.

Harman, G. (1973). *Thought*. Princeton: Princeton University Press.

Hetherington, S. (1998). Actually knowing. *Philosophical Quarterly, 48*, 453–469.

Hetherington, S. (2001). *Good knowledge, bad knowledge: On two dogmas of epistemology*. Oxford: Oxford University Press.

Hetherington, S. (2005). Gettier problems. *Internet Encyclopedia of Philosophy*. http://www.iep.utm.edu/gettier/

Jenkins-Ichikawa, J. & Steup, M. (2014). The analysis of knowledge. In E. N. Zalta (Ed.), *The Stanford encyclopedia of philosophy* (Spring 2014 Edition). http://plato.stanford.edu/archives/spr2014/entries/knowledge-analysis/

Kampourakis, K. (2014). *Understanding evolution*. Cambridge, UK: Cambridge University Press.

Kaplan, M. (1985). It's not what you know that counts. *Journal of Philosophy, 82*, 350–363.

Kvanvig, J. (1998). Why should enquiring minds want to know? Meno problems and epistemological axiology. *Monist, 81*, 426–452.

Kvanvig, J. (2003). *The value of knowledge and the pursuit of understanding*. Cambridge, UK: Cambridge University Press.

Lehrer, K. (1971). Why not scepticism? *Philosophical Forum, 2*, 283–298.

Lehrer, K. (1979). The Gettier Problem and the analysis of knowledge. In G. Pappas (Ed.), *Justification and knowledge* (pp. 65–78). Dordrecht: Reidel.

Lehrer, K. (2000). *Theory of knowledge* (2nd ed.). Boulder: Westview Press.

Ludwig, K. (2007). The epistemology of thought experiments: First vs. third person approaches. *Midwest Studies in Philosophy, 31*, 128–159.

Lycan, W. G. (2006). On the Gettier Problem problem. In S. Hetherington (Ed.), *Epistemology futures* (pp. 148–168). Oxford: Oxford University Press.

McCain, K., & Weslake, B. (2013). Evolutionary theory and the epistemology of science. In K. Kampourakis (Ed.), *The philosophy of biology: A companion for educators* (pp. 101–119). Dordrecht: Springer.

Russell, B. (1912). *The problems of philosophy*. London: Williams and Norgate.

Shope, R. K. (1983). *The analysis of knowing: A decade of research*. Princeton: Princeton University Press.

Sosa, E. (2007). Experimental philosophy and philosophical intuitions. *Philosophical Studies, 132*, 99–107.

Sosa, E. (2009). A defense of the use of intuitions in philosophy. In D. Murphy & M. Bishop (Eds.), *Stich and his critics* (pp. 101–112). Malden: Blackwell.

Steup, M. (1996). *An introduction to contemporary epistemology*. Upper Saddle River: Prentice Hall.

Turri, J. (2012a). In Gettier's wake. In S. Hetherington (Ed.), *Epistemology: The key thinkers* (pp. 214–229). London: Continuum.

Turri, J. (2012b). Is knowledge justified true belief? *Synthese, 184*, 247–259.

Turri, J. (2013). A conspicuous art: Putting Gettier to the test. *Philosophers' Imprint, 13*. http://quod.lib.umich.edu/p/phimp/3521354.0013.010/1

Unger, P. (1971). A defense of skepticism. *Philosophical Review, 30*, 198–218.

Weinberg, J., Nichols, S., & Stich, S. (2001). Normativity and epistemic intuitions. *Philosophical Topics, 29*, 429–460.

Williamson, T. (2000). *Knowledge and its limits*. Oxford: Oxford University Press.

Williamson, T. (2007). *The philosophy of philosophy*. Malden: Blackwell.

Zagzebski, L. (1994). The inescapability of Gettier problems. *Philosophical Quarterly, 44*, 65–73.

Part II
Knowledge of Scientific Claims

Chapter 9
Explanation and Understanding

Abstract This chapter begins the transition from talking about knowledge in general to talking about scientific knowledge in particular. Although scientific knowledge is itself something that falls under the general account of knowledge, it has special features that are worth exploring. In this chapter the importance of explanation to scientific knowledge is brought to the forefront of the discussion. The nature of scientific explanation itself as well as its relation to understanding is explored in this chapter. It is made clear that good explanations are those which provide understanding of particular phenomena. In addition to examining the relationship between explanation and understanding, this chapter also examines what can make one explanation superior to another. This examination of explanatory virtues is very important because it is common in scientific practice to adopt a particular theory as a result of its being more virtuous than its competitors. Not only is such a practice common in science, it is something that we do routinely in our everyday life.

The two chapters in this part of the book shift from focusing on the nature of knowledge in general to exploring the nature of scientific knowledge in particular. In this first chapter we will take a close look at the primary aim of science and at how we go about achieving that aim. Afterward, in Chap. 10, we will explore how the vehicle for achieving the aim of science helps shed light on how we gain knowledge in science as well as how we gain knowledge in general. Let us get to work on the tasks of this chapter.

It is generally accepted that science has three primary aims: the prediction, control, and explanation of phenomena. Various branches of science emphasize some of these aims more than others. Theoretical physicists exploring the merits of String Theory do not spend a lot of time focusing on controlling phenomena whereas biochemists researching new medicines do focus a lot on controlling various infections and diseases. Despite these differences, it is commonly held that *explanation* is the most important aim of science (Strevens 2006).[1] As the

[1] Although this is widely held, it is not universally so. Theorists sympathetic to constructive empiricism, such as van Fraassen (1980, 1989), are apt to maintain that the primary aim of science is to construct theories that simply fit the observable phenomena—explanations that go beyond

K. McCain, *The Nature of Scientific Knowledge*, Springer Undergraduate Texts in Philosophy, DOI 10.1007/978-3-319-33405-9_9

National Research Council (2012, p. 52) puts the point, "the goal of science is the construction of theories that can provide explanatory accounts of features of the world." Why is this commonly held to be true? That is, what makes explanation more important than prediction or control?

Before answering the question of the source of explanation's importance it is worth noting the reason we are using the term "explanation" rather than "scientific explanation" here. Despite the fact that most philosophical theories of explanation are described as theories of "scientific explanation", it is widely held that explanations in everyday life are roughly continuous with scientific explanations. The latter tend to be more precise and rigorous than our explanations in ordinary non-scientific contexts, but the differences between everyday explanations and scientific explanations are a matter of degree rather than differences in kind (Wilkenfeld 2014; Woodward 2003, 2014).

The close connection between everyday explanations and scientific explanations is not all that surprising when we consider a couple of facts. First, the practice of giving explanations is ubiquitous in our everyday lives. We often explain others' behaviors, we explain our own behaviors, we explain why various things happened, and so on. For example, we explain why our friend went to the store by pointing out that she wanted some milk and believed the store to be the best place to get milk. We explain why we read a particular article by citing the facts that we found the title interesting and that we had the time to read it. And so on. Second, commonsense reasoning and the methods of scientists are quite similar. Of course, scientific methods tend to be much more exact and sometimes lead to rather surprising conclusions from the standpoint of commonsense, yet the two share many of the same commitments to views of "rationality, truth, objectivity, and realism" (Gauch 2012, p. 33). "Science inherits indispensible presuppositions about the world" from our ordinary, commonsense view of the world (Gauch 2012, p. 33). In fact, there are a number of clinical studies which suggest that "the very methodology that a mature science uses to identify dependencies in the world . . . is in some sense the built-in method that we use in our attempts to understand the world" (Grimm 2009, p. 88).[2] Science is in the business of giving explanations, and it seems that in our ordinary lives we are too.

what is observable are superfluous at best; epistemically unacceptable at worst. Two points are worth keeping in mind here though. First, the primacy of explanation in science and science education is widely endorsed in science education reform documents. Take the National Research Council's (2012) framework for K-12 science education, for example. This framework consists of three dimensions: scientific and engineering practices, crosscutting concepts, and disciplinary core ideas. Explanation figures prominently in the first two of these three dimensions. It is similarly emphasized in other education reform documents such as AAAS (1993), National Research Council (1996, 2007), and NGSS Lead States (2013). Second, constructive empiricists themselves recognize that explanation is important—they simply claim that the goal of such explanations is only to adequately describe the observable world, not to give us knowledge of theoretical entities. So, even constructive empiricists, who deny that explanation is the primary aim of science, can agree that explanation is important.

[2]For the results of some of these clinical studies see Gopnik (1998), Gopnik and Glymour (2002), Gopnik, et al (2004), Gopnik and Sobel (2000), and Steyvers et al. (2003).

Not only is science and ordinary life in the same business when it comes to giving explanations, but it also seems that they conduct their business in a similar fashion. The similarity between explanations in scientific contexts and ordinary life is such that "the tendency in much of the recent philosophical literature has been to assume that there is a substantial continuity between the sorts of explanations found in science and at least some forms of explanation found in more ordinary non-scientific contexts, with the latter embodying in a more or less inchoate way features that are present in a more detailed, precise, rigorous etc. form in the former" (Woodward 2014, p. 1). What is more, in the philosophy of science it is assumed that a successful theory of explanation will be one that expresses what scientific and ordinary explanations have in common. "These assumptions help to explain... why, ...discussions of scientific explanation so often move back and forth between examples drawn from bona-fide science (e.g., explanations of the trajectories of the planets that appeal to Newtonian mechanics) and more homey examples involving the tipping over of inkwells" (Woodward 2014, p. 1). So, a proper account of explanation should capture both the everyday explanations that we employ as well as the more precise explanations we find in science and advocated for in science education reform documents, (e.g. National Research Council 1996, 2007, 2012). In light of the similarities between scientific and ordinary explanations and the assumptions commonly made by those theorizing about the nature of scientific explanation, the present discussion should be understood to be a discussion of explanation in general—so that both everyday and scientific explanations are included.

Now to the question of why explanation is so important to science. Explanation's purpose, i.e. the goal of explanation, is what makes it the most important aim of science. Successful explanations increase our understanding of the world around us.[3] It is in virtue of this understanding that we are able to predict and control phenomena. So, it is not that prediction and control are not important goals of science. They are! In fact, some are apt to doubt whether a theory is really scientific if it does not yield new predictions.[4] The reason that explanation is primary is that it allows for prediction and control.[5] Consider the phenomenon of a piece

[3]Philosophers with very diverse views on the nature of explanation and understanding agree that the two are closely linked with explanation providing a means to achieving understanding. See for example, Achinstein (1983), de Regt (2009, 2013), de Regt and Dieks (2005), Friedman (1974), Harman (1986), Khalifa (2012), Khalifa and Gadomski (2013), Kim (1994), Kitcher (1981, 2002), Kvanvig (2003), Lewis (1986), Lipton (2004), Moser (1989), Railton (1993), Salmon (1984, 1998), Sober (1983), Strevens (2006), (2013), Trout (2002), van Fraassen (1980), von Wright (1971), Wilkenfeld (2013, 2014), and Woodward (2003). Even Hempel (1965) and Hempel and Oppenheim (1948), who thought that understanding was too subject dependent to be a proper focus of philosophical study, agree that explanations provide understanding.

[4]This is one of the major criticisms that some press against String Theory. For discussion of this criticism see Dawid (2013).

[5]In some ways this view may be a bit simplistic. As we will see later, one might think that explanation and prediction are very similar (perhaps even the same—one is simply backward looking in time and the other forward looking). Also, it is reasonable to think that making predictions is a factor that can make one explanation better than another. Hence, the relationship

of iron rusting. An explanation of this phenomenon is that iron and oxygen have a redox reaction when brought together in the presence of H_2O (either in liquid form or as moisture in the air). This explanation facilitates understanding of why the particular piece of iron rusted as well as why iron rusts in general. This understanding allows us to predict various phenomena. For instance, we can predict that another piece of iron will begin to rust after a time of being left outside in moist air. The understanding gained via this explanation can also help us to control the rusting of pieces of iron. Since we understand that in order for iron to rust there must be moisture, we could control whether a piece of iron rusts by keeping it in a completely dehydrated environment. The understanding provided by the explanation of the piece of iron's rusting allows us to both make predictions and to control phenomena. Without such understanding it is difficult to see how we could manage either of these feats—particularly controlling whether a piece of iron rusts. In light of the role that understanding plays in both prediction and control, it is maybe a bit misleading to even characterize science as having *three* primary aims. As Erwin Schrödinger explains, our entire scientific worldview rests on the "hypothesis that *the display of Nature can be understood*" (1954, p. 90). Perhaps it is more accurate to say that the primary aim of science is to produce understanding via explanations.[6] Using the understanding gained via explanations to yield accurate predictions and to allow for increased control of phenomena are important secondary aims of science.[7]

Recognizing the centrality of understanding to science is an important insight as is the fact that the vehicle by which we typically gain understanding is explanation.[8] However, these insights raise questions. What exactly is understanding? How do explanations, when they are successful, provide us with understanding? Answers to both of these questions will be explored in this chapter. We will begin with the latter question because by coming to better understand the nature of explanation and the sort of information that explanations convey we will be better equipped to explore the nature of understanding.

between explanation and prediction is not completely clear-cut. However, for the present purpose we can treat them as separate in which case explanation is primary.

[6]Given the central importance of explanations in science, it is unsurprising that constructing and understanding explanations is a major goal of science education reform (Braaten and Windschitl 2011).

[7]According to de Regt et al. (2009, p. 1) "In the eyes of most scientists, and of educated laypeople, understanding is the central goal of science . . . it seems commonplace to state that the desire for understanding is a chief motivation for doing science."

[8]Strevens (2013) argues that when it comes to science, understanding can only be achieved via explanations, i.e. scientific understanding without explanation is impossible. de Regt et al. (2009), Gijsbers (2013), Hindriks (2013), and Lipton (2009) disagree. They each argue that there can be cases in which one comes to have scientific understanding without possessing an explanation. It is worth noting that even though they argue that it is possible to gain understanding without having an explanation, none of these philosophers contest the claim that we typically come to have understanding via explanations or the claim that providing understanding is the goal of producing explanations.

9.1 Explanation

As we have noted, explanation is the method by which we typically gain our understanding of the world around us, whether this is in a controlled scientific context or in our everyday lives. Given the importance of explanation to science it is not surprising that there has been much philosophical discussion of the nature of explanation. As a result of this philosophical discourse, there are many competing philosophical theories of the nature of explanation. There are far too many theories of explanation for us to survey here, in fact. So, we will proceed by exploring what Wesley Salmon (1990, p. 12) refers to as the "fountainhead" of contemporary discussions on the nature of explanation—Carl Hempel's Deductive-Nomological Model (D-N Model) of explanation—and some of its permutations and problems.[9] The importance of this account of explanation can hardly be overstated. It is commonly accepted that it shaped the contemporary discussion of explanation since its arrival on the scene in the mid-twentieth century. Further, it is fair to say that all contemporary theories of explanation are in some way or other reactions to the D-N Model and the problems that were exposed for this theory. We will also very briefly discuss some of the major theories of explanation that have arisen after the D-N Model's fall from dominance before settling on a general picture of explanation that will be useful as a working model for our purposes. However, before we begin our examination of various theories of explanation we need to first consider a very important distinction and some general features of explanation. Let us begin with the distinction.

While it might seem obvious that the two are different, it is important to explicitly distinguish between *explanation* and *explaining*. On the one hand, explaining is a particular action that we sometimes perform. Often this is construed as a speech act in which we verbally or non-verbally communicate an explanation to someone else (Harman 1986). On the other hand, an explanation is "something one grasps or understands that makes things more intelligible" (Harman 1986, p. 67). Typically, the thing that we grasp is a set of propositions—in other words, an explanation is a collection of propositions (de Regt 2009; Kim 1994; Moser 1989; Strevens 2006; Woodward 2003). One very rough way to understand this distinction is simply that explanations are the propositions we communicate to one another when we explain things. In other words, explanations are sets of propositions that provide answers to why-questions—the sort of questions we seek to answer when we are explaining things (Lehrer 2000; Lipton 2004, 2009; Moser 1989; Woodward 2014). Another

[9]There were other important supporters of this view of explanation, including Braithwaite (1953), Gardiner (1959), Nagel (1961), and Popper (1959). Nonetheless, as Woodward (2014) notes, "unquestionably the most detailed and influential statement is due to Carl Hempel". In fact, Hempel's work on this theory (1942, 1965) and Hempel and Oppenheim (1948) is so influential that it is not uncommon to refer to the D-N Model as "Hempel's theory of explanation". We will actually follow this tradition and refer to the D-N Model and the Inductive-Statistical Model (I-S Model) jointly as "Hempel's theories".

way to understand the distinction between explanations and explaining is in terms of cognitive outcomes versus cognitive processes. Explaining is a cognitive process which, when successful, yields a particular cognitive outcome. This cognitive outcome, explanation, is one that promotes understanding. Despite this connection between explanations and explaining, the vast majority of theorists hold that explanations and the explanatory facts which compose them are independent of our acts (or even intentions) of explaining (Strevens 2006).[10] It is the nature of explanation that we are concerned with here, not our acts of explaining (conveying explanations to others). The nature of explanation, the exact features a collection of propositions has to have in order to be an explanation, is the question that theories of explanation seek to answer. We will briefly explore some of the answers that have been offered in the subsections which follow.

9.1.1 Hempel's Theories of Explanation

In his vastly influential work on the nature of explanation, Carl Hempel proposes two theories of explanation, the D-N Model and the Inductive-Statistical (I-S) Model, though the D-N Model is the most widely discussed of the two (perhaps because Hempel considered the explanations it offered far superior to those of the I-S Model).[11] These two models share some features. First, both adhere to the majority view that explanations are sets of propositions. Second, both construe explanations in the form of an argument where a set of propositions (the *explanans*) provide support for a proposition that describes the phenomenon to be explained (the *explanandum*). Third, both require that the explanans include at least one proposition which describes a general law of nature as well as propositions that accurately describe the empirical conditions relevant to the phenomenon's occurrence (empirical conditions which fall under the domain of the law(s) given in the explanans).

Although the D-N Model and the I-S Model are similar in ways, they are importantly different. The general laws in a D-N explanation must be deterministic whereas those in the I-S Model are statistical (it is because some of our fundamental laws seem to be indeterministic that the I-S Model was proposed in the first place). Since the D-N Model involves general deterministic laws and conditions satisfying the conditions of those laws, the explanandum is shown to follow deductively from the explanans in this sort of explanation. In other words, a successful D-N Model explanation is a logically valid argument (an argument where the truth of

[10]Two notable exceptions to this majority opinion are Achinstein (1983) and van Fraassen (1980). Both of these philosophers hold that explanation cannot exist without acts of communication.

[11]Technically, Hempel (1965) discusses a third form of explanation, what he calls "Deductive-Statistical" explanations. These are explanations that derive a statistical uniformity from a more general statistical law. However, deductive-statistical explanations conform to the same pattern as the D-N Model so we will not treat them separately (Woodward 2014).

the premises guarantees the truth of the conclusion) where true explanans guarantee that the phenomenon described in the explanandum had to occur. For example, a D-N explanation of the position of Mars at some particular time, *t*, would include "Newton's laws of motion, the Newtonian inverse square law governing gravity, and information about the mass of the sun, the mass of Mars and the present position and velocity of each" in the explanans (Woodward 2014). These explanans (assuming that they are all true) would deductively entail that Mars has the particular position it does at *t*.

Successful I-S Model explanations are different. Since these explanations involve statistical (indeterministic) laws, the explanandum is not shown to deductively follow from the explanans in this sort of explanation. Instead, a successful I-S explanation shows that the explanans confer a high probability on the explanandum. For instance, if it is a statistical law that it is highly probable someone with disease X will recover after taking a particular drug, and Blake has taken the drug, then this information can be used (along with the law) to provide an I-S explanation of Blake's recovery from X (Woodward 2014).

There is much more that can be said about both of Hempel's theories of explanation including specific details about the I-S Model, which is very complex.[12] Nevertheless, we have a sufficient grasp of these theories for our purposes. Let us turn our attention to some of the objections to Hempel's theories.

9.1.2 Objections to Hempel's Theories

In spite of the fact that many believe Hempel's theories are largely correct about the nature of explanation and that these theories dominated the thinking on the topic of explanation for 20 years or more, numerous objections have been raised against Hempel's theories. Here we will consider two of the most prominent, and most decisive, objections—one for each model.[13]

Let us begin with perhaps the most famous objection to the D-N Model—Bromberger's flagpole counterexample.[14] Consider a situation where there is a flagpole of a particular height on a large patch of clear, level ground. The sun is shining brightly, and there is nothing to interfere with the sun's rays reaching the

[12]In addition to Hempel's own work see Salmon (1990) for an excellent discussion of Hempel's theories as well as the ensuing philosophical debate for the 40 years following the publication of Hempel and Oppenheim (1948).

[13]For discussion of these and several other objections to Hempel's theories see Salmon (1990), Strevens (2006), and Woodward (2014).

[14]Although this counterexample is attributed to Bromberger and often called "Bromberger's flagpole" (Salmon 1990), Bromberger's (1966) actual example involved a tower and its shadow. Nonetheless, the idea behind the counterexample is the same, and it was Bromberger who introduced this sort of counterexample. So, we will follow the literature and refer to this as "Bromberger's flagpole".

flagpole or with the flagpole's casting a shadow on the ground. Given the empirical facts about the flagpole's height, the position of the sun in the sky, and the law of rectilinear propagation of light, it is possible to deduce the length of the shadow that the flagpole casts on the ground. Hence, using the empirical information along with the law about the rectilinear propagation of light we can construct a successful D-N explanation of the length of the flagpole's shadow. This is all well and good for the D-N Model. There is a problem lurking here though. Using the empirical information about the sun's position in the sky and the length of the flagpole's shadow along with the law of rectilinear propagation of light we can construct an equally successful D-N explanation of the height of the flagpole. But, this cannot be right. While it seems we can explain the length of the shadow in this way it does not seem that the height of the flagpole can be explained like this. The shadow does not explain why the flagpole is the height that it is. Rather, a plausible explanation of the height of the flagpole will refer to the intentions of the person who erected the flagpole, the materials they had to use, and so on. The general problem is that for any explanandum for which we can give a successful D-N explanation we can turn around and use that explanandum to explain some of the explanans. The D-N Model does not properly capture explanatory asymmetries—the heights of flagpoles can explain the lengths of their shadows, but the length of flagpoles' shadows do not explain their heights.

Now, for a much discussed objection to the I-S Model.[15] This sort of objection raises a problem for the fact that in a successful I-S explanation the explanans *must* make the occurrence of the explanandum very probable. This is problematic in cases where the only cause of a particular kind of phenomena is *x*, but *x* does not *often* cause phenomena of that sort. The classic example of this sort of case involves syphilis and paresis (Scriven 1959). Only syphilis causes paresis. Yet, in most cases where someone has syphilis he does not contract paresis. So, if we were to find out that Joe has syphilis, we might think his odds of getting paresis are better than if he did not have syphilis, but his odds of contracting paresis are still very low. When we discover that Joe has paresis it seems like citing the fact that he has syphilis is an explanation of his having paresis though. After all, only people with syphilis contract paresis. Nevertheless, according to the I-S Model citing Joe's syphilis is not a good explanation of his having paresis because the fact that someone has syphilis does not make it very probable that he will contract paresis. It seems the restriction that the explanans must make the explanandum very probable is problematic.

Of course, there is much more that can be said about these objections and Hempel's theories in general. In fact, a lot has been said—the literature on explanation is vast and discussions of Hempel's theories dominated that literature for several years. However, for our purposes it is enough to have a handle on Hempel's theories and some of the objections which led the majority of theorists to look elsewhere for an account of explanation.

[15]See Hempel (1965), Lipton (2004), Scriven (1959), and van Fraassen (1980) for discussion of this objection.

9.1.3 Alternatives to Hempel's Theories and a Working Model of Explanation

In the mid-twentieth century there was largely a consensus about the nature of explanation—Hempel's theories, particular the D-N Model, were taken to be correct (Woodward 2003). In the wake of many purported counterexamples to Hempel's theories there is no longer a consensus concerning explanation. Rather, a number of competing accounts of explanation have been proposed. A complete, or even moderately complete, survey of these various alternatives to Hempel's theories is not possible here; nor is it necessary for our purposes. Consequently, we will simply very briefly consider some of the major alternatives that have been proposed and some of the challenges they face before settling on a working model of explanation. This working model of explanation will be fairly broad and ecumenical, which will work nicely for our purposes—we do not need to settle the intimate details of the nature of explanation in order to better understand NOS.

An early alternative to Hempel's theories is Wesley Salmon's (1971) Statistical Relevance (SR) account of explanation. According to the SR account, explanation requires describing the factors that are statistically relevant to the occurrence of the explanandum. A particular factor is statistically relevant to *E*'s occurring when the probability of *E*'s occurrence is greater when the factor is present than when it is not. In the ideal case an SR explanation would describe all of the factors that are statistically relevant to a particular explanandum's occurrence. For example, in the case of Joe's syphilis and paresis mentioned above Joe's syphilis is statistically relevant to his contracting paresis because the probability of contracting paresis when one has syphilis, while low, is higher than the probability of contracting paresis without having syphilis. In light of this, it seems that the SR account does not share the I-S Model's problem mentioned above. It also avoids the flagpole problem that plagues the D-N Model. While the height of the flagpole is statistically relevant to the length of the shadow, it does not seem that the length of the shadow is statistically relevant to the height of the flagpole. So far, so good.

The SR account is problematic, however. It has problems with cases where there is a common cause of two effects. Consider a case where a specific weather pattern typically leads to a storm, and the weather pattern also leads to a particular barometer reading. In such a case the barometer reading will be statistically relevant to the storm, when the barometer reading occurs it is more probable that the storm occurs than when the barometer reading does not occur. The storm will be similarly statistically relevant to the barometer reading though. The probability of the barometer reading is higher when the storm occurs than when it does not. But, intuitively the barometer reading does not explain the occurrence of the storm despite its statistical relevance to the storm's occurrence (Strevens 2006).[16]

[16]This sort of case is also a problem for the D-N Model.

Another alternative to Hempel's theories is the Unificationist account of explanation. The idea here is that explanation consists in subsuming the explanandum phenomenon under a system of laws which best unifies the relevant phenomena (Friedman 1974; Kitcher 1981, 1989). Very roughly, showing how phenomena or laws themselves can be taken to be instances of more fundamental laws creates a more unified picture of the world. The best explanations unify the most phenomena with the least number of fundamental principles. Unfortunately, the Unificationist account seems susceptible to the flagpole problem just like the D-N Model (Strevens 2006). After all, it seems that explaining the flagpole's height by appealing to the length of the shadow is just as unifying of a pattern as explaining the length of the shadow by appealing to the height of the flagpole. The reason is that, as we saw above, both of these "explanations" are apparently relying on the very same law and patterns to subsume the explanandum.[17]

The problems raised by the flagpole example and examples involving common causes, like the storm and the barometer, have led several theorists to think that the best way to approach explanation is by reference to causation. An early causal approach to explanation is Wesley Salmon's (1984) Causal Mechanical (CM) approach (adopted after his abandonment of the SR). On the CM account an explanation consists of a description of the causal influences of the explanandum. The sort of causal influences that are relevant to the CM approach are fundamental causes—close to how we would describe things in terms of the actions of fundamental particles in physics (Strevens 2006). There is a serious problem with this sort of account, however. As Strevens (2006) notes, in a normal case where a baseball shatters a window not only does the baseball count as part of the explanation of the window's shattering on the CM account, so do the shouts of the baseball players which cause very minute vibrations in the window. After all, these shouts do have a causal influence on the event of the window's shattering.[18] This seems implausible though because the fact that the baseball players are shouting seems irrelevant to the window's breaking.

The problem facing the CM account leads some causal theorists to focus on more nuanced accounts of causation. Some such accounts handle this sort of problem by allowing for multiple levels of causation. Often these accounts are cached out in terms of counterfactual views of causation (very roughly, accounts where x counts as a cause of y when it is the case that if x did not occur, y would not occur) (Lewis 1986; Woodward 2003). Other nuanced causal accounts agree with the CM account's focus on causal influence at a fundamental level while requiring the satisfaction of additional requirements in order for a particular fundamental causal influence to count as part of the explanation of the explanandum's occurrence (Salmon 1997; Strevens 2008).

[17]See Kitcher (1981) for attempts to address this issue.

[18]Strevens points out that Salmon appears to concede that this results from his theory.

Although such accounts of explanation hold promise, it is not clear that they can give us a completely general account of explanation.[19] Given their focus on causal relations it seems that some genuine explanations will not be properly accounted for under these theories. Assuming there can be genuine explanations in pure mathematics, it is very difficult to see how such explanations can be accounted for given any causal account of explanation.[20] Additionally, it seems like Philip can explain to his daughter why it is that Robert is her cousin by pointing out that Robert is the son of Philip's sister and that any children of one's parents' brothers or sisters are her cousins (Harman 1986). This seems like a perfectly good explanation, but it does not seem to be causal.

Obviously, our overview of alternatives to Hempel's theories is far from complete. There are many other accounts of the nature of explanation. There are also many other objections that have been raised for the theories we have discussed. Likewise, there are many responses on offer to the sorts of objections we have pointed out for these theories. There is no consensus on the exact nature of explanation because these objections and responses are still being raised and explored. Nevertheless, we now have a fairly good handle on the general issue of the nature of explanation and some of the surrounding debates. Before we turn our attention to the other major topic of this chapter, understanding, we will briefly discuss what we can take as a plausible working model of the nature of explanation for our purposes.

Our working model comes from Jaegwon Kim (1994). According to Kim, "*explanations track dependence relations*", that is to say, explanations are sets of propositions that provide information about dependence relations which hold between the explanans and the explanandum (1994, p. 68).[21] What makes Kim's proposal so useful as a working model of explanation is that there are a number of different kinds of dependence relations. Causal dependence is the sort of relation that holds when one thing causes another, so the sort of relations which causal accounts of explanation seek to capture fall within the purview of this model. The familial relations which constitute being a cousin are dependence relations too. The relation of constitution is itself a dependence relation. Mereological dependence, when "the properties of a whole, or the fact that a whole instantiates a certain property, may depend on the properties and relations had by its parts", is another sort of dependence relation (Kim 1994, p. 67). It is because of the fact that there are many kinds of dependence relations that our working model is ecumenical. It encompasses all of the sorts of relations that the prominent theories of explanation claim are of central importance. Of course, there are many details that would need to

[19]Of course, it might be a mistake to think that we can hope to find a theory of explanation that applies in all cases (Díez et al. 2013). Still, some argue that such a theory can be found (Nickel 2010). Until we have decisive reasons for thinking a fully general account of explanation is impossible, it seems to be a mark against a particular theory if it does not apply generally.

[20]For an overview of reasons for and against thinking there are genuine explanations in pure mathematics see Mancosu (2011).

[21]Strevens (2008) endorses this general conception of explanation as well.

be worked out in order for this dependence relation account to be a fully satisfactory final account of explanation. Fortunately, we do not need such an account for our purposes, and so, we do not need to trouble about these details. It is enough for us to have a working model of explanation, which we do.

Putting things together, we can understand an explanation to be a set of propositions where the explanans provides information about dependence relations which hold between phenomena described in the explanans and the phenomenon to be explained (the explanandum).[22] With this, undoubtedly abstract, working model in hand we are ready to turn our attention toward the aim of explanation: understanding.

9.2 Understanding

It is important to clarify the general sense of "understanding" that is our focus here before delving into more specific concerns. Since our primary focus throughout this book is scientific knowledge, we are especially interested in the sort of understanding which arises in science. For this reason, we will not concern ourselves with discussions of the myriad ways in which the word "understanding" might be used in English.[23] Also, the sense of "understanding" that concerns us is not merely the phenomenological sense in which an "explanation 'feels right'" or what is sometimes referred to as the "aha" experience of understanding (Trout 2002, p. 212). J.D. Trout (2002, 2005, 2007) argues, by appealing to psychological research concerning hindsight bias and overconfidence bias, that understanding construed as a phenomenological feeling of "rightness" is not a good indicator of the quality of an explanation. Although Trout tends to speak as if his arguments demonstrate that it is a mistake to think understanding is important for explanations or for science, they do not. Most philosophers who think providing understanding places constraints on what counts as an explanation are not thinking of understanding as a phenomenal feel, but as a kind of epistemic state. Although this epistemic state may be accompanied by the phenomenal feel that Trout targets, the two should not be conflated. It is the epistemic state we are concerned with, not merely the phenomenal "aha" feeling.[24]

---

[22]Admittedly, one might think that in order to have a "real" explanation one needs to provide information about causal dependence relations or information about natural laws governing the dependence relations in question. Two points about this concern are worth keeping in mind. First, insofar as it is plausible to think there are genuine explanations in pure mathematics it is unclear that such restrictions are necessary. Second, the addition of such restrictions would not affect the points made in this chapter. So, the reader is welcome to understand the account of explanation here as having such restrictions, if she believes they are necessary for genuine explanations.

[23]For a discussion of these issues see Franklin (1981).

[24]It is worth noting that most philosophers agree with Trout that this sort of "aha" experience is neither necessary nor sufficient for possessing genuine understanding. However, several

"Understanding" is "a cognitive success term", that is to say "understanding" denotes a mental state with positive epistemic status; it is a sort of cognitive achievement (Elgin 2007, p. 31).[25] There are various accounts of the nature of understanding construed as a positive epistemic state. For instance, Jonathan Kvanvig (2003, p. 192) says that "understanding requires the grasping of explanatory and other coherence-making relationships in a large and comprehensive body of information . . . understanding is achieved only when informational items are pieced together by the subject in question." Similarly, Linda Zagzebski (2001, p. 241) claims, "*understanding is the state of comprehension of nonpropositional structures of reality*." She goes on to identify three conditions of understanding:

1. It is acquired through mastering a *techne* [practical art/skill]
2. Its object is not a discrete proposition but involves seeing part/whole relations
3. It involves representing some portion of the world non-propositionally

Catherine Elgin (2007, p. 35) maintains, "understanding is primarily a cognitive relation to a fairly comprehensive, coherent body of information." The accounts of understanding that Kvanvig, Zagzebski, and Elgin offer, though different in important ways, seem to capture the general idea of what we have in mind when we claim that understanding is the goal of explanation. Since for our purposes a rough account of the nature of understanding will suffice, we will assume that understanding corresponds to something generally along the lines of what Kvanvig, Zagzebski, and Elgin propose.

Now that we have briefly outlined the relevant notion of "understanding" in broad strokes it is time to explore some of the details.

philosophers have argued Trout's arguments concerning this phenomenal sense of understanding do not undermine the importance of understanding as an aim of science or a goal of everyday explanations (de Regt 2004, 2009; Grimm 2009; Lipton 2009). Some (Lipton 2009; Grimm 2009) go so far as to argue that the "aha" experience may not be as misleading as Trout suggests, and in fact, it may be a reliable guide to the presence of genuine understanding in some cases.

[25]Although we tend to have an intuitive grasp of the notion, clearly defining "positive epistemic status" is difficult. Perhaps the best way to understand this notion is by noting other familiar states which have positive epistemic status such as justified/rational beliefs and knowledge on the one hand and states which lack positive epistemic status such as opinions, beliefs arising from wishful thinking, and guesses on the other hand. In this sense "understanding" is a term for a kind of mental state which belongs in the former group of mental states as opposed to the latter. The idea that understanding is a cognitive achievement (a mental state with positive epistemic status) is widespread. It is shared by those who hold very different views concerning other aspects of understanding, see, for example, Elgin (1996, 2007), Grimm (2001, 2006, 2014a, b), Khalifa (2011, 2012, 2013), Khalifa and Gadomski (2013), Kvanvig (2003, 2009), Pritchard (2009, 2014), Wilkenfeld (2013) and Zagzebski (2001).

9.2.1 Understanding Phenomena Versus Understanding Theories

Before exploring one of the central contemporary debates concerning understanding it will be helpful to first draw a further distinction between senses of understanding. We have already seen that for our discussion of understanding we are not concerned with the "aha" feeling of understanding, but rather with a cognitive achievement, a mental state with positive epistemic status. It is important to distinguish two sorts of cognitive achievements which are both kinds of understanding, and both relevant to our discussion: *understanding phenomena* (*UP*) and *understanding theories* (*UT*) (de Regt 2009).[26]

UP is the sort of understanding one has when she understands why or how a particular phenomenon occurs. This is the sort of understanding that is the aim of explanations, and so it is UP that is the primary aim of science. As we noted above, several philosophers claim that one gains UP by grasping a correct explanation of a particular phenomenon's occurrence. Thus, it is UP that has been our focus when discussing the relation between explanation and understanding.

At this point, one might wonder why we are bothering to distinguish UP and UT, or even discuss UT at all since UP is the sort of understanding that is the aim of science. The reason for this is simple. It is plausible that when it comes to scientific matters one can only truly possess UP, if she has UT (de Regt 2009; Strevens 2013). In other words, it is plausible that UT is necessary to have UP in scientific contexts.[27] So, what exactly is UT?

UT is the sort of understanding one has when she has the ability to use a theory.[28] The sense of "theory" that is relevant here is not the colloquial sense that occurs in ordinary English. Often in ordinary English the term "theory" is used to signify a claim or hypothesis that is still the subject of significant doubt. However, "among scientists, ... the term is often used to describe an established sub-discipline in which there are widely accepted laws, methods, applications, and foundations ... for

[26]Strevens (2013) draws a similar distinction between what he calls "understanding why" and "understanding with". The former corresponds to de Regt's UP while the latter corresponds to his UT.

[27]Plausibly, one way of understanding the distinction between UP and UT is in terms of their objects. The object of UP is natural phenomena; the object of UT is abstract content (theories).

[28]There is at least one other important sense of understanding, which is often characterized as "understanding-that". For example, Belle *understands that* the theory of relativity says X. While this sort of understanding is important, we will not focus on it here. The primary reason for this is that ascriptions of this sort of understanding are widely held to simply be knowledge ascriptions (see Kvanvig 2009; Pritchard 2009). For instance, when we say, "Belle understands that the theory of relativity says X" all we are saying is that "Belle knows that the theory of relativity says X". While this sort of understanding is important, it is fairly clear that it is simply a precondition of UT. One cannot have the ability to use a theory to construct explanations without knowing what the theory says. For present purposes we can assume individuals have this sort of understanding of the theories in question.

which there is strong empirical support" (Rosenberg 2012, p. 115).[29] It is the latter, scientific sense, of "theory" that is important for UT.[30] According to de Regt (2009, p. 37), UT "amounts to the ability of scientists to use relevant theories to construct explanations."[31] Whether or not you possess this ability with respect to a particular theory will depend on a variety of things. Of course, the virtues and nature of the theory itself will make a difference to whether you understand it in this sense. However, things like your "capacities, background knowledge, and background beliefs" (de Regt 2009, p. 33) will all make a difference to whether you have UT. This is extremely important because it is plausible that UT is necessary for UP. In light of this connection between the two kinds of understanding and the various conditions which affect one's having UT, it is possible that when "two people who possess exactly the same theories and background knowledge, one may achieve understanding of a phenomenon while the other does not" because the former might have the requisite capacities for the necessary UT while the latter does not (de Regt 2009, p. 37). [32]

It is worth emphasizing that the distinction between UP and UT along with their connection to one another helps make it clear how UP is a cognitive *achievement*. On this approach, gaining UP requires the exercising of various capacities, such as one's UT, in order to either come up with or properly appreciate an explanation which yields UP. It is not sufficient for understanding that you simply be informed

---

[29]These two very different understandings of the nature of theories, the ordinary and the scientific, may be partly to blame for some of the misguided objections to evolution such as the "it is just a theory" objection. For more on this and other misguided objections to evolution see McCain and Weslake (2013). See Kampourakis (2014) for an in-depth discussion of some of the common misunderstandings which lead to resistance against accepting evolutionary theory.

[30]Also, see National Research Council (2012) for a similar account of scientific theories. Of course, there are important issues concerning how we should understand the fundamental nature of scientific theories. It is not clear whether theories are best understood as ultimately collections of axiomatized sentences or nonlinguistic models, or both. Fortunately, for present purposes it is not necessary to settle the debate concerning the fundamental ontology of scientific theories. For more on this see Winther (2015).

[31]Plausibly, using a theory to make predictions about phenomena is simply an aspect of constructing explanations. After all, if explanations amount to information about dependence relations, then they allow one to make predictions about what will happen as well as to explain what has happened—explanation and prediction are simply backward looking (explanations) and forward looking (predictions) approaches to the same dependence relations. For this reason, we will focus primarily on explanation and the role that it plays in UT and UP. If one is convinced that explanation and prediction are not this closely connected, then she can construe the current discussion of UT in terms of the ability to construct explanations *and* the ability to make predictions.

[32]Both UT and UP come in degrees. One might have a greater or lesser degree of UT, and so understand a particular phenomenon that is explained by the relevant theory to a greater or lesser degree. In the text we are considering the high-end of the scale of UT—being able to construct explanations of phenomena requires a fairly significant degree of UT. This level of UT is something to strive for as an educational goal, but it is good to keep in mind that one can exhibit UT and UP without reaching this highest level. For example, one might have UT of a theory and UP of phenomena by being able to appreciate explanations of phenomena that are provided by a particular theory without being able to come up with such explanations on her own.

of an explanation of a particular phenomenon; you must be able to appreciate how the explanation provides an account of why or how the phenomenon occurs as it does.

9.2.2 *Understanding and Knowledge*

Now that we have gotten clearer on some of the important distinctions and basic points about the nature of understanding as well as its connection to explanation, it is time to briefly explore what is currently a major debate concerning explanation: the relationship between understanding and knowledge.

Interestingly, most philosophers of science hold that understanding is a kind of knowledge, but most epistemologists hold that understanding is not knowledge.[33] This debate is ongoing, and it does not need to be settled here. Nonetheless, we will briefly examine some of the considerations that have been adduced for thinking understanding (of either the UP or UT variety) is not a kind of knowledge. By doing so we will gain a sense of this debate.

To begin, one reason for thinking understanding is not a kind of knowledge is due to the purported transparency of understanding. Zagzebski (2001) argues that understanding is distinct from knowledge because while the former is transparent the latter is not. Zagzebski rightly notes that it is often the case that one knows that *p* without being aware of the factors which make it the case that she knows *p*. The reason for this is that a large number of the beliefs we have which are instances of knowledge depend upon other background evidence we possess for their justification. However, it is not the case that we are consciously aware of all of this background evidence. Consider your belief that you are reading this book. Setting aside skeptical worries, we can agree that your belief constitutes knowledge for you, i.e. you know that you are reading this book. The justification which you possess for this belief is dependent upon both your current perceptions and a large body of background evidence like the following: your knowledge of what a book is, your knowledge of what it is like to read a book, your knowledge of what experiences constitute your reading a book as opposed to your noting that someone else is reading a book, and so on. It is very probable that you are not consciously aware of all of the pieces of background evidence on which your knowledge that you are currently reading a book depends. So, your knowledge that you are reading a book is not transparent in the sense of your being aware of the factors which make it the case that you know it. Hence, Zagzebski is correct in claiming that knowledge is not always transparent because S can know that *p* without being aware of all of the factors which make it the case that she knows *p*.

[33]For a sampling of those who hold that understanding is a kind of knowledge see Achinstein (1983), Grimm (2006, 2009, 2014b), Khalifa (2011, 2013), Khalifa and Gadomski (2013), Kitcher (2002), Salmon (1990), Strevens (2008, 2013), and Woodward (2003). For a sampling of those who hold that understanding is not a kind of knowledge see Elgin (1996, 2007), Grimm (2001), Hills (2009, 2010), Kvanvig (2003, 2009), and Pritchard (2009, 2014).

Although knowledge is not transparent in this way, one might think that understanding is. According to Zagzebski, it is not possible for someone to have understanding of something and yet fail to be aware of the factors which make it the case that she understands it. This is an intuitive way of thinking of understanding, especially since understanding involves being aware of various relations between phenomena. In order to be said to understand phenomena one must be aware of the phenomena in some way and she has to be aware of the relations between the phenomena. Someone cannot lack awareness of this sort and still understand. Since it is the awareness of these relations which constitutes understanding, if someone understands something, she is aware of the factors which make it the case that she understands it. These sorts of considerations lead Zagzebski to plausibly claim that understanding and knowledge are two distinct kinds of things because understanding is transparent in this fashion, and knowledge is not.[34]

A further reason one might have for thinking understanding is not a kind of knowledge concerns factivity. As we have noted in earlier chapters, knowledge is factive—it is not possible to know that *p*, if *p* is false. Yet, some, such as Catherine Elgin (1996, 2007), argue that understanding is not factive.[35] Elgin (1996, p. 123) notes that the idea that "Objects in a vacuum fall toward the Earth at a rate of 32 ft/s" provides us with understanding of gravity and the rate of falling objects. Although this claim about the rate of falling objects provides understanding, it is not strictly speaking true. As Elgin observes, this claim ignores several factors which affect the rate at which an object will fall toward the earth, factors such as the force of the gravitational attractions of all other objects. Elgin even goes so far as to claim that the more complicated truth of this matter would be less likely to contribute to understanding of the phenomena than the simplified, but false, claim mentioned above.[36] Thus, Elgin claims that understanding is not a kind of knowledge because understanding is not factive.[37]

Perhaps the most persuasive reason to think that understanding is not a kind of knowledge comes from consideration of Gettier style cases. Jonathan Kvanvig (2003, pp. 197–98) makes a strong case for thinking understanding is not a kind of knowledge by way of the following example:

> Consider, say, someone's historical understanding of the Comanche dominance of the southern plains of North America from the late seventeenth until the late nineteenth centuries. Suppose that if you asked this person any question about this matter, she would

[34]See Grimm (2006) for criticisms of Zagzebski's claim that understanding is transparent in this way.

[35]It is worth noting that other philosophers who think understanding is not a kind of knowledge still maintain that understanding is factive, e.g. Kvanvig (2003, 2009) and Pritchard (2009, 2014).

[36]Elgin may be correct in this respect regarding the general person and her interests. However, it is highly dubious that the more complicated true statement would be less likely to contribute to the understanding of someone who comprehends this complicated statement as easily as the ordinary person does the simpler truth.

[37]The issue of how best to understand the role of idealizations in science is too complex to get into here. For now, it is simply worth mentioning that it is not clear that one must accept Elgin's view of how idealizations figure into our understanding of science.

> answer correctly. Assume further that the person is answering from stored information; she is not guessing or making up answers, but is honestly averring what she confidently believes the truth to be. Such an ability is surely constitutive of understanding, and the experience of query and answer, if sustained for a long enough period of time, would generate convincing evidence that the person in question understood the phenomenon of Comanche dominance of the southern plains.

Kvanvig adds to this example that the person gathered this information from an accurate history book, but that most all history books on the subject are inaccurate. Further, the person in this example had no reason to prefer the accurate history book that she did consult to any of the many inaccurate ones. So, it is a matter of luck that she came to possess accurate information, and her true beliefs about this subject are only accidentally connected to the truth. Thus, it seems the person in this example does not have knowledge of this subject because she is in a situation which matches the cases in the Gettier literature (notice, in particular, the similarity of this case to the fake barn case discussed in Chap. 8).

Although it seems the subject in this case does not have knowledge, Kvanvig argues that she does have understanding.[38] He points out that not only does she answer questions about this phenomenon correctly, but her "capacity for answering is counterfactual supporting . . . Ask anything about the phenomenon, and one would get a correct answer" (2003, p. 198). The person's awareness of the relations between the information she possesses concerning the Comanche dominance of the southern plains is not affected by how she came to be aware of those relations nor is her awareness of how the information she has about this phenomenon relates to her larger body of general historical information. Hence, it appears that the person in this example does not know because of the Gettier style situation she finds herself in, but she does understand. Therefore, it seems Kvanvig's example shows that understanding and knowledge are different in a significant way; understanding is not adversely affected by certain kinds of luck which keep one from having knowledge.[39]

Admittedly, the reasons we have examined in this section do not provide conclusive evidence that understanding is not a kind of knowledge, and the proposed

[38]See Hills (2009, 2010) and Pritchard (2009) for similar sorts of examples thought to show that understanding is distinct from knowledge. Grimm (2006, 2014b) and Khalifa (2013) contest whether such cases show that understanding is distinct from knowledge. It is worth noting, however, that Grimm does this by arguing that understanding is very similar to knowledge-how. As we noted in chapter two, knowledge-how seems to be importantly different than propositional knowledge (a difference which Grimm acknowledges). So, it may be that although Grimm argues that understanding is a kind of knowledge, his view is much closer to those who claim that understanding is not knowledge than it may seem at first glance.

[39]Pritchard (2009) helpfully distinguishes between two kinds of luck that arise in Gettier style cases. The first, which he terms "Gettier-style luck", is the sort of luck we find in Gettier's original cases and cases like the sheep in the field (all three of which were discussed in Chap. 8). The second, "environmental luck", is the sort of luck we find in fake barn cases. Pritchard argues that understanding is incompatible with Gettier-style luck, but it is compatible with environmental luck. Whereas, he claims that knowledge is incompatible with both kinds of luck.

distinction between understanding and knowledge is not without its detractors. Nonetheless, the fact that these issues are still up for debate should not trouble us. The purpose of this section was not to establish definitively that understanding is not a kind of knowledge, or that it is. Instead, the purpose of this section was merely to provide a sampling of some of the major points of contention in this debate. One final thought is worth mentioning. Even if it turns out that understanding is distinct from knowledge, this should not trouble us despite the fact that understanding is a primary aim of science. The reason for this is, as we have mentioned in Chap. 8 and earlier, the term "knowledge" with respect to scientific knowledge may perhaps be better understood as simply a general term for the sort of cognitive achievement we possess when we have understanding or simply a term for a strongly justified belief. At any rate, our conception of NOS may benefit from deeper appreciation of this debate, but it is not held hostage to its outcome.

9.3 Conclusion

In this chapter we have explored the nature of the fundamental aim of science, understanding. We have also examined the primary method by which we attain understanding in science, explanations. There are many ongoing debates about a number of issues concerning both understanding and explanation. However, we have enough here to help further our appreciation of the goals of science as well as how it is that we go about achieving those goals. One final point about understanding and explanation bears mentioning here. It seems plausible that since explanations aim at providing understanding, one very important criterion for determining whether one explanation is superior to another is the amount of understanding the explanation would provide if it were true (Lipton 2004). In the next chapter we will consider how we might compare explanations in order to gain the justification for accepting explanations and theories which is necessary for scientific knowledge. We will also see how the general explanatory method employed in science might be extended to provide an account of how our beliefs are justified more generally. Thus, we will explore the extent to which scientific knowledge is continuous with knowledge in general.

References

AAAS (American Association for the Advancement of Science). (1993). *Benchmarks for science literacy*. New York: Oxford University Press.

Achinstein, P. (1983). *The nature of explanation*. New York: Oxford University Press.

Braaten, M., & Windschitl, M. (2011). Working toward a stronger conceptualization of scientific explanation for science education. *Science Education, 95*, 639–669.

Braithwaite, R. (1953). *Scientific explanation*. Cambridge, UK: Cambridge University Press.

Bromberger, S. (1966). Why-questions. In R. G. Colodny (Ed.), *Mind and cosmos* (pp. 86–111). Pittsburgh: University of Pittsburgh Press.

Dawid, R. (2013). *String theory and the scientific method*. Cambridge, UK: Cambridge University Press.
de Regt, H. W. (2004). Discussion note: Making sense of understanding. *Philosophy of Science, 71*, 98–109.
de Regt, H. W. (2009). Understanding and scientific explanation. In H. W. de Regt, S. Leonelli, & K. Eigner (Eds.), *Scientific understanding: Philosophical perspectives* (pp. 21–42). Pittsburgh: University of Pittsburgh Press.
de Regt, H. W. (2013). Understanding and explanation: Living apart together? *Studies in History and Philosophy of Science, 44*, 505–509.
de Regt, H. W., & Dieks, D. (2005). A contextual approach to scientific understanding. *Synthese, 144*, 137–170.
de Regt, H. W., Leonelli, S., & Eigner, K. (2009). Focusing on scientific understanding. In H. W. de Regt, S. Leonelli, & K. Eigner (Eds.), *Scientific understanding: Philosophical perspectives* (pp. 1–17). Pittsburgh: University of Pittsburgh Press.
Díez, J., Khalifa, K., & Leuridan, B. (2013). General theories of explanation: Buyer beware. *Synthese, 190*, 379–396.
Elgin, C. Z. (1996). *Considered judgment*. Princeton: Princeton University Press.
Elgin, C. Z. (2007). Understanding and the facts. *Philosophical Studies, 132*, 33–42.
Feldman, R. (1974). An alleged defect in Gettier counter-examples. *Australasian Journal of Philosophy, 52*, 68–69.
Franklin, R. L. (1981). Knowledge, belief and understanding. *Philosophical Quarterly, 31*, 193–208.
Friedman, M. (1974). Explanation and scientific understanding. *Journal of Philosophy, 71*, 5–19.
Gardiner, P. (1959). *The nature of historical explanation*. Oxford: Oxford University Press.
Gauch, H. G., Jr. (2012). *Scientific method in brief*. Cambridge, UK: Cambridge University Press.
Gijsbers, V. (2013). Understanding, explanation, and unification. *Studies in History and Philosophy of Science, 44*, 516–522.
Gopnik, A. (1998). Explanation as orgasm. *Minds and Machines, 8*, 101–118.
Gopnik, A., & Glymour, C. (2002). Causal maps and Bayes nets: A cognitive and computational account of theory formation. In P. Carruthers, S. Stich, & M. Siegal (Eds.), *The cognitive basis of science* (pp. 117–132). Cambridge, UK: Cambridge University Press.
Gopnik, A., & Sobel, D. (2000). Detecting blickets: How young children use information about novel causal powers in categorization and induction. *Child Development, 71*, 1205–1222.
Gopnik, A., Glymour, C., Sobel, D., Schulz, L., Kushnir, T., & Danks, D. (2004). A theory of causal learning in children: Causal maps and Bayes nets. *Psychology Review, 111*, 1–31.
Grimm, S. (2001). Ernest Sosa, knowledge, and understanding. *Philosophical Studies, 106*, 171–191.
Grimm, S. (2006). Is understanding a species of knowledge? *British Journal for the Philosophy of Science, 57*, 515–535.
Grimm, S. (2009). Reliability and the sense of understanding. In H. W. de Regt, S. Leonelli, & K. Eigner (Eds.), *Scientific understanding: Philosophical perspectives* (pp. 83–99). Pittsburgh: University of Pittsburgh Press.
Grimm, S. (2014a). Understanding. In S. Bernecker & D. Pritchard (Eds.), *The Routledge companion to epistemology* (pp. 84–94). New York: Routledge.
Grimm, S. (2014b). Understanding as knowledge of causes. In A. Fairweather (Ed.), *Virtue epistemology naturalized: Bridges between virtue epistemology and philosophy of science* (pp. 329–346). Dordrecht: Springer.
Harman, G. (1986). *Change in view*. Cambridge, MA: MIT Press.
Hempel, C. G. (1942). The function of general laws in history. *Journal of Philosophy, 39*, 35–48.
Hempel, C. G. (1965). *Aspects of scientific explanation and other essays in the philosophy of science*. New York: The Free Press.
Hempel, C. G., & Oppenheim, P. (1948). Studies in the logic of explanation. *Philosophy of Science, 15*, 135–175.
Hills, A. (2009). Moral testimony and moral epistemology. *Ethics, 120*, 94–127.

Hills, A. (2010). *The beloved self: Morality and the challenge from egoism*. New York: Oxford University Press.
Hindriks, F. (2013). Explanation, understanding, and unrealistic models. *Studies in History and Philosophy of Science, 44*, 523–531.
Kampourakis, K. (2014). *Understanding evolution*. Cambridge, UK: Cambridge University Press.
Khalifa, K. (2011). Understanding, knowledge, and scientific antirealism. *Grazer Philosophische Studien, 83*, 93–112.
Khalifa, K. (2012). Inaugurating understanding or repackaging explanation? *Philosophy of Science, 79*, 15–37.
Khalifa, K. (2013). Understanding, grasping, and luck. *Episteme, 10*, 1–17.
Khalifa, K., & Gadomski, M. (2013). Understanding as explanatory knowledge: The case of Bjorken scaling. *Studies in History and Philosophy of Science, 44*, 384–392.
Kim, J. (1994). Explanatory knowledge and metaphysical dependence. *Philosophical Issues, 5*, 51–69.
Kitcher, P. (1981). Explanatory unification. *Philosophy of Science, 48*, 507–531.
Kitcher, P. (1989). Explanatory unification and the causal structure of the world. *Minnesota Studies in the Philosophy of Science, 13*, 410–505.
Kitcher, P. (2002). Scientific knowledge. In P. Moser (Ed.), *The Oxford handbook of epistemology* (pp. 385–407). New York: Oxford University Press.
Kvanvig, J. (2003). *The value of knowledge and the pursuit of understanding*. Cambridge, UK: Cambridge University Press.
Kvanvig, J. (2009). The value of understanding. In A. Haddock, A. Millar, & D. Pritchard (Eds.), *Epistemic value* (pp. 95–111). Oxford: Oxford University Press.
Lehrer, K. (2000). *Theory of knowledge* (2nd ed.). Boulder: Westview Press.
Lewis, D. (1986). Causal explanation. In *Philosophical papers, volume II* (pp. 214–240). Oxford: Oxford University Press.
Lipton, P. (2004). *Inference to the best explanation* (2nd ed.). New York: Routledge.
Lipton, P. (2009). Understanding without explanation. In H. W. de Regt, S. Leonelli, & K. Eigner (Eds.), *Scientific understanding: Philosophical perspectives* (pp. 43–63). Pittsburgh: University of Pittsburgh Press.
Mancosu, P. (2011). Explanation in mathematics. In E. N. Zalta (Ed.), *The Stanford encyclopedia of philosophy* (Summer 2011 Edition). http://plato.stanford.edu/archives/sum2011/entries/mathematics-explanation/
McCain, K., & Weslake, B. (2013). Evolutionary theory and the epistemology of science. In K. Kampourakis (Ed.), *The philosophy of biology: A companion for educators* (pp. 101–119). Dordrecht: Springer.
Moser, P. K. (1989). *Knowledge and evidence*. Cambridge, UK: Cambridge University Press.
Nagel, E. (1961). *The structure of science: Problems in the logic of scientific explanation*. New York: Harcourt, Brace and World.
National Research Council. (1996). *National science education standards*. Washington, DC: The National Academies Press.
National Research Council. (2007). *Taking science to school: Learning and teaching science in grades K-8*. Committee on Science Learning, Kindergarten through Eighth Grade. R. A. Duschl, H. A. Schweingruber, & A. W. Shouse (Eds.), Board on Science Education, Center for Science Education, Division of Behavioral and Social Sciences and Education. Washington, DC: The National Academies Press.
National Research Council. (2012). *A framework for K-12 science education: Practices, crosscutting concepts, and core ideas*. Committee on a Conceptual Framework for New K-12 Science Education Standards. Board on Science Education, Division of Behavioral and Social Sciences and Education. Washington, DC: The National Academies Press.
NGSS Lead States. (2013). *Next generation science standards: For states, by states*. Washington, DC: The National Academies Press.
Nickel, B. (2010). How general do theories of explanation need to be? *Nous, 44*, 305–328.
Popper, K. R. (1959). *The logic of scientific discovery*. London: Hutchinson.

Pritchard, D. (2009). Knowledge, understanding and epistemic value. In A. O'Hear (Ed.), *Epistemology: Royal Institute of Philosophy Supplement 64* (pp. 19–44). Cambridge, UK: Cambridge University Press.
Pritchard, D. (2014). Knowledge and understanding. In A. Fairweather (Ed.), *Virtue epistemology naturalized: Bridges between virtue epistemology and philosophy of science* (pp. 315–328). Dordrecht: Springer.
Railton, P. (1993). Probability, explanation, and information. In D.-H. Ruben (Ed.), *Explanation* (pp. 160–181). New York: Oxford University Press.
Rosenberg, A. (2012). *Philosophy of science: A contemporary introduction* (3rd ed.). New York: Routledge.
Salmon, W. (1971). Statistical explanation. In W. Salmon (Ed.), *Statistical explanation and statistical relevance* (pp. 29–87). Pittsburgh: University of Pittsburgh Press.
Salmon, W. (1984). *Explanation and the causal structure of the world*. Princeton: Princeton University Press.
Salmon, W. (1990). *Four decades of scientific explanation*. Minneapolis: University of Minnesota Press.
Salmon, W. (1997). Causality and explanation: A reply to two critiques. *Philosophy of Science, 64*, 461–477.
Salmon, W. (1998). *Causality and explanation*. New York: Oxford University Press.
Schrödinger, E. (1954). *Nature and the Greeks*. Cambridge, UK: Cambridge University Press.
Scriven, M. (1959). Explanation and prediction in evolutionary theory. *Science, 130*, 477–482.
Sober, E. (1983). Equilibrium explanation. *Philosophical Studies, 43*, 201–210.
Steyvers, M., Tenenbaum, J., Wagenmakers, E., & Blum, B. (2003). Inferring causal networks from observations and interventions. *Cognitive Science, 27*, 453–489.
Strevens, M. (2006). Scientific explanation. In D. M. Borchert (Ed.), *Encyclopedia of philosophy* (2nd ed.). Detroit: Macmillan Reference USA.
Strevens, M. (2008). *Depth: An account of scientific explanation*. Cambridge, MA: Harvard University Press.
Strevens, M. (2013). No understanding without explanation. *Studies in History and Philosophy of Science, 44*, 510–515.
Trout, J. D. (2002). Scientific explanation and the sense of understanding. *Philosophy of Science, 69*, 212–233.
Trout, J. D. (2005). Paying the price for a theory of explanation: De Regt's discussion of Trout. *Philosophy of Science, 72*, 198–208.
Trout, J. D. (2007). The psychology of scientific explanation. *Philosophy Compass, 2*, 564–591.
van Fraassen, B. C. (1980). *The scientific image*. Oxford: Oxford University Press.
van Fraassen, B. C. (1989). *Laws and symmetry*. Oxford: Oxford University Press.
von Wright, G. H. (1971). *Explanation and understanding*. Ithaca: Cornell University Press.
Wilkenfeld, D. (2013). Understanding as representation manipulability. *Synthese, 190*, 997–1016.
Wilkenfeld, D. (2014). Functional explaining: A new approach to the philosophy of explanation. *Synthese, 191*, 3367–3391.
Winther, R. G. (2015). The structure of scientific theories. In E. N. Zalta (Ed.), *The Stanford encyclopedia of philosophy* (Spring 2015 Edition). http://plato.stanford.edu/archives/spr2015/entries/structure-scientific-theories/
Woodward, J. (2003). *Making things happen: A theory of causal explanation*. New York: Oxford University Press.
Woodward, J. (2014). Scientific explanation. In E. N. Zalta (Ed.), *The Stanford encyclopedia of philosophy* (Winter 2014 Edition). http://plato.stanford.edu/archives/win2014/entries/scientific-explanation/
Zagzebski, L. (2001). Recovering understanding. In M. Steup (Ed.), *Knowledge, truth, and duty: Essays on epistemic justification, responsibility, and virtue* (pp. 235–252). New York: Oxford University Press.

Chapter 10
From Explanation to Knowledge

Abstract Building upon the insights of the previous chapter concerning the nature of explanation and its relation to understanding this chapter argues for a close connection between explanation and evidential support. That is to say, this chapter argues that the degree to which a given body of evidence supports believing that a particular proposition is true depends upon how well that proposition explains the evidence or is explained by the best explanation of that evidence. The upshot of this explanationist view of evidential support is that explanation is an integral component of epistemic justification. As a result of detailing this explanationist view of evidential support, this chapter offers a clear conception of when we should accept claims in science as well as an account of epistemic justification more generally. Thus, the chapter establishes a very close connection between scientific inference and the justification we might have for any of our beliefs.

In the previous chapter we explored the nature of explanation and its aim, understanding. We noted it is plausible that in order to have scientific understanding of a phenomenon (UP) it is necessary to have understanding of a particular theory, or theories (UT). One must understand a theory before she can use that theory to generate, and possibly even to truly understand, explanations (explanatory hypotheses) of phenomena.

Although the connection between UT, explanations, and UP is very important, a key component of scientific knowledge is missing. We need to have a grasp of what is required to know, or at least justifiedly believe (recall our discussion from Chap. 8 that justification may be what we really care about rather than knowledge), that the theories for which we have UT are true. After all, having UT, understanding how a theory, *T*, works and being able to generate possible explanations of a particular phenomenon from *T* will not help us to truly understand the phenomenon unless we know that *T* is true. For example, understanding my theory of gnomes and being able to use this theory to generate an explanation of why iron rusts (because gnomes love the taste of iron and their saliva causes rust, of course!) does not provide one with genuine understanding of why iron rusts. Why not? My theory about gnomes is simply not true. Even though you can generate "explanations" using my theory, you cannot gain understanding of the phenomena "explained" by those explanations. You need to know that the theory is true. This sort of knowledge is an important part

K. McCain, *The Nature of Scientific Knowledge*, Springer Undergraduate Texts in Philosophy, DOI 10.1007/978-3-319-33405-9_10

of our scientific knowledge. But, how do we come to have this scientific knowledge? We will see in this chapter that such knowledge comes by way of explanatory inferences.[1] In this chapter we will also explore the connection between this sort of explanatory reasoning and the justification of our beliefs more generally.

10.1 Knowledge of Scientific Theories

When it comes to our knowledge of scientific theories part of the picture is quite mundane and unsurprising. Often we base our beliefs concerning which theories are true, at least in part, on observations. How do we gain knowledge of the observational data that we use to support these theories? Just like we gain knowledge of any other empirical facts about the world around us—through observation or by learning of the observations others have made. We use our senses either directly such as when we look outside our window to see if the dog is in the yard, or indirectly by using measuring instruments such as when we determine the temperature by looking at a thermometer or when we receive reports of someone else's observations of deviations in Uranus' orbit. We will discuss how our sensory experiences might be understood to support our observational beliefs below. For now it is sufficient to recognize that this piece of our scientific knowledge, knowledge of observations, is the same as any other knowledge we gain by using our five senses or via the testimony of others as to what they have observed.

The issue that is of interest to us is how we move from observations to knowledge of a particular scientific theory.[2] We tend to do so via a particular method of inference: inference to the best explanation (*IBE*).[3] In simplest terms the idea behind IBE is that explanatory virtues are a guide to truth. That is to say, "the explanation that would, if true, provide the deepest understanding is the explanation that is likeliest to be true" (Lipton 2004, p. 61). IBE is extremely widespread in the sciences. As Clark Glymour (1984, p. 173) aptly notes "One can find such arguments [inferences to the best explanation] in sociology, in psychometrics, in chemistry and astronomy, in the time of Copernicus, and in the most recent of our scientific journals." Here is an historical example to help illustrate the use of IBE in science:

[1]Similar considerations apply to our knowledge of the explanatory hypotheses generated from theories and our knowledge of laws of nature.

[2]Here we are concerned with the sort of knowledge that is gained in a particular scientific context—how a theorist can come to know that a particular theory is true. Later, in Chap. 15, we will discuss how one can gain scientific knowledge via testimony from others whether this is through studying the written works of others or being told directly about the theories and their claims, supporting evidence, and so on.

[3]IBE is sometimes referred to as "abduction". It is best to use separate terminology because there are good reasons for thinking IBE and abduction are not the same. See Hintikka (1998) and Minnameier (2004) for arguments to this effect.

> At the beginning of the nineteenth century, it was discovered that the orbit of Uranus, one of the seven planets known at the time, departed from the orbit as predicted on the basis of Isaac Newton's theory of universal gravitation and the auxiliary assumption that there were no further planets in the solar system. One possible explanation was, of course, that Newton's theory is false. Given its great empirical successes for (then) more than two centuries, that did not appear to be a very good explanation. Two astronomers, John Couch Adams and Urbain Leverrier, instead suggested (independently of each other but almost simultaneously) that there was an eighth, as yet undiscovered planet in the solar system; that, they thought, provided the best explanation of Uranus' deviating orbit. Not much later, this planet, which is now known as "Neptune," was discovered. (Douven 2011)

There are many other examples of IBE being used in the history of science. Copernicus' main argument in support of the heliocentric model of the solar system over the geocentric model of Ptolemaic theory was that his theory provided the best explanation of the observational data (Gauch 2012). Galileo's arguments in support of the Copernican view of the solar system were IBE arguments too (Pitt 1988). Joseph John Thomson's discovery of the electron was an IBE concerning the behavior of cathode rays (Achinstein 2001). Joseph Priestley's discovery of photosynthesis involved employing IBE (Matthews 2015). The reason that we are justified in accepting Einstein's Special Relativity instead of Lorentz's version of aether theory, which also fits the empirical data, is that the former is the best explanation of the data (Janssen 2002). There are several other instances in the history of science where IBE has been successfully employed, for example Antoine Lavoisier's argument against phlogiston theory in favor of the oxygen theory of combustion, Christiaan Huygens' argument in support of the wave theory of light, and, of course, Charles Darwin's argument in support of natural selection (Thagard 1978). As Darwin (1859/1962, p. 476) said about his theory, "it can hardly be supposed that a false theory would explain, in so satisfactory a manner as does the theory of natural selection, the several large classes of facts above specified. It has recently been objected that this is an unsafe method of arguing; but it is a method used in judging of the common events of life, and has often been used by the greatest natural philosophers." Isaac Newton seemed to support IBE as well. In his *Principia* Newton offered several rules for scientific reasoning. The very first rule of reasoning that he offered is an appeal to the sort of explanatory virtues used in IBE. "No more causes of natural things should be admitted than are both true and sufficient to explain their phenomena...For nature is simple and does not indulge in the luxury of superfluous causes" (Newton 1687/1999, p. 794). Clearly, Newton believed that simplicity is a very important factor in choosing which theory to accept. There are numerous additional examples of IBE employed both historically and currently in the sciences. The many instances where Ockham's razor (considerations of parsimony/simplicity) is employed to support one theory over another are all examples of IBE.[4] In fact, IBE is so pervasive in the sciences that Ernan McMullin (1992) refers to it as "the inference that makes science."

[4]See Sober (2015) for discussion of the many historical uses and defenses of Ockham's razor as a method of theory selection.

It is clear that IBE plays a major role in the sciences (Boyd 1981, 1984; Douven 2011; Harré 1986; Lipton 2004; McMullin 1992; Psillos 1999). It accounts for how we gain scientific knowledge of theories.[5] When a particular theory generates specific hypotheses that best explain a sufficiently large set of data and those explanations are sufficiently better than their rivals, we can infer that those hypotheses are true. When the best explanation of why a theory produces true hypotheses is that the theory is true, such as in cases where the hypotheses of the theory offer novel predictions, we can infer that the theory itself is true. This is how we gain knowledge of scientific theories. Of course, such knowledge is tentative in the sense that we may revise what we think about a particular theory in the light of new evidence or new rival theories. After all, additional evidence, or new rival theories, may make it so that the hypotheses provided by the original theory are no longer the best explanations, or the new rival may provide the same explanations as the original while providing additional explanatory hypotheses of even more data.

Now that we have had a glimpse of the pervasiveness of IBE in the sciences and we have seen that IBE is the common method by which we gain knowledge of scientific hypotheses and theories, it is worth exploring the structure of such inferences in more detail.[6] Here is the general structure of IBE:

> $F_1, F_2, \ldots, F_n$ are facts in need of explanation.
> Hypothesis H explains the F_i.
> No available competing hypotheses would explain the F_i as well as H does.
>
> ---
>
> Therefore, H is true (Lycan 2002, p. 413).[7]

Something like this formulation seems to be what most people have in mind when they speak of IBE.

There are a number of points about the above formulation of IBE that need to be clarified. First, the line "$F_1, F_2, \ldots, F_n$ are facts in need of explanation" needs to be made clearer. Specifically, it is important to consider what counts as a "fact". For our purposes, facts are simply evidence we have and which is taken as given. For example, in a scientific setting the "facts" are typically the experimental observations that have been made.

Second, recall from the previous chapter that to say H explains F is to say that H (perhaps along with various other auxiliary hypotheses or further information) provides information about dependence relations which hold between various

---

[5]Despite its ubiquity in the sciences, some question the veracity of IBE. We will consider some of the primary objections that have been leveled at IBE in Chap. 14. We will see that these objections are not persuasive.

[6]Presumably, we come to have knowledge of laws of nature by either coming to know a theory of which they are a part or by inferring them from numerous known theories which are themselves best explained by the truth of the law. Hence, IBE is integral to this aspect of our scientific knowledge as well.

[7]Also, see Lycan (1988) for a similar, but different formulation. The primary difference is that Lycan's (1988) formulation has the conclusion that "[probably] H is true" instead of "H is true".

phenomena and F. For instance, the hypothesis that a baseball hit the window (H) explains the window's breaking (F) because, along with auxiliary information about windows and baseballs, H provides information about the dependence of F on other phenomena.

Third, it is important to note a distinction we glossed over in the previous chapter—the distinction between *actual* explanations and *potential* explanations. An actual explanation of a phenomenon is true. Potential explanations are such that they *would* provide understanding of the phenomenon to be explained *if* they were true. The explanations referred to in the above formulation should be understood as potential explanations. Thus, when making an inference to the best explanation one is inferring that the best potential explanation of F is the actual explanation of F.

Fourth, the line "No available competing hypotheses would explain the F_i as well as H does" contains two points that need clarification. One point is what is meant by "available" hypotheses. Admittedly, there are some difficult issues in trying to say when exactly an explanation is available to someone. For instance, it might be thought that for hypothesis H to be available to S, S must be consciously aware of H. Alternatively, one might think that H is available to S so long as S possesses the requisite concepts and cognitive abilities to understand H. Though giving precise availability conditions for an explanation is difficult, one thing seems indisputable; if S is consciously aware of H and she understands H (in the UT sense we explored in the previous chapter), then H is available to her. Plausibly, we might hold that S has H available anytime she has UT of H or she has UT of the theory from which she can generate H, even if she is not consciously aware of H at the moment. We will say a bit more about availability below.

Another point that needs to be clarified is what it means for hypotheses to be "competing". Competing hypotheses are simply contrary hypotheses which offer explanations of a fact or set of facts. On the one hand, the hypothesis that a baseball broke the window competes with the hypothesis that Piper broke the window by kicking it and then she placed a baseball by the window. At most one of these hypotheses is true. On the other hand, the hypothesis which purports to explain why ice melts in water by appealing to modern chemistry is not a competitor for the hypothesis which purports to explain this phenomenon by appealing to quantum mechanics. It is possible that both of these hypotheses are true. So, two hypotheses are competing when it comes to a particular claim just in case they disagree about that claim—when they cannot both be correct about the claim.

Fifth, it is important to understand what it means to say that no available hypothesis explains F *as well as* H does. This is essentially the claim that the explanation inferred to be true must be the *best* available explanation. Before continuing it is worth very briefly considering what might make one explanation better than another. A host of explanatory virtues have been identified and appealed to in various scientific contexts: empirical adequacy (actually explaining the data in question), various kinds of simplicity, explanatory power (the range of phenomena explained and/or how illuminating the explanation is), consistency with currently accepted theories, non-ad hocness, predictive power (making novel predictions), and raising fewer unanswered questions are just a few.[8] In general, inferences to the best explanation involve evaluating the various available hypotheses in terms of their explanatory virtues and inferring that the most explanatory virtuous hypothesis is (likely) true. Peter Lipton (2004, p. 59) offers the following straightforward construal of what it means for something to be the best explanation: "we may characterize the best explanation as the one which would, if correct, be the most explanatory or provide the most understanding." So, according to Lipton, even if various competing hypotheses each explain all of the relevant facts, there can still be a best explanation among these competitors. The best explanation is the one that, if true, would provide the most understanding of the dependence relations of the phenomena to be explained.

Sixth, another key point which needs to be clarified is that the mere fact that a potential explanation is the best available explanation is not enough for the truth of the explanation to be inferred. The best must also be a good (enough) explanation of the phenomenon in question. That is to say, in order for an explanation to be legitimately inferred as true because it is the best available explanation the explanation needs to meet certain minimal standards.[9] Admittedly, it is a difficult task to make these minimal standards completely explicit. Nevertheless, for the current purposes an intuitive understanding should suffice. To help make the general idea clearer consider the following sort of example:

> Monte has three available explanations for a particular phenomenon, which has ten points of data that require explaining. H_1, H_2 and H_3 each individually explain one of these data points, but offer no explanation of the other nine.

Intuitively, in this example Monte should not infer that the best of the three explanations, H_1 say, is true because it does not sufficiently explain the relevant data. Alternatively, consider the following example:

> Martha has three available explanations for a particular phenomenon, which has ten points of data that require explaining. H_1, H_2 and H_3 each individually explain all ten of these data points.

Whichever explanation is best H_1, H_2 or H_3 clearly meets the standards of being a good (enough) explanation of the data. Certainly, there are difficult cases that lie between these two examples; however, the intuitive idea behind the restriction is

[8]See Beebe (2009), Lacey (2005), Lipton (2004), Longino (1990), Lycan (1988), Kuhn (1977), McAllister (1996), McMullin (1982), Quine and Ullian (1978), Thagard (1978), and Vogel (1990) for a sampling of the explanatory virtues that have been proposed in various scientific contexts and the literature on the nature of explanation. Some might question whether all of the virtues listed are distinct—for example, some claim that predictive power is what separates ad hoc theories from those that are not (Popper 1959; Psillos 1999). As a result, they might question whether predictive power and non-ad hocness are actually two virtues rather than one. Fortunately, for our purposes it is sufficient to simply have a grasp of what some of the most commonly cited explanatory virtues are.

[9]For more on this see Lipton (2004).

fairly clear. Thus, an important caveat for our understanding of IBE is that it is not enough to be the best available explanation; the inferred explanation must be the best available explanation, *and* it must be a sufficiently good explanation of the relevant data.

Now that we have a good handle on the nature of IBE it is worth briefly sketching a picture of our scientific knowledge. Inferring to the best explanation allows us to gain knowledge of explanatory hypotheses which are generated by our scientific theories on the basis of their explaining our observational data. From our knowledge of explanatory hypotheses we are able to infer that our scientific theories are true. If one then has UT of that theory, she can use the theory she knows to be true to construct (or at least appreciate) explanations of further phenomena. These explanations can in turn lead to her having UP of the phenomena in question. The exact process by which we gain knowledge of any particular theory is apt to be much more complex than what we have here, but we do have a good handle on how the various pieces of our puzzle fit together to yield scientific knowledge and understanding.

10.2 IBE Everywhere

In addition to being widely used in the sciences, inference to the best explanation is ubiquitously employed in everyday life. We use IBE to infer that a patient has a particular illness from the fact that she has particular symptoms. In fact, some claim that IBE is the primary method of medical diagnosis (Josephson and Josephson 1994).[10] IBE is used in agriculture to construct models which help increase crop yields (Gauch 1992, 2012). Many philosophers argue that we use IBE when we gain information via the testimony of others. Our explanatory reasoning is key to whether or not we trust the information we have been given by a testifier (Adler 1994; Fricker 1994; Harman 1965; Lipton 1998). Some claim that the use of IBE is relevant to our comprehension of language because it is required to even determine what speakers mean when they tell us things (Dascal 1979; Hobbs 2004). We use IBE so often in our lives many philosophers and psychologists claim that it may be "so routine and automatic that it easily goes unnoticed" (Douven 2011).

Explanatory reasoning more broadly, whether it occurs in explicit instances of IBE or implicit ones, is pervasive in our everyday lives. There are large amounts of cross-cultural research which suggests explanations are pervasive in "our activities from the most simple and mundane... to the most sophisticated and unusual"

[10]This claim is somewhat controversial because some think appeal to likelihood ratios alone may be the key to medical diagnosis. Although it is plausible that likelihood ratios can be important tools in medical diagnosis (see Grimes and Schulz 2005), it is not clear that even their use cannot be accounted for under the umbrella of IBE (see Chap. 12). For present purposes, it is enough to note that it has been claimed that IBE is the primary method of medical diagnosis, and this claim has some plausibility.

(Wilson and Keil 2000, p. 87). As William Brewer et al. (2000, p. 281) point out "when one is given an explanation of a phenomenon there appears to be a natural human tendency to evaluate the quality of the explanation." This tendency seems so natural that many think explanatory reasoning may be hard-wired into us. It may be that we are simply born equipped to engage in explanatory reasoning. Even if we are not born ready to engage in this sort of reasoning, it seems clear that the capabilities for explanatory reasoning appear very early in our development. Evidence suggests that infants may have "at least a rudimentary form of explanatory understanding" (Keil and Wilson 2000, p. 4). Brewer et al. (2000, p. 103) assert that their cognitive science research "suggests that, qualitatively, children show competence with most aspects of everyday explanations at an early age." Importantly, the "aspects of everyday explanations" which Brewer et al. refer to are what are typically understood as the sort of explanatory virtues we mentioned above, things like: consistency, explanatory power, simplicity, etc. Further support for the idea that children are competent at evaluating explanations comes from the work of Alison Gopnik (1998, p. 103), who claims that children have "powerful and flexible theory-formation abilities", which seems to suggest they are good at coming up with their own explanations. The empirical evidence clearly suggests that IBE is ubiquitous throughout our entire lives. It is so ubiquitous that it is not uncommon to think that IBE is a basic belief forming method for humans (Enoch and Schechter 2008).

Explanation, understanding, and knowledge all come together in IBE. IBE involves using our understanding of theories to generate explanations which are evaluated in terms of their explanatory virtues. The most virtuous explanation is inferred to be true (or at least likely to be true). Once we infer the truth of the best explanation of a given set of data we arrive at further understanding and knowledge of the phenomena it explains. This method of reasoning, IBE, is a powerful mode of arriving at knowledge in both science and our everyday lives.[11]

Given how integral IBE is to both our scientific and everyday reasoning, we have some reason to think that perhaps the way to understand what is required for a given body of evidence to justify believing a particular proposition (evidential support) is really a matter of explanatory relations. This is an insight that has been noted by many "explanationists" (those who understand epistemic justification to be a matter

[11]Despite its widespread use in science and everyday life, IBE is not without its critics. See van Fraassen (1989), Ladyman et al. (1997), Roche and Sober (2013), and Wray (2008). One of the lines of criticism many find particularly troubling is the claim that IBE leads to probabilistic incoherence. In other words, critics charge that IBE is inconsistent with accepted theories of probabilistic reasoning such as Bayesianism. For a survey of responses to objections to IBE see Douven (2011). For responses to the claim that IBE runs afoul of probabilistic reasoning see Lipton (2004), McCain and Poston (2014), McGrew (2003), Okasha (2000), Psillos (1999), and Weisberg (2009). Some (Huemer 2009; Poston 2014) even go so far as to argue that without IBE probabilistic reasoning, including Bayesian confirmation theory, straightforwardly falls prey to the skeptical problem of induction. We will explore criticisms of IBE as well as responses to those criticisms more fully in Chap. 14.

of explanatory considerations).[12] We will turn our attention to the prospects for such an explanationist account of evidential support now.

10.3 Explanatory Reasoning and Evidential Support

Consideration of IBE and explanatory reasoning more generally puts us in a position to answer a promissory note given in Chap. 5. We are now in a position to sketch an account of what it takes for a body of evidence to provide justification for believing a proposition (evidential support).[13] As we have noted, a plausible place to look for an account of evidential support is in our practices of explanatory reasoning. As Earl Conee and Richard Feldman (2008, p. 97) explain, explanationists hold that "fundamental epistemic principles are principles of best explanation." Here is a plausible way of spelling out evidential support in terms of explanatory considerations:

Explanationism

A person, S, with evidence *e* at time *t* is justified in believing *p* at *t* if and only if at *t* S has considered *p*, and:

(i) *p* is part of the best explanation available to S at *t* for why S has *e*

or

(ii) *p* is available to S as an explanatory consequence of the best explanation available to S at *t* for why S has *e*.[14]

It is worth very briefly explicating a few points about *Explanationism* before continuing.

First, when we discuss whether S is justified in believing that *p*, "S's evidence" should be understood to mean S's total evidence. This is important to keep in mind because making judgments about justification while construing *Explanationism* in terms of only part of S's evidence would have the obvious flaw of ignoring the potential justificatory impact of defeating evidence. For instance, the best explanation of why Jeff seems to see an elephant is that there is an elephant in his vicinity. However, the best explanation for why Jeff seems to see an elephant when he is on a commercial airline flight, has taken a medication which has the side-effect of causing elephant hallucinations, and several trustworthy companions are assuring him that there are no elephants around is that Jeff is suffering one of the side-effects of his medication and hallucinating an elephant. So, <there is an elephant in the vicinity> is the best explanation of part of Jeff's evidence, but it

---

[12] Conee and Feldman (2008), Goodman (1965, 1978), Harman (1973, 1986), Lycan (1988, 2012), McCain (2013, 2014), Moser (1989), Poston (2014), and Sellars (1963) each defend explanationist theories of justification.

[13] The account of evidential support that we will sketch, and the subsequent discussion of it, is based on the account defended in McCain (2013, 2014, 2015).

[14] McCain (2015, p. 339) This is also very similar to McCain's (2014) "*Ex-EJ*".

is not the best explanation when we consider all of the evidence that he has. This makes clear that it is very important to take account of the total evidence one has when determining whether a particular proposition is justified for her. For simplicity, the examples below only focus on a portion of S's evidence; however, it is assumed in the examples that S's total evidence does not include defeating evidence relevant to the proposition being discussed.

Second, in order for *p* to be part of the "best" explanation available to S at *t* it must be the case that there is no competing explanation with respect to *p* available to S at *t* which is as good of an explanation as, or better than, the best explanation of which *p* is a part. Importantly, this does not require S to have a unique best explanation available to her. It could be that S has two explanations available to her, H_1 and H_2. H_1 and H_2 are both equally good explanations of why S has the evidence that she does. Consequently, S could not justifiedly accept H_1 over H_2 or vice versa. In such a case S may still be justified in believing that *p* because it is part of both H_1 and H_2. If *p* is not a part of one of the two explanations, H_1 say, then *p* is not part of the best explanation available to S because there is an equally good explanation available to her which does not have *p* as a part—H_1 and H_2 are equally good after all. In order to be justified in believing that *p* there cannot be an explanation which does not contain *p* that is as good as the best explanation S has which contains *p*.

Relatedly, as we noted above with respect to IBE, the mere fact that a potential explanation is the best available explanation is not enough for the truth of that explanation to be inferred. The best must also be a good (enough) explanation of the phenomenon in question. Similar considerations apply to *Explanationism*. In order for S to be justified in believing that *p* it must not only be the best available explanation of S's evidence, it must also be a sufficiently good explanation of S's evidence.

Finally, it is important to say a bit about what it means for S to have an explanation available or have *p* available as an explanatory consequence of the best explanation available to S. The simplest way to explain what "available" means here is in similar terms to how we construed UT in the previous chapter. Recall, UT requires being able to use a theory to construct or to appreciate explanatory hypotheses. Similarly, we can roughly characterize when S has an explanation available as when she can construct that explanation or at least appreciate it as an account of why she has the evidence that she does without having to first gather additional evidence. Likewise, we can roughly characterize S's having *p* available as an explanatory consequence of the best explanation available to S as her being in a position to recognize that the best available explanation of her evidence would explain *p* significantly better than it would $\sim p$.[15]

---

[15]This approach is influenced by earlier explanationist views such as Harman (1973) where *p* is justified when it explains or is explained by one's evidence. Notably, the approach here does not say that *p* is justified when it is explained by one's evidence though. Rather, it holds that *p* is justified when it best explains S's evidence or when it would be explained by the best explanation of S's evidence. The difference here is subtle, but important.

Of course, we are glossing over many details that an epistemologist would require before accepting *Explanationism* as the correct account of evidential support. For our purposes though we do not need to go into all of these details. We are not attempting to defend a complete account of evidential support and epistemic justification here.[16] Instead, we are simply providing a sketch of how a plausible general account of evidential support and epistemic justification can be constructed from the sort of explanatory reasoning we engage in routinely in our everyday lives. In light of this we are better served by briefly considering how such an account might apply in a number of cases than delving into the minutia of *Explanationism*, or similar accounts.

10.3.1 Applying Explanationism

In order to appreciate the plausibility of our rough sketch of evidential support, it will be instructive to consider some cases to see how *Explanationism* fares when compared with our intuitive judgments.

Let us begin with a simple case of perceptual experience. Intuitively, in normal circumstances (when one does not have the sort of defeating evidence that he had above) Jeff's visual experience as of an elephant provides him with justification for believing <there is an elephant nearby>. *Explanationism* delivers this result. Part of the best explanation of why Jeff has an experience as of an elephant is that there is an elephant. It is exceedingly plausible that an ordinary person, like Jeff, would have the proposition <there is an elephant nearby> as part of the best explanation of his evidence (which includes his current visual experience as of an elephant). Furthermore, in normal situations Jeff will not have an equally good or better explanation of his experience, which fails to include the proposition <there is an elephant nearby>, available to him. Thus, according to *Explanationism*, Jeff is justified in believing that there is an elephant nearby in this case. This is the intuitively correct judgment for such a case.[17]

Explanationism also seems to work well in cases where the proposition we have justification for believing is a truth of reasoning. Consider this sort of situation, Florence considers the proposition <every second grade teacher is a teacher>. Plausibly, since Florence understands what this proposition means, she is justified

[16]For a full development and sustained defense of the sort of account of evidential support we are sketching here see McCain (2014).

[17]One might worry whether this proposition is really part of the best explanation of Jeff's evidence because of concerns having to do with external world skepticism (the view that we cannot know or have good reason to believe propositions about the world around us). There are good reasons for thinking the ordinary propositions that we believe are better explanations than their skeptical rivals. Detailing these reasons is outside of the scope of this chapter, but we will return to this issue in Chap. 11 when we discuss, and respond to, the threat that external world skepticism poses for our scientific knowledge.

in believing it to be true. How can we account for Florence's justification in this case? Presumably, when she reflects on this proposition and understands its content Florence has an experience. Likely, Florence has some sort of awareness of the proposition and of the relations which hold among the conceptual components of the proposition.[18] That is, she is aware that the predicate term "teacher" is contained in the subject term "second grade teacher". It is reasonable to think that in such a situation the truth of the proposition <every second grade teacher is a teacher> is part of the best explanation of Florence's awareness of the relations between the conceptual components of the proposition. After all, when the predicate term of a proposition is contained in the subject term, and the predicate term is affirmed of the subject, there is no way for the proposition to be false.[19] Again, *Explanationism* yields the intuitively correct judgment that Florence's belief is justified.

In both of the previous cases the person was not making an explicit inference. In this next kind of case the person is making an explicit, inductive inference. Bert is aware that all of the vast number of observed emeralds are green. Intuitively, his possession of this evidence puts Bert in a position to justifiedly infer that the next observed emerald will also be green.[20] Again, *Explanationism* yields the correct result. Plausibly, in such a case part of the best explanation available to Bert for his evidence concerning the vast number of observed emeralds is that all emeralds are green.[21] <The next observed emerald will be green> is better explained by the best explanation of Bert's evidence in this case than <the next observed emerald will not be green>. After all, <all emeralds are green> provides a very good explanation of the first proposition, but no explanation at all of the second. And so, <the next observed emerald will be green> is available to Bert as an explanatory consequence of the best explanation available to him for his evidence concerning the observed emeralds.

A more complex case of inductive inference arises when Bert has made many varied observations of some kind of object and most, but not all, of them have had a particular characteristic. Perhaps he has observed a large number of swans in a variety of situations, and the vast majority of observed swans are white, but Bert has seen a few black swans too. In such a case <all swans are white> is not part of the

[18]See Conee (1998) for an explanation and defense of this view. See Markie (2013) for criticism.

[19]One might worry about how a proposition like <every second grade teacher is a teacher> can explain any feature of someone's evidence. While it is true that <every second grade teacher is a teacher> cannot offer much by way of a *causal* explanation of Florence's evidence, explanation should not be restricted to causal explanations when it comes to *Explanationism*. The relevant notion of explanation here is the sort we appealed to in our working model in the previous chapter—it is a matter of providing information about dependency relations. It is not implausible to think that the truth of <every second grade teacher is a teacher> does help explain the dependency relation Florence finds herself aware of when she recognizes that the predicate term of this proposition is contained in the subject term.

[20]We can assume here that Bert knows there will be more emeralds observed in addition to those which have been observed so far.

[21]Whether this regularity is itself a law of nature or some other, perhaps contingent, regularity does not matter for the present purpose.

best available explanation of Bert's evidence. Instead, something like <most swans are white> is part of the best available explanation of Bert's evidence. Often in such cases we still think Bert would be justified in believing <the next observed swan will be white>, or more guardedly, <the next observed swan is likely to be white>. Of course, Bert is not as justified in believing either of these propositions in this case as he would be if all the swans he had observed had been white.[22]

Again, *Explanationism* yields the intuitive result in this sort of case. Bert is justified in believing <the next observed swan will be white> because the best explanation of his evidence, which includes <most swans are white>, better explains that proposition than its denial, namely, <the next observed swan will not be white>. The reason for this is that large probabilities explain better than smaller ones. That is to say, if we are considering two hypotheses and, for example, one hypothesis says that the probability of A occurring is X and the other hypothesis says that the probability of A occurring is less than X, although both hypotheses might offer potential explanations of A's occurrence, all other things being equal, the first hypothesis is a better explanation of A.[23] Likewise, if a particular hypothesis says that the probability of A occurring is X and the probability of B occurring is less than X, then, all other things being equal, the hypothesis provides a better explanation of A than it does of B. When we say, "most swans are white" we mean the probability of observing a swan that is white in a random sampling is X (in this case X is greater than 50%), and the probability of observing a swan that is not white is less than X. Hence, <most swans are white> would better explain <the next observed swan will be white> than it would explain <the next observed swan will not be white> because it offers a higher probability explanation of the first proposition than it does for the second. Thus, *Explanationism* coupled with the very plausible claim that large probabilities explain better than smaller ones yields the intuitively correct results in these sorts of cases of inductive inference.

We have seen that there is at least some plausibility to *Explanationism*. As we have already noted, much more would need to be said in order to provide a full elaboration and defense of this sort of account of evidential support and justification. Nonetheless, we have seen enough to recognize that such an account is a live option. Furthermore, we have now seen that it is not implausible to think that the very method we utilize to gain scientific knowledge can be extended to an account of how we gain the justification required for knowledge in general. The fact that IBE is widely used in most every area of science and the fact that we utilize the very same method, in admittedly a less precise fashion, in our everyday lives gives reason to think that there may be more unification to the sciences on a deep level than may be apparent at first. Additionally, consideration of explanationist accounts of evidential

---

[22]We are assuming that in this case Bert will be making his next observation of a swan from a random sample. Things would be different if he were making an observation from a sample he has reason to believe is biased in some way.

[23]For further articulation and defense of why large probabilities explain better than smaller ones see Strevens (2000).

support, such as *Explanationism*, suggest that scientific knowledge may not be a different kind of thing than general knowledge. If something like *Explanationism* is correct, there may differences between knowledge in everyday contexts and scientific knowledge in terms of precision and the amount of supporting evidence we require before we believe the known proposition, but the two do not differ in any deep, conceptual ways.

10.4 Conclusion

In this chapter we have considered how it is that we employ explanatory reasoning in order to come to have scientific knowledge—particularly, knowledge of theories and the hypotheses constructed from those theories. We also saw that there are grounds for thinking the sort of explanatory reasoning that is ubiquitous in science, IBE, also permeates our everyday lives. As a result of the extremely widespread use of IBE in our lives it is plausible that evidential support and epistemic justification themselves may be a matter of explanatory considerations. Although the evidence of the ubiquity of IBE and of the plausibility of an explanationist account of evidential support adduced here is far from complete, it does help us to better understand the nature of scientific knowledge and to appreciate its contiguity with knowledge in general. Both may be important for deepening our understanding of NOS.

We have now completed Part II of this book. In these two chapters we have narrowed our focus from the nature of knowledge in general to scientific knowledge in particular. Despite the fact that there are numerous points which are currently unsettled when it comes to the nature of scientific knowledge we have continued to develop a philosophical foundation for understanding NOS. By appreciating the various aspects of scientific knowledge and the debates which are still raging over various components we are in a position to better appreciate the components of scientific knowledge that enjoy widespread support and to understand the relevant issues that remain unsettled. This puts us in a position from which we can continue to deepen our understanding of NOS.

References

Achinstein, P. (2001). *The book of evidence*. New York: Oxford University Press.
Adler, J. (1994). Testimony, trust, knowing. *Journal of Philosophy, 91*, 264–275.
Beebe, J. (2009). The abductivist reply to skepticism. *Philosophy and Phenomenological Research, 79*, 605–636.
Boyd, R. (1981). Scientific realism and naturalistic epistemology. *PSA, 1980*(2), 613–662.
Boyd, R. (1984). The current status of scientific realism. In J. Leplin (Ed.), *Scientific realism* (pp. 41–82). Berkeley: University of California Press.
Brewer, W. F., Chinn, C. A., & Samarapungavan, A. (2000). Explanation in scientists and children. In F. Keil & R. A. Wilson (Eds.), *Explanation and cognition* (pp. 279–298). Cambridge, MA: MIT Press.

Conee, E. (1998). Seeing the truth. *Philosophy and Phenomenological Research, 58*, 847–857.
Conee, E., & Feldman, R. (2008). Evidence. In Q. Smith (Ed.), *Epistemology: New essays* (pp. 83–104). Oxford: Oxford University Press.
Darwin, C. (1859/1962). *The origin of species*. New York: Collier.
Dascal, M. (1979). Conversational relevance. In A. Margalit (Ed.), *Meaning and use* (pp. 153–174). Dordrecht: Kluwer.
Douven, I. (2011). Abduction. In E. N. Zalta (Ed.), *The Stanford encyclopedia of philosophy* (Spring 2011 Edition). http://plato.stanford.edu/archives/spr2011/entries/abduction/
Enoch, D., & Schechter, J. (2008). How are basic belief-forming methods justified? *Philosophy and Phenomenological Research, 76*, 547–579.
Fricker, E. (1994). Against gullibility. In B. K. Matilal & A. Chakrabarti (Eds.), *Knowing from words* (pp. 125–161). Dordrecht: Kluwer.
Gauch, H. G., Jr. (1992). *Statistical analysis of regional yield trials: AMMI analysis of factorial designs*. New York: Elsevier.
Gauch, H. G., Jr. (2012). *Scientific method in brief*. Cambridge, UK: Cambridge University Press.
Glymour, C. (1984). Explanation and realism. In J. Leplin (Ed.), *Scientific realism* (pp. 173–192). Berkeley: University of California Press.
Goodman, N. (1965). *Fact, fiction, and forecast*. Cambridge, MA: Harvard University Press.
Goodman, N. (1978). *Ways of worldmaking*. Cambridge, MA: Harvard University Press.
Gopnik, A. (1998). Explanation as orgasm. *Minds and Machines, 8*, 101–118.
Grimes, D. A., & Schulz, K. F. (2005). Refining clinical diagnosis with likelihood ratios. *The Lancet, 365*, 1500–1505.
Harman, G. (1965). The inference to the best explanation. *Philosophical Review, 74*, 88–95.
Harman, G. (1973). *Thought*. Princeton: Princeton University Press.
Harman, G. (1986). *Change in view*. Cambridge, MA: MIT Press.
Harré, R. (1986). *Varieties of realism*. Oxford: Blackwell.
Hintikka, J. (1998). What is abduction? The fundamental problem of contemporary epistemology. *Transactions of the Charles S. Peirce Society: A Quarterly Journal in American Philosophy, 34*, 503–533.
Hobbs, J. R. (2004). Abduction in natural language understanding. In L. Horn & G. Ward (Eds.), *The handbook of pragmatics* (pp. 724–741). Oxford: Blackwell.
Huemer, M. (2009). Explanationist aid for the theory of inductive logic. *British Journal of Philosophy of Science, 60*, 345–375.
Janssen, M. (2002). Reconsidering a scientific revolution: The case of Einstein *versus* Lorentz. *Physics in Perspective, 4*, 421–446.
Josephson, J. R., & Josephson, G. S. (Eds.). (1994). *Abductive inference*. Cambridge, UK: Cambridge University Press.
Keil, F. C., & Wilson, R. A. (2000). Explaining explanation. In F. C. Keil & R. A. Wilson (Eds.), *Explanation and cognition* (pp. 1–18). Cambridge, MA: MIT Press.
Kuhn, T. S. (1977). *The essential tension: Selected studies in scientific tradition and change*. Chicago: University of Chicago Press.
Lacey, H. (2005). *Is science value free? Values and scientific understanding* (2nd ed.). London: Routledge.
Ladyman, J., Douven, I., Horsten, L., & van Fraassen, B. C. (1997). A defence of van Fraassen's critique of abductive inference: Reply to Psillos. *Philosophical Quarterly, 47*, 305–321.
Lipton, P. (1998). The epistemology of testimony. *Studies in History and Philosophy of Science, 29*, 1–31.
Lipton, P. (2004). *Inference to the best explanation* (2nd ed.). New York: Routledge.
Longino, H. (1990). *Science as social knowledge*. Princeton: Princeton University Press.
Lycan, W. G. (1988). *Judgement and justification*. Cambridge, UK: Cambridge University Press.
Lycan, W. G. (2002). Explanation and epistemology. In P. Moser (Ed.), *The Oxford handbook of epistemology* (pp. 408–433). Oxford: Oxford University Press.
Lycan, W. G. (2012). Explanationist rebuttals (coherentism defended again). *Southern Journal of Philosophy, 50*, 5–20.

Markie, P. J. (2013). Rational intuition and understanding. *Philosophical Studies, 163*, 271–290.
Matthews, M. (2015). *Science teaching: The contribution of history and philosophy of science* (2nd ed.). New York: Routledge.
McAllister, J. W. (1996). *Beauty and revolution in science*. Ithaca: Cornell University Press.
McCain, K. (2013). Explanationist evidentialism. *Episteme, 10*, 299–315.
McCain, K. (2014). *Evidentialism and epistemic justification*. New York: Routledge.
McCain, K. (2015). Explanationism defended on all sides. *Logos & Episteme, 6*, 333–349.
McCain, K., & Poston, T. (2014). Why explanatoriness is evidentially relevant. *Thought, 3*, 145–153.
McGrew, T. (2003). Confirmation, heuristics, and explanatory reasoning. *British, Journal for the Philosophy of Science, 54*, 553–567.
McMullin, E. (1982). Values in science. *PSA 1982, 2*, 3–28.
McMullin, E. (1992). *The inference that makes science*. Milwaukee: Marquette University Press.
Minnameier, G. (2004). Peirce-suit of truth—Why inference to the best explanation and abduction ought not to be confused. *Erkenntnis, 60*, 75–105.
Moser, P. K. (1989). *Knowledge and evidence*. Cambridge, UK: Cambridge University Press.
Newton, I. (1687/1999). *The Principia: Mathematical principles of natural philosophy* (trans: Cohen, I. B. & Whitman, A.). Berkeley: University of California Press.
Okasha, S. (2000). Van Fraassen's critique of inference to the best explanation. *Studies in History and Philosophy of Science, 31*, 691–710.
Pitt, J. C. (1988). Galileo, rationality and explanation. *Philosophy of Science, 55*, 87–103.
Popper, K. R. (1959). *The logic of scientific discovery*. London: Hutchinson.
Poston, T. (2014). *Reason & explanation: A defense of explanatory coherentism*. New York: Palgrave-MacMillan.
Psillos, S. (1999). *Scientific realism: How science tracks truth*. London: Routledge.
Quine, W. V. O., & Ullian, J. S. (1978). *The web of belief* (2nd ed.). New York: Random House.
Roche, W., & Sober, E. (2013). Explanatoriness is evidentially irrelevant, or inference to the best explanation meets Bayesian confirmation theory. *Analysis, 73*, 659–668.
Sellars, W. (1963). *Science, perception and reality*. Atascadero: Ridgeview Publishing.
Sober, E. (2015). *Ockham's razors: A user's manual*. Cambridge, UK: Cambridge University Press.
Strevens, M. (2000). Do large probabilities explain better? *Philosophy of Science, 67*, 366–390.
Thagard, P. (1978). The best explanation: Criteria for theory choice. *Journal of Philosophy, 75*, 76–92.
van Fraassen, B. C. (1989). *Laws and symmetry*. Oxford: Oxford University Press.
Vogel, J. (1990). Cartesian skepticism and inference to the best explanation. *Journal of Philosophy, 87*, 658–666.
Weisberg, J. (2009). Locating IBE in the Bayesian framework. *Synthese, 167*, 125–143.
Wilson, R. A., & Keil, F. C. (2000). The shadows and shallows of explanation. In F. C. Keil & R. A. Wilson (Eds.), *Explanation and cognition* (pp. 87–114). Cambridge, MA: MIT Press.
Wray, K. B. (2008). The argument from underconsideration as grounds for anti-realism: A defence. *International Studies in the Philosophy of Science, 22*, 317–326.

Part III
Challenges to Scientific Knowledge

Chapter 11
Skepticism About the External World

Abstract One way to challenge our scientific knowledge is to challenge all of our knowledge of the world around us. This chapter explores the classic philosophical problem of external world skepticism. This philosophical problem challenges whether we can know anything at all about the world outside of our own minds. In the process of examining and responding to arguments for external world skepticism important insights about the nature of scientific knowledge are revealed. One of the foremost of these insights is that knowledge in general does not require evidence that makes the believed proposition absolutely certain—beyond all *possible* doubt. Instead, significant, yet fallible, evidence is all that is required for knowledge, both scientific and mundane. Another insight is that the explanationist account of evidential support developed in the previous chapter helps to show that, despite initial appearances, external world skepticism is not a significant threat to our knowledge after all. We do have good evidence for believing the world is roughly the way we commonsensically take it to be.

We have discussed many features of knowledge as well as what is required for us to come to possess knowledge, both scientific and mundane. Throughout this discussion we have been working under the assumption that we do in fact know a great many things. Despite the fact that this assumption is exceedingly plausible, our knowledge of scientific claims faces many challenges. In this part of the book we will examine some of the major threats to our scientific knowledge. We will begin by exploring threats which challenge our scientific knowledge by way of challenging some broader category under which our scientific knowledge falls. Afterward, we will examine threats arising from the debate between realists and anti-realists which specifically target our scientific knowledge.

This chapter begins our consideration of the more sweeping challenges to our scientific knowledge. The first such challenge we will think through is one that threatens all of our knowledge of the world around us. It is particularly worrisome when we realize that included in this body of knowledge is not just scientific knowledge, but all sorts of things we typically take ourselves to know: there are trees, other people exist, we have bodies, and so on. This challenge comes from a classic philosophical problem: external world skepticism. Although the threat of external world skepticism is one that can be overcome, consideration of this

K. McCain, *The Nature of Scientific Knowledge*, Springer Undergraduate Texts in Philosophy, DOI 10.1007/978-3-319-33405-9_11

philosophical problem is worthwhile because it illuminates important features of the nature of scientific knowledge and knowledge more generally.

11.1 Challenge I: Lack of Certainty

One common way skeptics challenge our scientific knowledge is by pointing out that we cannot be *certain* that scientific claims are true. The sense of "certain" that is relevant here is not merely a psychological state of being completely sure or exceedingly confident that something is true. Some people are completely sure that humans never landed on the moon. They believe the moon landing is a hoax, and they believe this without a shadow of a doubt. Nevertheless, despite their confidence the claim that humans never landed on the moon is definitely not certain for these people! The relevant sense of "certain" here is the sense in which you are certain that *p* just in case your evidence for *p* guarantees *p*'s truth. In other words, to be certain in this sense requires that your evidence for a particular claim is so strong that it is impossible for you to have that evidence when the claim is false.

Of course, it is at least possible that despite the evidence we have in support of any particular scientific claim we could be mistaken. It is often awareness of this tentativeness that leads some to the misguided objection that evolutionary theory is "merely a theory".[1] The thought being that the evidence in support of evolution does not make it absolutely certain that it is true, so it is merely a theory and not something we know to be true. This is sometimes a bit disconcerting because it is partly correct. We do not have sufficient evidence to be absolutely certain of the truth of evolution, or any other scientific claim for that matter. Our evidence is always such that it is possible for us to have that evidence and yet the scientific claim in question turn out to be false. Recognition of this fact is one of the reasons why it is commonly accepted that science is tentative—since we lack certainty for scientific claims, the scientific claims we accept as true are always revisable in the light of new evidence.

The fact that our evidence does not make the truth of scientific claims certain for us should not be surprising. After all, as we noted in Chap. 5, the same is true of any claim about the world around us—in a scientific setting or not. For example, it is possible for you to have the same evidence you do now and yet not be reading this book. After all, you could be having a particularly vivid hallucination. So, your evidence for thinking that you are currently reading this book, while outstandingly strong evidence, is not so strong that it makes it absolutely certain for you that you are reading this book. This fact about our situation with respect to propositions about

[1]For more on this sort of objection to evolutionary theory and why it is misguided see McCain and Weslake (2013). For extensive discussion of various misconceptions concerning evolutionary theory and analysis of why some resist evolution in spite of the evidence see Kampourakis (2014).

the external world leads us to the first argument for external world skepticism we will examine:

CERTAINTY

C1) If you know some proposition about the external world, then your evidence makes the truth of that proposition certain for you.
C2) Your evidence does not make the truth of any proposition about the external world certain for you.
C3) Therefore, you do not know any proposition about the external world.

A couple points about CERTAINTY should be noted immediately. First, its second premise is clearly true. For any proposition about the external world your evidence in support of that proposition could be misleading. As noted above, it is at least possible that you could have all of the same evidence you do now, but you are not in fact reading this book. It is possible that you are really being deceived by a malicious demon of the sort philosopher René Descartes (1641/1988) described. It is possible you are really simply having an especially vivid dream. It is possible that unbeknownst to you someone has injected you with a hallucinogenic drug. Admittedly, these possibilities are far-fetched. Nevertheless, they are genuine possibilities—these scenarios are not impossible. Since they are possible, there is a very slight chance that your evidence for thinking you are reading a book is misleading, and you are not reading a book at all. But, if your evidence could be misleading with regard to whether you are reading a book, you cannot be *certain* that you are doing so. Second, C2 clearly applies to our scientific knowledge. Most (perhaps all) scientific claims are to some degree based upon observations we have made of the world around us. If we cannot be certain that any of those observations are accurate, we cannot be certain that the scientific claims based on those observations are true.

At this point you may be getting slightly worried. CERTAINTY concludes that we cannot know anything about the external world, which entails that not only do we lack scientific knowledge, we cannot know mundane things like that we have bodies, that other people exist, and so on. CERTAINTY has a very threatening conclusion. To make matters worse we have seen that one of its two premises is clearly true! Fortunately, it is easy to adequately respond to CERTAINTY because its other premise is clearly false. Knowledge simply does not require certainty. This is something we noted in Chap. 5, and it is something that nearly all epistemologists accept (Cohen 1988). This is why the dominant view of the evidence required for knowledge is fallibilism. Fallibilism is the idea that knowledge does not require evidence which guarantees truth (evidence that makes the proposition in question certain). According to fallibilism, it is possible for you to know that you are reading this book even though there is a remote chance that your evidence is misleading. Accepting fallibilism allows us to reasonably deny C1, and so, given fallibilism, CERTAINTY fails to commit us to denying that we have knowledge of the external world.

Although CERTAINTY does not pose a significant threat to our scientific knowledge or our mundane knowledge of the world around us, consideration of this argument helps illuminate important facts. First, we have to acknowledge that if knowledge really does require evidence which makes propositions certain, then we have very little (if any) knowledge. Perhaps we can have evidence which is this strong for thinking that we exist when we are currently thinking. Perhaps we can even have this sort of evidence for believing things like we are in pain when we are currently experiencing an intense pain. Regrettably, that is pretty much the full extent of the propositions about whose truth we can be absolutely certain. We definitely cannot have this sort of evidence for propositions about the external world. Thus, although we may wish for certainty when it comes to the things we believe, this is something that we simply cannot have for nearly any of our beliefs.

Second, consideration of CERTAINTY reveals why we should be fallibilists about knowledge. As we noted in Chap. 5, it is difficult to say precisely how much evidence is required for knowledge, but we can clearly recognize that it is something less than what is required for certainty. Additionally, as we noted in that chapter, a plausible answer to the question of how much evidence is required for knowledge is the criminal standard. Recall, the idea with the criminal standard is that in order to know that *p* your evidence must make the truth of *p* beyond a reasonable doubt for you. This means knowledge requires a significant amount of evidence, but nowhere near what is needed for certainty. By examining CERTAINTY we have not only taken an important first step in defending our scientific knowledge, we have furthered our understanding of the nature of knowledge in general. Unfortunately, there are more threatening skeptical challenges to our knowledge of the external world.

11.2 Challenge II: Underdetermination

Consider our commonsense view of the world, what we might call the "Real World Hypothesis" (RWH).[2] The RWH will include various claims, perhaps things like: "The earth is more than 3 minutes old", "Fire is hot", "Water is wet", and so on. The exact claims making up the RWH are not important for our purposes here. Rather, all that matters is that the claims of the RWH have two significant features. First, the truth of these claims entails an external world which is independent of our minds. That is, the truth of these claims requires that there is a world which is a certain way regardless of what we believe or take ourselves to know about that world. Second, if the claims of the RWH are known to be true, they constitute empirical knowledge. That is to say, we cannot come to know all of the claims of the RHW without investigating the world around us—not all of them are things we can know a priori (by reasoning alone). One of the classic ways of attacking our

[2]This name for our commonsense view of the world was put forward by Vogel (1990).

knowledge of the external world is to present skeptical alternatives to the RWH and argue that our evidence fails to determine which of these hypotheses is correct.

This is a broader threat than the underdetermination challenge that often arises in the context of debates between realists and anti-realists about science, which we will discuss in Chap. 14. This broader underdetermination challenge is particularly threatening for two reasons. One reason is that the skeptical argument which claims our evidence underdetermines whether the RWH or some rival skeptical alternative is true threatens all of our knowledge of the external world, not just our scientific knowledge. The other reason this challenge is particularly threatening is that it seems at least initially plausible that such rivals to the RWH do exist. For example, the skeptical alternative which claims that all of our experiences are caused by a manipulating demon who wants us to erroneously believe the RWH is consistent with all of our sensory experiences. The skeptical alternative that we are really brains in vats being deceived by the workings of a supercomputer is consistent with all of our sensory experiences too. A host of other skeptical alternatives fit with our sensory experiences as well. So, we seem to have genuine rivals to the RWH that at least initially seem to fit our evidence just as well as the RWH does. This is worrisome.

The skeptical argument from underdetermination can be formulated more precisely as follows:

UNDERDETERMINATION

U1) If you know the RWH, then you have evidence that favors the RWH over all rival skeptical alternatives.
U2) You do not have evidence that favors the RWH over all rival skeptical alternatives.
U3) Therefore, you do not know the RWH.

Three points about UNDERDETERMINATION are worth keeping in mind. First, knowing the RWH should be understood as knowing the claims of the RWH are true. This does not mean that one has ever thought that there is such a hypothesis or that she thinks of the claims of the RWH as hypotheses. Instead, the idea is simply that when one "knows the RWH" she knows (most of) the claims which make up the RWH such as "The earth is more than 3 minutes old", "Fire is hot", "Water is wet", and so on to be true. Second, to say that the evidence favors the RWH, over its rivals simply means that the evidence supports the RWH more than it does the rival skeptical alternatives to our commonsense picture of the world. Third, it should be noted that UNDERDETERMINATION cannot be dismissed simply by accepting fallibilism like CERTAINTY can. UNDERDETERMINATION claims that we do not even meet the fallibilist standard of evidence for knowledge—it holds that our evidence does not even make the RWH beyond a reasonable doubt for us. This makes UNDERDETERMINATION a much more worrisome threat to our knowledge of the external world. Fortunately, we have already equipped ourselves with the tools to respond to this skeptical threat.

11.3 The Explanationist Response

The first step to responding to UNDERDETERMINATION is to recognize that premise U1 is true. If your evidence does not support the RWH over its skeptical rivals, then you are not justified in believing the claims of the RWH. And, of course, if you are not justified in believing these claims, you do not know that the RWH is true. Since we recognize that U1 is true, we can focus all of our attention on showing that U2 is false.

The explanatory account of evidential support that we explored in Chap. 10 provides a very promising way of attacking premise U2. Recall, according to the explanationist account of evidential support discussed in that chapter, *Explanationism*, the evidence supports believing a particular proposition when that proposition is part of the best explanation of the evidence, or when that proposition is an explanatory consequence of the best explanation of the evidence. A similar approach can be used to respond to UNDERDETERMINATION. This *Explanationist Response* involves arguing that the truth of the RWH is a better explanation of relevant features of our sensory experiences than any of the available competing explanations (the various rival skeptical alternatives). Since the RWH is the best explanation of relevant features of our sensory experiences, the evidence we gain from those sensory experiences supports it.[3] As a result, the combination of explanatory considerations and our sensory experiences provides us with evidence that favors the RWH over all skeptical rivals. Thus, the Explanationist Response shows that premise U2 is false.[4]

Of course, before we can be confident that the Explanationist Response satisfactorily defends our knowledge from the threat of UNDERDETERMINATION we must first get clear on a couple things. We have to be clear about what exactly the relevant features of our sensory experiences are, and we have to get clear about why

[3]Putting things more precisely, inferring to the best explanation of our sensory experiences supports the RWH because the RWH is not only the best available explanation of the relevant features of those experiences, it is also a very good explanation in its own right. Though it can be difficult to spell out what exactly is required for an explanation to be good enough to be legitimately inferred, it is plausible that the RWH satisfies this requirement because the RWH accounts for all of the relevant features of our sensory experiences in a highly unified manner. See the previous chapter as well as Lipton (2004) for an explanation of why it is necessary to limit legitimate inferences to the best explanation to only those where the inferred explanation is of sufficient quality.

[4]There are other ways of responding to UNDERDETERMINATION. However, we will only focus on the Explanationist Response for a couple reasons. The Explanationist Response fits very nicely with the explanatory account which we developed in the previous chapter. Additionally, this response has the most promise of providing a satisfactory response to the threat of external world skepticism. Readers interested in other responses to the threat of external world skepticism are encouraged to see Greco (2000) for a response that is particularly appealing to those who like externalist views of epistemic justification, Huemer (2001) for the sort of response to this argument provided by the internalist account of epistemic justification known as "phenomenal conservatism", as well as Moore (1939), Pryor (2004), and Willenken (2011) for what has come to be known as the "Moorean" response to external world skepticism.

we should think the RWH really does explain these features better than its skeptical rivals. We will briefly consider the first point in the next section and the second point in the section after that.

11.3.1 Relevant Features of Sensory Experience

There are a number of features of our sensory experiences that are relevant to evaluating the explanatory merits of the RWH and its skeptical rivals. First of all, our sensory experiences come to us in an involuntary, spontaneous fashion (BonJour 1999; BonJour and Sosa 2003; Vogel 2008). Assuming that you do not have problems with your vision, when you open your eyes in normal lighting conditions you spontaneously have various visual experiences. These spontaneous visual experiences will also be involuntary. After all, while your eyes are open, the lighting is good, and you are looking at a tree you cannot make it so that you no longer have the experience of seeing a tree by just willing to no longer see it.

Additionally, our sensory experiences exhibit a high degree of regularity and coherence with one another. For example, our sensory experiences of one kind, say vision, fit with one another coherently, and they also fit coherently with our other kinds of sensory experiences. When you have a visual experience of one car crashing into another car nearby, your auditory sensations of a loud sound correspond to your visual sensations of the cars colliding.[5] Finally, there is coherence between our sensory experiences and our volitional activities. For instance, when you decide to grab (what you take to be) a glass of water your visual experiences usually change in a regular way which the hypothesis that you have limbs and are using one of them to voluntarily reach for a glass explains very well.

The RWH explains each of these facts about our sensory experiences. Given the RWH, our sensory experiences come to us in an involuntary and spontaneous fashion because they are caused by the interaction of mind-independent objects in the external world and our sense organs, and this interaction is not directly under our voluntary control. The reason our sensory experiences fit together coherently with one another is that the various kinds of sensory experiences are caused by the same mind-independent objects in the external world affecting our sense organs. In order to see how the RWH works it will be helpful to consider how it provides explanations of our sensory experiences in a particular example.

Consider the sensory experiences you are having right now. You are having visual sensations as of a roughly rectangular object as well as tactile sensations as of an object with edges some of which are longer than others (assuming that

---

[5]This is not to assume that we have different sense organs responsible for different kinds of sensory experiences. Such an assumption would surely stack the deck in favor of the RWH. Instead, here it is simply important to pay attention to the fact that we have sensory experiences of different kinds (visual, auditory, tactile, etc.), and these sensory experiences fit together coherently.

you are holding a hardcopy of this book in your hand rather than reading on an e-reader or your computer). You do not have to put forth effort to have these sensory experiences—they come to you spontaneously and involuntarily. The RWH explains these sensory experiences by positing a three-dimensional, rectangular object (the book), as well as your sense organs, and an appropriate light source. It is these same external objects which account for both kinds of sensations you are having.

The RWH also explains why our sensory experiences are coherent over time. Consider again your sensory experiences while reading this book. Your visual experiences from a few moments ago cohere with those you are having now. They are not exactly the same because you are reading different words, but they are very similar, as one would expect, if the same external object caused your sensations now as before. Likewise, your tactile sensations now are similar to those you had just a short while ago. You have a very similar tactile sensation of the weight of the book throughout this period (assuming that you have not set the book down, of course). According to the RWH, you have similar tactile sensations throughout this period of time because there is an actual three-dimensional object in your hand the whole time.[6]

Clearly, the RWH explains the relevant features of our sensory experiences quite well. Unfortunately, the mere fact that the RWH provides a good explanation of these features is not enough for the Explanationist Response to successfully defeat UNDERDETERMINATION. In order to provide a reasonable defense of our knowledge of the external world the RWH needs to not only provide a good explanation of these features, it needs to be the *best* available explanation of them. In other words, the RWH must be a better explanation of these features of our sensory experiences than its skeptical rivals. We will now see that it is reasonable to think the RWH is in fact the best explanation of these features of our sensory experiences.

11.3.2 The Superiority of the RWH

The RWH offers a very good explanation of why our sensory experiences have the features that they do. By comparison the classic skeptical alternatives put forward as rivals of the RWH are clearly inadequate. These skeptical alternatives claim that our sensory experiences are caused by a deceptive demon or by our brains being directly stimulated by a supercomputer without providing a story as to how the demon or the supercomputer produces our sensory experiences. Classic skeptical alternatives leave us without much of an explanation at all for why our sensory experiences have the features that they do. They simply tell us that whatever features our sensory experiences do have are the result of the actions of the demon or the supercomputer.

[6]For further discussion of how the RWH explains various features of our sensory experiences see McCain (2014).

Additionally, the RWH allows us to make accurate predictions concerning what our future sensory experiences will be like. For example, the RWH predicts that in a few seconds you will have similar visual sensations to those you are having now because you will be reading the same book. That is, the same external object will be interacting with your sense organs in a very similar way to how it is right now. Classic skeptical alternatives cannot make such predictions though. These skeptical alternatives simply claim that a demon or a supercomputer or whatever causes your sensory experiences; this does not allow for predictions at all (let alone accurate ones). Instead, these skeptical alternatives can only account for your sensory experiences after the fact by claiming that your sensory experiences a few seconds from now are also caused by the demon or the supercomputer. These classic skeptical alternatives cannot match the explanatory power of the RWH. Thus, the RWH is clearly a better explanation of the relevant features of our sensory experiences than the classic skeptical alternatives.[7]

Since the classic skeptical alternatives are clearly inferior to the RWH, it seems the only rivals to the RWH that might be thought to pose a genuine threat to the RWH's superiority are what Jonathan Vogel (1990, p. 660) calls "improved skeptical hypotheses" (ISHs).[8] ISHs are improved versions of the classic skeptical alternatives. These versions of classic skeptical alternatives to the RWH satisfy two constraints. First, ISHs "should invoke items corresponding to the elements" of the RWH (Vogel 1990, p. 660). Second, they "should also posit, as holding of these items, a pattern of properties, relations, and explanatory generalizations mirroring" the RWH's (Vogel 1990, p. 660). Essentially, to generate an ISH we simply "extract the explanatory skeleton or core from the RWH—that there are some entities bearing some properties that are related in ways exactly analogous to those specified by the RWH—and then to add that the entities and their properties are somehow different from the ones mentioned in the RWH" (Vogel 1990, p. 661). ISHs seem to provide some explanation of how the objects of our sensory experiences are related to one another and why they behave as they do because ISHs are isomorphic to the RWH. We will focus on ISHs when evaluating whether the RWH is the best available explanation of the relevant features of our sensory experiences because they are the only skeptical rivals that seem to have any hope of matching the RWH's explanatory power.

Although ISHs perhaps seem to be a match for the RWH when it comes to explanatory power, the RWH is a better explanation of the features of our sensory experiences than these skeptical rivals. Despite the fact that the RWH and ISHs are isomorphic and capable of explaining the same facts, the RWH offers a *simpler* explanation of those facts. All theories, the RWH included, must posit some fundamental explanatory regularities, which cannot be reduced to some other

[7] See McCain (2012) for further elaboration of why these sorts of skeptical alternatives are inferior to the RWH.

[8] See Vogel (1990, 2005) for further reason to think that ISHs are the only skeptical rivals to the RWH that have the potential to match the RWH's explanatory power.

regularity or set of regularities. In any theory some facts have to be taken as brute or else we will be caught in an infinite regress of regularities that can be reduced to other regularities. As we will see, however, the RWH can make use of necessary truths in its explanations in ways that ISHs can only mimic by positing additional contingent fundamental regularities. That is to say, we will see that ISHs can only truly match the explanatory power of the RWH by positing fundamental regularities beyond those required by the RWH. This is very important because, all other things being equal, an explanation which posits fewer fundamental regularities is better than an explanation that posits more. Thus, we will see that the RWH is superior to ISHs because the RWH is simpler in this important way.[9] Since it is superior to its skeptical rivals, the RWH is the best available explanation of the relevant features of our sensory experiences.

In order to see how it is that the RWH is simpler than the ISHs it will be helpful to consider a particular ISH. One commonly presented skeptical rival to the RWH is the "Brain in a Vat" hypothesis. According to this skeptical hypothesis, it may be that we are brains floating in vats with all of our sensory experiences simply resulting from the stimulation of parts of our brains by electrodes connected to a supercomputer. Of course, as it is, this barebones skeptical alternative is clearly inferior to the RWH—it does not offer any explanation of why our sensory experiences have the features they do beyond claiming that the supercomputer causes them to have those features. However, it is possible to construct an improved version of this skeptical hypothesis that at least has a chance of matching the RWH's explanatory power. Let us call the ISH version of the Brain in a Vat hypothesis "BIV".[10] In order to be an ISH, BIV must "invoke items corresponding to the elements" of the RWH, and it must "posit, as holding of these items, a pattern of properties, relations, and explanatory generalizations mirroring" the RWH's (Vogel 1990, p. 660). For example, the RWH posits a number of external objects and causal relations between them and your sense organs to explain your current sensory experiences while you are reading this book. BIV explains those same sensory experiences by positing sequences of code in the supercomputer's operating program corresponding to the various external objects of the RWH and relations holding between these sequences of code which mirror the causal relations of the objects posited by the RWH. With

[9]There are other explanatory virtues we might use to compare the RWH and ISHs. Our focus is only on explanatory simplicity because it is an explanatory virtue that is clearly relevant to determining which theory is best (there is controversy concerning the importance of many other purported explanatory virtues). Additionally, comparing the RWH and ISHs in this manner will be sufficient for us to see that the RWH has features which make it superior to these skeptical rivals. For discussion of other explanatory virtues, see Beebe (2009), Lycan (1988), Quine and Ullian (1978), Thagard (1978), and Vogel (1990). For further discussion of the superiority of the RWH over ISHs and comparisons between the two with respect to additional explanatory virtues see McCain (2014).

[10]The conclusions drawn from the comparison of the RWH and BIV are equally applicable to comparisons of the RWH and other ISHs.

our understanding of roughly how BIV explains various sensory experiences in hand we are ready to compare it to the RWH.

The RWH is simpler than BIV because it can make use of necessary truths in explanations in a way that BIV can only mimic by positing additional contingent fundamental regularities.[11] Since both the RWH and BIV are committed to the necessary truths (these truths hold regardless of whether the RWH or some ISH is true) and BIV has to posit additional contingent fundamental regularities that the RWH does not, the RWH offers the simplest explanation of the features of our sensory experiences.

In an effort to illustrate how the RWH can make use of necessary truths in explanations in a way that BIV cannot it will be helpful to consider an example. Consider a case where you have sensory experiences as of a baseball field in normal conditions. In particular, consider the sensory experiences of first base, third base, and home plate. These three bases visually appear to be arranged in a triangular pattern.[12] According to the RWH, three external world objects which occupy three physical locations in a triangular pattern cause your sensory experiences of these three bases. Given the RWH, a particular necessary truth can help explain why when you have sensations of moving at the same speed your sensation of walking lasts longer when you have the sensation of walking from first base to third base to home plate than when you have the sensation of walking directly from third base to home plate. The relevant necessary truth here is the triangle inequality theorem. The triangle inequality theory states that the sum of the lengths of any two sides of a triangle is greater than the length of the remaining side.[13] According to the RWH, part of the explanation of why your sensations of walking differ in duration is that, according to the triangle inequality theorem, the distance from first base to third base to home plate in this case *must* be greater than the distance from third base directly to home plate.

BIV cannot make use of the triangle inequality theorem in its corresponding explanation of your sensations. The counterparts of the three external objects (the bases) in the RWH that BIV posits do not have genuine locations because they are sequences of code in the supercomputer's software. Since these "BIV-bases" do not have genuine locations, the triangle inequality theorem does not entail that the distance from BIV-first base to BIV-third base to BIV-home plate is greater than the distance from BIV-third base directly to BIV-home plate. Consequently, in order for BIV to explain the differing duration of your sensations when you seem to take different routes between what appear to you to be bases it must posit some contingent regularity which governs the relations between BIV-first

[11] See BonJour (1999), BonJour and Sosa (2003), and Vogel (1990, 2005, 2008) for further support of this claim.

[12] This example is adapted from Vogel (2008, pp. 547–548).

[13] The triangle inequality theorem is normally regarded as axiomatic for Euclidean geometry as well as many non-Euclidean geometries. In fact, in any geometry with a well-defined distance function the triangle inequality theorem holds. That is, the triangle inequality theorem is a theorem in any geometry that is a metric space.

base, BIV-third base, and BIV-home plate in a way that is similar to the triangle inequality theorem. For example, BIV will need to posit something like a regularity that when the sequence of code which produces sensations of walking from base to base is ran that code is paired with specific codes which produce various sensations of duration. This pairing will need to be such that the sequence of code which produces sensations of walking from first base to third base to home plate is paired with code which produces sensations of duration X. The sequence of code which produces sensations of walking directly from third base to home plate will need to be paired with code which produces sensations of duration less than X. Of course, this additional contingent regularity of the code pairings in BIV does not have a counterpart in the RWH. Thus, BIV will have to posit contingent fundamental regularities that the RWH does not in order to explain your sensory experiences in this situation. There are numerous other situations where the triangle inequality theorem (or other necessary truths) does explanatory work in the RWH's explanations that it cannot do in BIV's explanations. Since the set of necessary truths are fundamental regularities of both the RWH and BIV, and BIV has to posit more contingent fundamental regularities than the RWH, the RWH is simpler than BIV in an important way.

In light of the fact that they have the same explanatory power while the RWH is simpler, the RWH is a better explanation of the features of our sensory experiences than BIV. Further, all other ISHs face this same problem. Thus, the RWH is the best explanation of the relevant features of our sensory experiences. Since it is the best explanation of the relevant features of our sensory experiences, our evidence does favor the RWH over the skeptical alternatives. Hence, U2 of UNDERDETERMINATION is false. Thus, although UNDETERMINATION poses more of a challenge to our knowledge of the external world than CERTAINTY, it is a challenge that we can overcome.

11.4 Conclusion

In this chapter we have seen that one way our scientific knowledge can be challenged is by threatening all of our knowledge of the external world. We have also seen that both of the major arguments for external world skepticism are unsound. Consideration of these arguments has revealed some important points about the nature of knowledge though. First, knowledge does not require certainty. Since fallibilism is true, we only need evidence that makes a proposition beyond a reasonable doubt for us in order to know it. Second, the challenge posed by underdetermination of theory by evidence can be expanded into a seemingly powerful attack on all of our knowledge of the external world. Third, our knowledge can be successfully defended from the threat of underdetermination by recognizing that the RWH better explains our experiences than any of its skeptical rivals. Although we have seen that our knowledge, both scientific and mundane, withstands the threat of external world skepticism, other challenges to our knowledge await in the remaining chapters of this part of the book.

References

Beebe, J. (2009). The abductivist reply to skepticism. *Philosophy and Phenomenological Research, 79*, 605–636.

BonJour, L. (1999). Foundationalism and the external world. *Philosophical Perspectives, 13*, 229–249.

BonJour, L., & Sosa, E. (2003). *Epistemic justification: Internalism vs. externalism, foundations vs. virtues*. Malden: Blackwell Publishing.

Cohen, S. (1988). How to be a fallibilist. *Philosophical Perspectives, 2*, 91–123.

Descartes, R. (1641/1988). *Meditations on first philosophy*. In J. Cottingham, R. Stoothoff, & D. Murdoch (Trans.), *Descartes: Selected philosophical writings* (pp. 73–123). Cambridge, UK: Cambridge University Press.

Greco, J. (2000). *Putting skeptics in their place: The nature of skeptical arguments and their role in philosophical inquiry*. Cambridge, UK: Cambridge University Press.

Huemer, M. (2001). *Skepticism and the veil of perception*. Lanham: Rowman & Littlefield.

Kampourakis, K. (2014). *Understanding evolution*. Cambridge, UK: Cambridge University Press.

Lipton, P. (2004). *Inference to the best explanation* (2nd ed.). New York: Routledge.

Lycan, W. G. (1988). *Judgement and justification*. Cambridge, UK: Cambridge University Press.

McCain, K. (2012). A predictivist argument against scepticism. *Analysis, 72*, 660–665.

McCain, K. (2014). *Evidentialism and epistemic justification*. New York: Routledge.

McCain, K., & Weslake, B. (2013). Evolutionary theory and the epistemology of science. In K. Kampourakis (Ed.), *The philosophy of biology: A companion for educators* (pp. 101–119). Dordrecht: Springer.

Moore, G. E. (1939). Proof of an external world. *Proceedings of the British Academy, 25*, 273–300.

Pryor, J. (2004). What's wrong with Moore's argument? *Philosophical Issues, 15*, 349–378.

Quine, W. V. O., & Ullian, J. S. (1978). *The web of belief* (2nd ed.). New York: Random House.

Thagard, P. (1978). The best explanation: Criteria for theory choice. *Journal of Philosophy, 75*, 76–92.

Vogel, J. (1990). Cartesian skepticism and inference to the best explanation. *Journal of Philosophy, 87*, 658–666.

Vogel, J. (2005). The refutation of skepticism. In M. Steup & E. Sosa (Eds.), *Contemporary debates in epistemology* (pp. 72–84). Malden: Blackwell.

Vogel, J. (2008). Internalist responses to skepticism. In J. Greco (Ed.), *The Oxford handbook of skepticism* (pp. 533–556). Oxford: Oxford University Press.

Willenken, T. (2011). Moorean responses to skepticism: A defense. *Philosophical Studies, 154*, 1–25.

Chapter 12
Skepticism About Induction

Abstract Another way of challenging our scientific knowledge is to challenge our knowledge of all unobserved cases and our ability to make justified predictions about what will happen on the basis of previous observations. The skeptic about induction claims that while we might observe many, many instances of As that are Bs this does not allow us to know that the next A we observe will be a B (or even reasonably believe that it will be or is likely to be a B). This sort of inductive skepticism poses a major threat to our scientific knowledge as well as our commonsense knowledge of the world around us. After all, we depend on this sort of inference (from observed cases to what we expect to observe in the future) every day. This chapter argues that again the skeptical challenge can be overcome by carefully understanding the sort of explanationist account of evidential support which has been developed in earlier chapters.

As we saw in the previous chapter, one way to challenge our scientific knowledge involves challenging all of our knowledge of the world around us. Another sweeping skeptical challenge comes by way of attacking our practices of inductive reasoning. This sort of reasoning is ubiquitous in our everyday lives and in science. Roughly, this is the sort of reasoning that we employ when we infer what will happen on the basis of previous observations. Our justification for thinking that the sun will rise tomorrow, that water will boil when heated to a sufficiently high temperature, that a pet goldfish will not survive without being fed regularly, and so on depends upon inductive reasoning. Although this sort of reasoning is "at the heart of science and is crucial to common-sense reasoning", it has long been a target of skeptical attack (Feldman 2003, p. 130). Of course, a successful skeptical attack on our inductive reasoning would undermine a vast swath of our knowledge, including our scientific knowledge.

Skepticism about induction is a significant philosophical problem—one that cannot be easily set aside. On the one hand, as we will see, the sort of argument that is marshaled in support of inductive skepticism seems to employ pretty uncontroversial premises. On the other hand, the great strides which science has made in helping us to learn about our universe makes it clear that the sort of reasoning at the core of scientific practice cannot be rotten. Yet, the premises of arguments for inductive skepticism seem true, and they seem to lead to the clearly

K. McCain, *The Nature of Scientific Knowledge*, Springer Undergraduate Texts in Philosophy, DOI 10.1007/978-3-319-33405-9_12

unacceptable conclusion that our practices of inductive reasoning do not give us justified beliefs or knowledge. As C.D. Broad (1952, pp. 142–143) put it, it appears that inductive reasoning is "the glory of science and the scandal of philosophy." So, we face yet another significant threat to our scientific knowledge.

Despite the fact that the challenge posed by inductive skepticism is a difficult one which cannot be easily put to rest, we will see that there are promising ways of dealing with this scandal. In this chapter we will explore the nature of this challenge and one of the more promising responses to inductive skepticism. Again, the response to this skeptical challenge we will examine is an outflowing of the explanatory account we have been developing in the previous chapters. We shall see that while it is not easy to dismiss the threat of inductive skepticism, we should not despair. Our scientific knowledge remains untarnished.

12.1 Examples of Inductive Reasoning

Before assessing the threat to our scientific knowledge and our knowledge more generally posed by skepticism about induction, it will be helpful to briefly clarify some of the more common varieties of inductive inference. The most widely discussed pattern of this sort of reasoning is when we infer something about the next instance on the basis of previous observations (Feldman 2003). Here is an example:

1. There have been many observations of emeralds in a variety of circumstances and conditions.
2. All of the observed emeralds have been green.
3. Therefore, the next emerald to be observed will be green.

It is also common to reason from observations to general conclusions like in the following example:

1. There have been many observations of emeralds in a variety of circumstances and conditions.
2. All of the observed emeralds have been green.
3. Therefore, all emeralds are green.

Both of the above examples of inductive reasoning are ones in which all of the observations are the same in important ways. In both of these examples *all* of the observed emeralds have been green. Of course, not all of our inductive reasoning proceeds from perfectly uniform sampling. Often we find that only a portion of our observations has the characteristic we are interested in. For instance, if controlled studies have been performed to determine the effectiveness of a particular drug in treating an illness, and it has been found that 75 % of those given the drug recover from the illness, we tend to reasonably conclude that (roughly) 75 % of the untreated people with the illness would recover after receiving the drug. More precisely, we reason in the following way:

1. There have been many observations of the effects of drug X on illness Y in a variety of circumstances and conditions.
2. 75 % of all patients suffering from illness Y have recovered after being given drug X.
3. Therefore, 75 % of patients suffering from illness Y will recover after given drug X.

Certainly, there are complications when it comes to inductive reasoning. If illness Y is something like a cold, say, that is known to end after a particular duration, then we may not infer that drug X leads to recovery 75 % of the time. The reason we do not infer the conclusion in this case is that we have a better explanation of the patients' recoveries—namely, the cold has run its course. Similarly, if we have good reason to think that our observations of objects of kind *A* are skewed in some way, perhaps because the observations have only come from one particular lab and cannot be replicated in other labs, then despite the fact that all of the observed *A*s have had a particular property we may not reasonably infer that they all do or that the next *A* will have that property. As we will see below, this fact about inductive reasoning can be readily explained by the sort of account of inductive reasoning our response to inductive skepticism builds upon. We can set aside these sorts of worries for now. It is enough to recognize that we often justifiedly reason in the ways described above.

There are other examples of inductive reasoning, but for our purposes this gives us a sufficient grasp of what is targeted by the inductive skeptic. It is now time to turn our attention to examining the nature of this skeptical challenge.

12.2 The Challenge of Inductive Skepticism

Like many philosophical problems the challenge of inductive skepticism arises from questioning something that we often take for granted. Our commonsense reasoning and scientific practices take it for granted that inductive reasoning provides us with knowledge (or at the very least with justified beliefs). But, this is something which might be questioned. P.F. Strawson (1952, p. 249) characterizes the challenge of inductive skepticism as the challenge of answering the following question: "Why should we suppose that the accumulation of instances of *A*s which are *B*s, however various the conditions in which they are observed, gives any good reason for expecting the next *A* we encounter to be a *B*?" The challenge begins with this sort of question, but as a question it does not yet provide any reason to doubt our knowledge, scientific or otherwise. The reason that this sort of question poses a threat to our knowledge is that one can argue that the answer to Strawson's question is "we should not suppose this at all!"

The challenge of inductive skepticism in its modern form is often attributed to the work of David Hume.[1] Although the argument first appears in Hume's *A Treatise of Human Nature* (1739–1740/1978), it is given a particular clear expression in his *Enquiries Concerning Human Understanding* (1748/1975, pp. 35–36):

> Concerning matter of fact and existence... there are no demonstrative arguments... it implies no contradiction that the course of nature may change... All of our experimental conclusions proceed upon the supposition that the future will be conformable to the past. To endeavor, therefore, the proof of this last supposition by probable arguments, or arguments regarding existence, must be evidently going in a circle, and taking that for granted, which is the very point in question.

In essence, Hume claims that all of our inductive reasoning relies upon the assumption that the future will be like the past. He then goes on to point out that such an assumption cannot be established demonstratively—we cannot derive this assumption from self-evidently true principles via deductive inference. Unfortunately, Hume claims that this assumption cannot be established by using our experiences and inductive reasoning ("probable arguments") either because any such argument must itself rely on the very assumption that we are trying to establish as reasonable—such arguments must rely upon the assumption that the future will resemble the past. So, he claims the assumption is groundless.

Let us make the skeptical challenge to inductive reasoning a bit more precise before turning our attention to how we can overcome this challenge. The first thing we need to do is to recognize that Hume construes the relevant principle too narrowly. It is not just that the future will resemble the past, but instead, the relevant principle is "unobserved instances are like observed instances". After all, it could be that we are making inductive inferences not from past to future, but about things that are all in the past. For example, suppose you have observed random drawings without replacement from an urn of marbles. All 10,000 draws have produced a red marble. You have drawn a marble without looking at it. You infer from the marbles you have observed that the other marble is red too. Notice in this case you are not inferring that the marble *will be red*; you are inferring that *it is red*. Your drawing of the marble is in the past. Although for the discussion that follows the distinction between the principle that "the future will resemble the past" and the broader principle (because it includes the claim that the future will resemble the past) "unobserved instances are like observed instances" will not make much of a difference, it is useful to keep this in mind so as to avoid confusion.

With our clarified understanding of Hume's argument in hand we are ready to formulate the skeptical argument against induction more precisely:

HUME'S ARGUMENT

1. Inductive reasoning is justified only if the principle that unobserved instances are like observed instances (ULO) can be justified.

[1] Weintraub (1995) offers a nice discussion of how Hume's argument is a refinement of an ancient skeptical argument put forward by Sextus Empiricus.

2. If ULO can be justified, then ULO can be justified by either demonstrative argument or by inductive reasoning.
3. ULO cannot be justified by demonstrative argument.
4. ULO cannot be justified by inductive reasoning.
5. Therefore, ULO cannot be justified.
6. Therefore, inductive reasoning is not justified.

(1) is simply expressing the fact that inductive reasoning seems to rely on the assumption that unobserved cases are like observed cases (ULO). (2) is supported by the idea that if we justify a principle (provide good reasons in support of relying upon the principle) we only have two options—we can provide reasons that come from self-evident truths along with deductive reasoning (demonstrative arguments) alone or we can provide reasons by way of experience and inductive reasoning. The thought here is that we either support principles like ULO through the use of reasoning alone or we do so at least in part on the basis of experience. Hume supports (3) by pointing out that there is nothing contradictory in thinking ULO is false. Since the denial of ULO is not a contradiction, it is plausible that ULO could be false. So, Hume concludes that ULO cannot be established via demonstrative argument because its truth is not a matter which can be determined by reason alone. With respect to (4) Hume claims that supporting ULO with inductive reasoning "must be evidently going in a circle, and taking that for granted, which is the very point in question", so he claims that ULO cannot be justified by inductive reasoning. In light of these considerations, it seems that ULO cannot be justified (5). However, if ULO cannot be justified, it seems that inductive reasoning cannot be justified because it relies upon ULO. Thus, HUME'S ARGUMENT yields the skeptical conclusion that inductive reasoning is not justified (6). As we noted above, if this conclusion is true, then our scientific knowledge, as well as an enormous portion of our everyday knowledge, is undermined. Inductive skepticism is a significant threat indeed.

12.3 Responding to the Challenge of Inductive Skepticism

As might be expected from the fact that there are four premises in HUME'S ARGUMENT for inductive skepticism (5 and 6 are conclusions), there are four broad ways to respond to the challenge of inductive skepticism. Each way of responding to the threat of inductive skepticism involves denying one of premises (1)–(4). One might deny premise (1), but doing so seems extremely implausible. It does seem that inductive reasoning rests on ULO or something very similar. After all, if we had no good reason to think that ULO is true, then it is hard to see how we could be justified in relying on inductive reasoning. Consequently, it seems that there are really only three broad responses to this challenge which may be tenable. Let us briefly take a look at some of the more prominent versions of these broad responses before turning to the explanatory response.

First of all, one might deny (2) of HUME'S ARGUMENT on the grounds that there is an additional way ULO might be justified. Some philosophers argue that we can justify inductive reasoning pragmatically.[2] The thrust of this pragmatic justification of induction involves showing that if any method of reasoning about the world will be successful, then our inductive practices will be. Such a response is considered a pragmatic justification because it does not actually involve arguing that our inductive practices are successful or epistemically justified. Instead, this response argues that our inductive reasoning practices are the best we can do. So, supporters of this view argue that there is a way to justify induction which does not rely upon demonstrative arguments or providing inductive arguments in support of inductive reasoning.

The second broad approach to responding to this challenge is to deny (3) of HUME'S ARGUMENT. Typically, this is done by arguing that induction can be justified without any appeal to experience, but instead, it can be justified through considerations of pure reasoning alone. One popular way of doing this is to argue that we can simply tell by rationally reflecting on our methods of reasoning that inductive reasoning is justified.[3] Often the thought here is that when we are discussing something as fundamental as inductive reasoning we will have no recourse but to appeal to our intuitions about what is rational, and these intuitions support thinking that inductive reasoning is rational. We will discuss this issue a bit more fully when discussing the explanatory response.

The third sort of broad approach one might take in responding to the challenge of inductive skepticism is to deny (4) of HUME'S ARGUMENT and maintain that we can justify inductive reasoning via inductive reasoning.[4] Roughly, the idea here is that we can justify ULO by noting that when we have used inductive reasoning before it has led us to correct (justified) conclusions, and then arguing that since our observed cases of inductive reasoning have been successful, the unobserved cases are/will be too. Essentially, this response seeks to show that it is possible to defend the rationality of inductive reasoning by using that very sort of reasoning. This sort of response is circular in some sense, but supporters maintain that the kind of circularity involved is not vicious. We will examine this idea further in sections that follow.

There is much more that can be, and has been, said both for and against each of these approaches to responding to the challenge of inductive skepticism. For our purposes, however, simply noting some of these primary ways of responding to HUME'S ARGUMENT is enough. Accordingly, we will not delve into the debates over the acceptability of these various responses. Some of the details of these responses will be fleshed out further when discussing the explanatory response, but

[2]See Reichenbach (1938, 1949), Salmon (1957, 1974), and Skyrms (1975).

[3]See Carnap (1968), Kyburg (1956, 1965), Lycan (1988), Psillos (1999), and Strawson (1952) for this sort of view.

[4]Prominent defenders of this sort of response to the challenge of inductive skepticism include Black (1954, 1958, 1963), Braithwaite (1953), Papineau (1993), and Van Cleve (1984).

these details can wait until this response has been explained. Briefly explaining these methods of responding to the challenge of inductive skepticism as we have here is worthwhile for at least two reasons. First, doing so helps us to get clear on some of the variation in the ways in which we might seek to respond to this skeptical challenge. Second, even if one were unsatisfied with the explanatory response that we will develop in this chapter, we can see that there are still many other avenues for addressing the challenge of inductive skepticism which might be explored. It remains an open possibility that our scientific knowledge can be defended from this skeptical threat in a variety of distinct ways.

12.4 The Explanatory Response

The first step to developing the explanatory response to inductive skepticism is to get clear on how we should understand the nature of inductive reasoning. It is plausible that the inductive argument forms which we considered in Sect. 12.1, and all other plausible instances of inductive reasoning, should be understood in terms of inference to the best explanation. Recall, from Chap. 10 that inferences to the best explanation are roughly of this form:

$F_1, F_2, \ldots, F_n$ are facts in need of explanation.
Hypothesis H explains the F_i.
No available competing hypotheses would explain the F_i as well as H does.

Therefore, H is true

Admittedly, at first glance the arguments we examined above do not seem to fit this form. Initial appearances notwithstanding, the best way to understand inductive reasoning, when it is appropriate, is as inference to the best explanation/explanatory reasoning. Consider, why is it reasonable to infer that all emeralds are green on the basis of having observed only green emeralds? The reason is that the best explanation of our observations all being the way they are is that there is something about emeralds which makes them all green. Hence, the claim that all emeralds are green is part of the best explanation of our evidence. This answer to why it is reasonable to infer this conclusion is a natural consequence of the account of evidential support we developed in Chap. 10— *Explanationism*. According to *Explanationism*, we have justification for accepting propositions that are part of the best explanation of our evidence or that are better explained by the best explanation of our evidence than their denials.

The case for understanding all good inductive reasoning as explanatory reasoning is clearer when we consider arguments like this one:

1. There have been many observations of the effects of drug X on illness Y in a variety of circumstances and conditions.

2. 75 % of all patients suffering from illness Y have recovered after being given drug X.
3. Therefore, 75 % of patients suffering from illness Y will recover when given drug X.

As we noted above, the conclusion that 75 % of patients suffering from illness Y will recover when given drug X does not seem to follow from the premises when we know that the illness is something like a cold which will go away on its own (without drug X) in the amount of time elapsed before the observation is to be made. What explains the difference between instances where we can reasonably accept the conclusion of this sort of argument and instances where we cannot? In the cases where we can reasonably accept the conclusion the truth of that conclusion is part of the best explanation of our observations, or it is better explained by the best explanation of our evidence than its denial. The best explanation of the fact that 75 % of all patients suffering from illness Y have recovered after being given drug X, when this is all the information we have, is that there is something about drug X (a property governed by a law-like regularity, some mechanism, etc.) which makes it so that 75 % of the patients given drug X will recover. The claim that (roughly) 75 % of patients suffering from illness Y will recover when given drug X is much better explained by the best explanation of our observations than its rival (it is not the case that (roughly) 75 % of patients suffering from illness Y will recover when given drug X) is. So, according to *Explanationism* that conclusion is justified for us.

In a case where we also know that illness Y is something which tends to disappear on its own in the length of time that we have observed patients given drug X things are different. In this case there being something about drug X (a property governed by a law-like regularity, some mechanism, etc.) which makes it so that 75 % of the patients given drug X will recover is not the best explanation of our observations. Instead, we have an equally good rival explanation—illness Y is such that it naturally (without the drug treatment) goes away in the amount of time elapsed prior to our observations. In this case we are not justified in accepting the conclusion of the argument above.

Additionally, once we understand induction as a kind of explanatory reasoning we can explain why certain methods of inductive inference are fallacious (Lycan 1988). It is widely recognized that inductive inferences drawn from an insufficient sample size and those drawn from biased samples do not provide good reason to accept their conclusions. But, why is that? If we understand inductive reasoning as nothing over and above the sorts of inferences we considered in Sect. 12.1 above (with no underlying explanatory connection being necessary), then it seems we do not have an explanation of why it is fallacious to drawn conclusions in these ways. However, construing induction as a kind of explanatory inference explains why these are fallacies.

Consider a simple example of drawing a conclusion from an insufficient sample size. Saundra is trying to determine the color of marbles in an urn. Saundra knows there are a million marbles in the urn and that they can be any color. She draws a single marble and notices that it is red. Saundra infers on the basis of that single observation that all the marbles are red. In such a case it is clear that Saundra has reasoned poorly. She has committed the fallacy of reasoning from an insufficient sample size. The flaw in Saundra's reasoning is that her conclusion is not the best explanation of why she has the observation that she does. An equally good (or better) explanation is that some, but not all, of the marbles are red. Without understanding induction in terms of explanatory reasoning it is not clear what is wrong with Saundra's inference in this case.

When it comes to reasoning from biased samples things are even clearer. Consider a simple example of this fallacy. Steve wants to know which college football team is the most popular team in the United States. He surveys students at the University of Alabama at Tuscaloosa and finds that almost all of them like the University of Alabama's football team better than any other college football team. On the basis of his data, Steve concludes that the University of Alabama's football team is by far the most popular college football team in the United States. Like Saundra, Steve has reasoned poorly. The flaw in Steve's reasoning is perhaps even clearer than the flaw in Saundra's. The best explanation of Steve's data is not that the University of Alabama's football team is by far the most popular college football team in the United States. A better explanation of Steve's data is that students at a particular university prefer their university's football team to the teams at other universities. In light of considerations about these fallacies, and the other considerations raised above, it is reasonable to conclude that inductive reasoning, when it is justified, is an instance of explanatory reasoning (IBE).[5]

At this point we have seen it is plausible that good cases of inductive reasoning should be understood as an explanatory reasoning and that when inductive reasoning goes wrong it is because it fails to form a good IBE. It is now time to explore how recognizing this fact helps with the challenge of inductive skepticism. Recall P.F. Strawson's (1952, p. 249) expression of this challenge in terms of the following question: "Why should we suppose that the accumulation of instances of *A*s which are *B*s, however various the conditions in which they are observed, gives any good reason for expecting the next *A* we encounter to be a *B*?" We now have a good answer to this question. We have good reason for expecting the next *A* we encounter to be a *B* when our accumulated observations of *A*s that are *B*s are best explained by some regularity connecting *A*-ness to *B*-ness.[6] For example, why is it that we have good reason to think the next unsupported object relatively close to the surface of the earth will be pulled toward the earth? The reason is that the best explanation of our previous observations of unsupported objects relatively close to the surface of the earth being pulled toward the earth is that it is a law of nature, gravity, that

---

[5]See Harman (1965, 1968, 1973, 1986), Weintraub (2013), and White (2005) for further considerations in support of this claim. Fumerton (1980) demurs, but see Weintraub (2013) for convincing responses to Fumerton's objections.

[6]Some, such as Armstrong (1983) and Foster (1982–1983, 2004), hold that this regularity must be a law of nature. Others, such as White (2005), argue that it is not necessary that the regularities appealed to are natural laws. Fortunately, we do not need to settle this issue here. Either way things turn out the explanatory response to inductive skepticism is still compelling.

objects with mass are drawn toward one another with the less massive object moving more toward the more massive object. Now, we do not have to know the exact regularity which is operative in order to reasonably draw an inference on the basis of our previous observations, it is enough for it simply to be the best explanation of our observations that there is some regularity or other that connects A-ness (like unsupported mass) to B-ness (falling).

Putting this response in terms of HUME'S ARGUMENT from Sect. 12.2, we have good reason to think that premise (4) is false. Once we recognize that inductive reasoning is best understood in terms of explanatory reasoning we can see that there is good inductive (in the broad sense of being non-deductive/non-demonstrative) support for ULO—it is justified for us when the best explanation of why the observed instances have a feature that they do is that there is some regularity which makes it so that they have that feature. When such a regularity is the best explanation for why the observed *A*s are *B* we have good reason to think the unobserved *A*s will be *B* as well. So, we do have a sort of inductive reasoning which can legitimately support ULO. Thus, premise (4) is false, and HUME'S ARGUMENT fails to establish the conclusion that our inductive reasoning is not justified. Therefore, the threat of inductive skepticism is not all that threatening.[7]

12.5 The Challenge Returns?

Although we have seen that appealing to explanatory reasoning provides us with a satisfactory way of overcoming the challenge of inductive skepticism, one might worry that the problem simply returns as a problem for IBE (Weintraub 2013). That is, one might question what reason we have for thinking explanatory reasoning will get us to the truth of things. In other words, rather than questioning why we should think ULO is true, one might instead question why we should think the best explanation of some phenomenon is likely to be true.

This objection has been termed the "Truth Demand" because it demands that in order for IBE to be acceptable we have to have reason for thinking that the features which make one explanation better than another, explanatory virtues, are linked to the truth (Vogel manuscript). The Truth Demand has been pressed in numerous forms. James Beebe (2009, p. 619) says, "the satisfaction of the explanatory criteria cannot provide us with an epistemic reason to believe . . . if those criteria are not themselves truth-linked." Peter Lipton (2004, p. 144) expresses the Truth Demand as follows:

> Why should the explanation that would provide the most understanding if it were true be the explanation that is most likely to be true? Why should we live in the loveliest of all possible worlds? Voltaire's objection is that, while loveliness may be as objective as you like, the

[7]For further considerations in support of this sort of response to inductive skepticism see Armstrong (1983), BonJour (2010), Feldman (2003), Foster (1982–1983, 2004), and Lycan (1988).

> coincidence of loveliness and likeliness is too good to be true. It would be a miracle if using explanatory considerations as a guide to inference were reliably to take us to the truth.

Bas van Fraassen (1980, p. 90) also appears to be pressing the Truth Demand when he says, "some writings on the subject of induction suggest that simpler theories are more likely to be true. But it is surely absurd to think that the world is more likely to be simple than complicated." All of these versions of the Truth Demand present the same challenge to the acceptableness of relying on explanatory reasoning. The Truth Demand challenges the supporter of IBE to give reasons for thinking that explanatory virtues, which make one explanation better than another, are connected to the truth. That is, the Truth Demand is a demand for reasons to think that the best explanation of a given phenomenon is likely to be true.

Jonathan Vogel (manuscript) aptly notes that the various versions of the Truth Demand can be understood as pressing an argument that beliefs licensed by IBE are not justified. Essentially the Truth Demand is the challenge of inductive skepticism, which was overcome above, returning as an attack on explanatory reasoning. Here is Vogel's formulation of the Truth Demand argument:

1. A belief licensed by inference to the best explanation will be justified only if we are justified in believing that such a belief is likely to be true.
2. We are justified in believing that a belief licensed by inference to the best explanation is likely to be true only if we are justified in believing that the world is lovely.
3. We aren't justified in believing that the world is lovely.
4. Therefore, we aren't justified in believing that a belief licensed by inference to the best explanation is likely to be true.
5. Therefore, a belief licensed by inference to the best explanation isn't justified. (manuscript, p. 4)[8]

In light of this, one might worry that we have overcome the challenge of inductive skepticism only to find ourselves facing an analogous challenge for explanatory reasoning.

12.6 Responding to the Returned Challenge

Fortunately, we have plausible ways of responding to the challenge presented by the Truth Demand too. One way of responding to the Truth Demand involves appealing to rational reflection. As Henry Kyburg Jr. (1974, p. 65) says:

> I think that in some sense our justification of inductive rules must rest on an ineradicable element of inductive intuition—just as I would say that our justification of deductive rules

[8]With respect to an explanation "loveliness" refers to the explanation's possessing explanatory virtues. Hence, the lovelier of two explanations is the explanation which is more explanatorily virtuous, i.e. the better explanation of the two. The claim that "the world is lovely" refers to the idea that explanatory virtues are correlated to the truth in the sense that explanations which are explanatorily virtuous are likely to be true.

> must ultimately rest, in part, on an element of deductive intuition: we *see* that *modus ponens* is truth-preserving—that is simply the same as to reflect on it and fail to see how it can lead us astray. In the same way, we *see* that if all we know about in all the world is that all the A's we've seen have been B's, it is *rational* to *expect* that the next A will be a B.[9]

Kyburg appeals to rational reflection as a way to justify inductive reasoning because he takes inductive reasoning to be a fundamental form of reasoning just as deductive reasoning is. Since inductive reasoning is fundamental, it does not make sense to attempt to justify it in terms of some other more basic forms of reasoning. Thus, rational reflection provides the only non-circular way of justifying inductive reasoning.

As we have discussed above, Kyburg's claim about inductive reasoning being fundamental does not seem correct when we think of inductive reasoning as simply enumerative induction—the sort of inductive reasoning that says there is nothing more to the inferences in Sect. 12.1 than moving directly from observations to conclusions about the unobserved. This is why we could not simply employ Kyburg's method to respond to our initial challenge of inductive skepticism. Yet, we have seen that it *is* reasonable to think that inductive reasoning of that sort can really be reduced to, or subsumed under, explanatory reasoning. It is plausible that explanatory reasoning is a basic form of reasoning.[10]

If explanatory reasoning, like IBE, is a fundamental form of reasoning, it is to be expected that it will be justified in the same way as other fundamental forms of reasoning, that is, by rational reflection. So, mirroring Kyburg, one might claim that we *see* that if H is the best explanation of our evidence, it is *rational* to *expect* H to be true. This is the best that can be expected when one is justifying a fundamental form of reasoning; to ask for justification in terms of some other more fundamental truth relation is to misunderstand what is involved in being a fundamental form of reasoning.

Given the plausible assumption that in order for it to be rational for one to expect H to be true it must be that H is likely to be true given one's evidence, appeal to rational reflection offers a plausible way of meeting the Truth Demand. In other words, one might attempt to satisfy the Truth Demand by claiming that when we reflect on various explanations we see that the one which is the

[9]Carnap (1968), Kyburg (1956), Lycan (1988), and Psillos (1999) all claim that both basic forms of deductive reasoning and basic forms of inductive reasoning are justified by rational reflection. In a similar vein, Goodman (1965, p. 64) claims that all inference rules are justified by a sort of reflective equilibrium, "The process of justification is the delicate one of making mutual adjustments between rules and accepted inferences; and in the agreement achieved lies the only justification needed for either." Of course, Goodman's reflective equilibrium just is an example of employing explanatory reasoning of the sort that we are attempting to defend. White (2005) suggests something similar to Goodman. After arguing that explanatory considerations can be used to sort good and bad non-deductive inferences, White claims that inference to the best explanation may be justified by seeking reflective equilibrium with respect to our judgments concerning instances of non-deductive inference. We will discuss this sort of response below.

[10]For reasons in addition to those we discussed above see Enoch and Schechter (2008) for strong grounds for thinking that explanatory reasoning is a fundamental form of reasoning.

most explanatorily virtuous is likely to be true. Thus, one might attempt to meet the Truth Demand by making the plausible claim that explanatory reasoning is a fundamental form of reasoning, and it is justified through rational reflection.

Of course, one might question what justifies us in using rational reflection as a way of justifying forms of reasoning. According to our account of evidential support, *Explanationism*, this must be because the best explanation of the observations we make through rational reflection is that those observations are true. As a result, it seems that we may end up relying on explanatory reasoning in order to justify explanatory reasoning even if we go via the route of rational reflection. This is not necessarily a bad thing though.

This brings us to our second way of responding to the Truth Demand. We can plausibly maintain that the best explanation for why most of our inferences to the best explanation have been successful is that explanatory reasoning is a justifying form of reasoning. That is to say, we might appeal to explanatory reasoning as a way of justifying the claim that explanatory reasoning gets us to truth. We might express this in the following sort of way:

Argument IBE

1. Most of our observed instances of (good) IBE have true conclusions.
2. The best explanation of (1) is that explanatory virtues are positively correlated to the truth.
3. Therefore, explanatory virtues are positively correlated to the truth.

Of course, there are details concerning *Argument IBE* which could use further development, but for our purposes this sketch of our reasoning in response to the Truth Demand is sufficient.

At this point one might object that we are doing the very thing which Hume (1748/1975, p. 36) complained occurs when you try to justify induction inductively—we "must be evidently going in a circle, and taking that for granted, which is the very point in question." It is true that there is circularity here, however, it is not vicious. In order to appreciate this point it is important to distinguish two kinds of circularity: *premise* and *rule*.

Premise-circularity occurs when the conclusion of an argument occurs as one of the premises of that argument. Here is a blatant example of this sort of circularity:

1. Explanatory virtues are positively correlated with truth.
2. Therefore, explanatory virtues are positively correlated with truth.

Clearly, this sort of circularity is vicious because its conclusion is the same as its premise. The very thing that is trying to be established is accepted as a premise—the argument gives you no reason to accept the conclusion which you did not have prior to formulating the argument.

Things are different with rule-circularity. Rule-circularity occurs when the rule relied upon in the argument for moving from the premises to the conclusion is the very rule that is supported in the conclusion. This sort of circularity does not seem to be vicious. The reason for this is that a rule-circular argument can give you reason to accept the conclusion that you did not have prior to formulating the argument. You can gain such reasons because in some cases, those involving fundamental methods of reasoning, you might reasonably employ a method of reasoning without explicitly endorsing that method as you would have to if it were a premise of your argument. This is a very important difference between premise-circular and rule-circular arguments. *Argument IBE* is rule-circular, but not premise-circular.

Quite plausibly, when we reach a fundamental method of reasoning it must be justified in a rule-circular fashion (Matheson 2012). By definition, a fundamental method of reasoning is one which cannot be reduced to or supported by some more fundamental method. After all, if the method in question can be reduced to, or supported by something more fundamental, then the method being reduced/supported is not fundamental. Hence, it is not surprising that explanatory reasoning is justified in a rule-circular fashion. Further, it seems like a perfectly legitimate way to justify this fundamental method of reasoning. After all, there is no more fundamental method of reasoning which we can appeal to in justifying explanatory reasoning—any attempt will end up employing explanatory reasoning, either explicitly or implicitly, at some point.

12.6.1 Residual Concerns

Before we set aside the Truth Demand and consider the challenge it presents to be overcome, there are two concerns that need to be addressed. The first concern is simply the question of why we cannot employ this sort of rule-circular response from the start—when responding to the challenge of inductive skepticism. As we noted above, this is something that several philosophers have attempted to do. Nevertheless, we have seen that the plausibility of this sort of response depends upon whether the principle of reasoning being defended is fundamental. We have considered good reasons for thinking that inductive reasoning, construed as distinct from explanatory reasoning, is not fundamental. Thus, it is doubtful that this sort of response can be successfully employed against the original challenge of inductive skepticism.

The second concern is that if rule-circular reasoning can support explanatory reasoning as it does in *Argument IBE*, then it can just as easily be employed to support clearly false principles of reasoning. The worry here is that it seems reading tea leaves or consulting one's horoscope or employing counter-induction (the rule that since all observed *A*s have been *B*, the next observed *A* will not be *B*) and other clearly poor ways of reasoning can also be defended in a rule-circular fashion. After all, one might point out that it seems the counter-inductivist can argue in the following fashion:

1. The previously observed instances of counter-induction have failed to lead to true conclusions.
2. Therefore, (by counter-induction) the next instance of counter-induction will lead to a true conclusion.[11]

Similar rule-circular defenses of other clearly poor methods of reasoning seem to be easily generated. Consequently, one might claim that it is trivial to point out that a method of reasoning can be supported by employing that very method of reasoning. Thus, one might worry that the sort of self-support which a rule-circular explanatory argument provides in support of explanatory reasoning fails to justify explanatory reasoning.

One thing to note immediately is that it is not a trivial matter that a self-supporting, rule-circular argument can be given for a particular method of reasoning. Consider the simple rule that calls for consulting tea leaves in order to determine how a particular unobserved situation will turn out. It is entirely possible that if one were to read tea leaves in order to determine whether tea leaf reading is a good way to reason, the result would be negative (Matheson 2012). The same applies to consulting the horoscope and numerous other flawed reasoning methods (Weintraub 1995). It is a live possibility that many of these methods of reasoning would fail to provide grounds for a rule-circular argument supporting their own reasonableness. So, it is not a trivial accomplishment that explanatory reasoning is self-supporting in the way that it is (Boghossian 2000).

Despite the fact that not all methods of reasoning can appeal to self-supporting rule-circular defenses of themselves, it is not enough to allay this concern by noting that explanatory reasoning's self-support is non-trivial. Fortunately, we have already noted a very important difference between explanatory reasoning and these other reasoning methods—explanatory reasoning is fundamental. In addition to the reasons we have adduced above for thinking that explanatory reasoning is fundamental there is a further reason to think explanatory reasoning is special in this way. Consider any of the other flawed methods of reasoning that we mentioned above. How would we determine whether that method is successful or not? Intuitively, in order to judge whether a method of reasoning is successful we would look at the observed instances when that method was employed and see what the results were. In cases where a particular method has been observed to yield true conclusions in the majority of the observed instances of its use we might reasonably conclude that the method is successful. That is to say, we might reasonably conclude that the method will be likely to lead us to true conclusions in future instances of its use. Of course, the means by which we are reasonably

[11] Salmon (1957) argues that if a rule-circular justification can be given for inductive reasoning, then it can be given for counter-induction too. Black (1958) argues that counter-induction cannot be supported in this way because counter-induction is actually internally inconsistent. We will not settle this issue here because there are other reasons for thinking that rule-circular reasoning is a viable way of supporting explanatory reasoning, but not illegitimate forms of reasoning like counter-induction.

concluding that the method is successful is by appealing to explanatory reasoning. We infer that the best explanation of the observed successes of the method is that it is the sort of method which is generally successful. So, we conclude that future uses of the method are likely to lead us to true conclusions too. Thus, in order to determine whether another method of reasoning is a successful method we will have to rely on explanatory reasoning.[12]

We now have a clear rationale for distinguishing between methods of reasoning that can be supported in a rule-circular fashion (such as explanatory reasoning) and those that cannot—the former are basic/fundamental methods of reasoning; the latter are not. If a method of reasoning is not fundamental then, if it can be supported at all, there will be some more basic method which supports it. However, if a reasoning method is fundamental, there is nothing more basic or fundamental which can be appealed to in order to justify it. To expect such a justification of a fundamental method of reasoning is to misunderstand what it means to be *fundamental*. Hence, it is not at all surprising that a rule-circular justification is required for fundamental methods of reasoning like explanatory reasoning. Once a fundamental method is reached there is nowhere else to look for justification than that method itself.[13]

Now, one might worry that while a rule-circular justification of explanatory reasoning is all that we can (or should) hope for because it is a fundamental method of reasoning, this will not satisfy the inductive skeptic. After all, the inductive skeptic doubts that it is reasonable to employ explanatory reasoning in the first place. Accordingly, utilizing the sort of reasoning being questioned by the skeptic will not assuage the skeptic's concerns.

The key to responding to this worry is to recall our distinction in Chap. 5 between justifying and being justified. It is true that a rule-circular defense of explanatory reasoning will not be dialectically satisfying to the skeptic. So, if we employ such a defense, we will fail in *justifying* our use of explanatory reasoning to the skeptic. Nevertheless, this does not mean that our use of explanatory reasoning fails to be *justified* in this rule-circular way (Boghossian 2000; Matheson 2012). It may very well be that if a skeptic does not grant that our most basic methods of reasoning are reasonable, we cannot persuade him otherwise. It is likely that if the skeptic refuses to acknowledge the reasonableness of our fundamental methods of reasoning, we cannot reason with him at all! This should not trouble us though. Without at least some common ground it is not possible to justify anything to the skeptic. That is not a problem for our fundamental methods of reasoning, but instead, it is a problem for the skeptic.

[12]Both Jones (1982) and Strawson (1952) emphasize this point in terms of inductive reasoning. However, as we have seen above, inductive reasoning is itself really a kind of explanatory reasoning.

[13]See Enoch and Schechter (2008), Feldman (2003), Feigl (1950), and Matheson (2012) for further discussion of fundamental methods of reasoning.

12.7 Conclusion

As we have seen, the challenge that inductive skepticism poses for our scientific knowledge has the potential to be quite threatening. Fortunately, we have also seen that the explanatory account that we have been developing throughout earlier chapters offers a compelling way of overcoming this challenge. Although there is much work to be done in spelling out the exact details of the amount of justification conferred upon claims via our inductive reasoning practices, and there are additional challenges to this sort of reasoning, we have reason to be optimistic with respect to our prospects for spelling out these details and meeting these challenges.[14] In light of this, our scientific knowledge seems secure despite challenges from inductive skepticism.

References

Armstrong, D. M. (1983). *What is a law of nature?* Cambridge, UK: Cambridge University Press.

Beebe, J. (2009). The abductivist reply to skepticism. *Philosophy and Phenomenological Research, 79*, 605–636.

Black, M. (1954). *Problems of analysis*. Ithaca: Cornell University Press.

Black, M. (1958). Self-supporting inductive arguments. *Journal of Philosophy, 55*, 718–725.

Black, M. (1963). Self-support and circularity: A reply to Mr. Achinstein. *Analysis, 23*, 43–44.

Boghossian, P. (2000). Knowledge of logic. In P. Boghossian & C. Peacocke (Eds.), *New essays on the a priori* (pp. 229–254). Oxford: Clarendon Press.

BonJour, L. (2010). *Epistemology: Classic problems and contemporary responses* (2nd ed.). Lanham: Rowman & Littlefield.

Boyce, K. (2014). On the equivalence of Goodman's and Hempel's paradoxes. *Studies in History and Philosophy of Science, 45*, 32–42.

Braithwaite, R. (1953). *Scientific explanation*. Cambridge, UK: Cambridge University Press.

Broad, C. D. (1952). *Ethics and the history of philosophy*. London: Routledge.

[14]One of the more famous of these challenges is Goodman's (1965) "New Riddle of Induction". Essentially, Goodman's riddle poses the challenge of explaining why observations support what we typically assume rather than some strange hypothesis. Goodman makes this point by noting that it seems the fact that all observed emeralds are green supports thinking that all emeralds are *grue* (where "grue" means that the object is observed before a particular time and green, or observed after that time and blue) just as well as it supports thinking that all emeralds are green. This is yet another inductive challenge which can be met with explanatory reasoning. Very roughly, the reason that our normal hypotheses are supported by our observations rather than strange grue-like hypotheses is that the former are better explanations of our data than the latter. For more detailed discussion of the New Riddle of Induction and convincing arguments that the solution to the riddle lies in appealing to explanatory reasoning, see Hesse (1969), Lycan (1988), Ward (2012), and White (2005). Interestingly, Boyce (2014) argues that the New Riddle of Induction is logically equivalent to the purported paradox of ravens illuminated by Hempel (1945). Consequently, if Boyce is correct, then explanatory reasoning can solve this paradox as well. Even if they are not logically equivalent, it is plausible that explanatory reasoning offers a solution to Hempel's paradox of ravens too (Lycan 1988).

Carnap, R. (1968). Inductive intuition and inductive logic. In I. Lakatos (Ed.), *The problem of inductive logic* (pp. 258–267). Amsterdam: North-Holland Publishing Company.
Enoch, D., & Schechter, J. (2008). How are basic belief-forming methods justified? *Philosophy and Phenomenological Research, 76*, 547–579.
Feigl, H. (1950). De principiis non disputandum ... ?: On the meaning and the limits of justification. In M. Black (Ed.), *Philosophical analysis: A collection of essays* (pp. 113–147). Freeport: Prentice Hall.
Feldman, R. (2003). *Epistemology*. Upper Saddle River: Prentice Hall.
Foster, J. (1982–1983). Induction, explanation and natural necessity. *Proceedings of the Aristotelian Society, 83*, 87–101.
Foster, J. (2004). *The divine lawmaker*. Oxford: Clarendon Press.
Fumerton, R. (1980). Induction and reasoning to the best explanation. *Philosophy of Science, 47*, 589–600.
Goodman, N. (1965). *Fact, fiction, and forecast*. Cambridge, MA: Harvard University Press.
Harman, G. (1965). The inference to the best explanation. *Philosophical Review, 74*, 88–95.
Harman, G. (1968). Enumerative induction as inference to the best explanation. *Journal of Philosophy, 64*, 529–533.
Harman, G. (1973). *Thought*. Princeton: Princeton University Press.
Harman, G. (1986). *Change in view*. Cambridge, MA: MIT Press.
Hempel, C. G. (1945). Studies in the logic of confirmation (I). *Mind, 54*, 1–26.
Hesse, M. (1969). Ramifications of 'grue'. *British Journal for Philosophy of Science, 20*, 13–25.
Hume, D. (1739–1740/1978). *A treatise of human nature*. Oxford: Clarendon Press.
Hume, D. (1748/1975). *Enquiries concerning human understanding*. Oxford: Clarendon Press.
Jones, G. E. (1982). Vindication, Hume, and induction. *Canadian Journal of Philosophy, 12*, 119–129.
Kyburg, H. E., Jr. (1956). The justification of induction. *Journal of Philosophy, 53*, 394–400.
Kyburg, H. E., Jr. (1965). Discussion: Salmon's paper. *Philosophy of Science, 32*, 147–151.
Kyburg, H. E., Jr. (1974). Comments on Salmon's 'inductive evidence'. In R. Swinburne (Ed.), *The justification of induction* (pp. 62–66). Oxford: Oxford University Press.
Lipton, P. (2004). *Inference to the best explanation* (2nd ed.). New York: Routledge.
Lycan, W. G. (1988). *Judgement and justification*. Cambridge, UK: Cambridge University Press.
Matheson, J. (2012). Epistemic relativism. In A. Cullison (Ed.), *Continuum companion to epistemology* (pp. 161–179). London: Continuum.
Papineau, D. (1993). *Philosophical naturalism*. Oxford: Blackwell.
Psillos, S. (1999). *Scientific realism: How science tracks truth*. London: Routledge.
Reichenbach, H. (1938). *Experience and prediction*. Chicago: University of Chicago Press.
Reichenbach, H. (1949). *The theory of probability*. Berkeley: University of California Press.
Salmon, W. (1957). Should we attempt to justify induction? *Philosophical Studies, 8*, 33–48.
Salmon, W. (1974). The pragmatic justification of induction. In R. Swinburne (Ed.), *The justification of induction* (pp. 85–97). Oxford: Oxford University Press.
Skyrms, B. (1975). *Choice and chance* (2nd ed.). Belmont: Wadsworth.
Strawson, P. F. (1952). *Introduction to logical theory*. London: Methuen.
Van Cleve, J. (1984). Reliability, justification, and the problem of induction. *Midwest Studies in Philosophy, 9*, 555–567.
van Fraassen, B. C. (1980). *The scientific image*. Oxford: Oxford University Press.
Vogel, J. (manuscript). Explanation, truth, and the external world. http://fitelson.org/current/vogel.pdf
Ward, B. (2012). Explanation and the new riddle of induction. *Philosophical Quarterly, 62*, 365–385.
Weintraub, R. (1995). What was Hume's contribution to the problem of induction? *Philosophical Quarterly, 45*, 460–470.
Weintraub, R. (2013). Induction and inference to the best explanation. *Philosophical Studies, 166*, 203–216.
White, R. (2005). Explanation as a guide to induction. *Philosopher's Imprint, 5*, 1–29.

Chapter 13
Empirical Evidence of Irrationality

Abstract Whereas the previous two chapters responded to philosophical challenges to our scientific knowledge this chapter explores a more practical threat to scientific knowledge. This challenge comes from research which suggests we are subject to a number of biases and irrational processes when forming our beliefs. Numerous studies have seemingly shown that people are prone to make systematic errors of reasoning in particular kinds of cases. Some take this evidence of human irrationality to undercut our knowledge in general, and hence, our scientific knowledge as well. This chapter argues that this challenge does not pose a significant threat to our scientific knowledge. Although there is evidence for human irrationality, we have ways of keeping this sort of irrationality contained so that it does not "infect" all of our beliefs. So, while we are prone to make systematic errors in certain cases, we are aware of our proclivities, and we can take steps to counteract our natural shortcomings.

In the two previous chapters we explored skeptical challenges to our scientific knowledge that come by way of philosophical arguments threatening a wide swath of our knowledge of the world around us. Although the arguments for these forms of skepticism are interesting, and they pose challenges to our having the knowledge we ordinarily take ourselves to have, we have the requisite tools for overcoming the challenges they present. Considering these sorts of challenges to our scientific knowledge is important because doing so helps illuminate various key features of the nature of knowledge in general, which in turn can help to deepen our understanding of scientific knowledge in particular.

Despite their importance, readers of a more practical disposition are apt to find the traditional skeptical arguments of the previous two chapters unimpressive and perhaps not worth taking all that seriously. In this chapter we will explore a challenge to our scientific knowledge which does not rest upon philosophical speculation. The challenge to our scientific knowledge we will consider in this chapter is more concrete than the previous challenges because it arises from empirical studies of human reasoning. Several studies seem to suggest that we systematically make mistakes in reasoning. On the whole these studies appear to paint a rather dismal picture of our reasoning abilities (Nisbett and Borgida 1975). The mistakes these studies reveal are common enough that they lead some to worry

K. McCain, *The Nature of Scientific Knowledge*, Springer Undergraduate Texts in Philosophy, DOI 10.1007/978-3-319-33405-9_13

we are systematically irrational in our reasoning. Given widespread systematic irrationality, one might worry that we lack the sort of knowledge we take ourselves to have. After all, if we are often wrong when we reason as we do, then it seems that when we get things right as a result of our reasoning it is largely a matter of luck. Consequently, these empirical studies appear to seriously threaten our scientific knowledge and, as will become clear shortly, a large amount of our ordinary (non-scientific) knowledge as well. While the threat posed here may not be as far-reaching as those of the skeptical arguments in the previous chapters, it is still very worrisome—especially, given the fact that this threat is grounded in empirical research not just philosophical speculation.

13.1 The Empirical Evidence

There have been numerous studies on human reasoning and the mistakes we sometimes make when given specific reasoning tasks.[1] Rather than conduct an exhaustive survey of this literature we will confine our examination of the issue to two of the most widely discussed experiments and the evidence they are purported to provide for our systematic irrationality.

13.1.1 Errors in Deductive Reasoning: The Selection Task

One of the most well known experiments concerning reasoning is Peter Wason's (1966) selection task experiment.[2] The purpose of this experiment is to test people's ability to accurately apply principles of deductive reasoning. In this experiment subjects are shown four cards. Each card has a visible side and a covered side. Subjects are told that each card has a letter on one side and a number on the other side. The visible sides are as follows:

Card 1	Card 2	Card 3	Card 4
A	K	4	7

Subjects are then asked which of the four cards they *need* to see the covered side of in order to determine if the following rule is true:

[1]For helpful surveys of this literature see Gilovich (1991), Kahneman et al. (1982), Nisbett and Ross (1980), Plous (1993), and Tweney et al. (1981).

[2]See Wason (1968) for a similar, but more extensive, experiment. See Wason and Johnson-Laird (1972) for further discussion of these sorts of experiments.

> If a card has a vowel on one side, then it has an even number on the other side. (Stein 1996, p. 80)

In order to correctly evaluate this rule subjects should follow this principle:

> *Conditional-Testing Principle*: To test the truth of a conditional, examine cases where the antecedent is true to make sure that the consequent is true and examine cases where the consequent is false to make sure the antecedent is false. (Stein 1996, p. 81)

The reason for this is that the only way a conditional statement, such as the rule subjects are asked to evaluate, can be false is if the antecedent (in this case "a card has a vowel on one side") is true and the consequent (in this case "it has an even number on the other side") is false. So, the only situations in which there is a violation of the rule are situations in which a card has a vowel on one side and an odd number on the other side. Any other combination will be consistent with the rule. Thus, if one is following the Conditional-Testing Principle in her reasoning about this task, she will rightly note that in order to test the rule you need to see the covered sides of card 1 and card 4.

Experimenters found that subjects overwhelmingly give the wrong answers in this sort of experiment. In one instance of the experiment only 5 out of 128 subjects correctly answered that one needs to see the covered sides of cards 1 and 4 in order to test the rule (Stich 1990). Unfortunately, these results are fairly typical. In the many iterations of the experiment it turns out that 33 % of subjects claim that just card 1 needs to be seen in order to test the rule, 46 % claim that card 1 and card 2 need to be uncovered, 17 % maintain that some other, incorrect, combination of cards needs to be examined—less than 5 % of subjects give the correct answer that cards 1 and 4 need to be seen in order to test the rule experimenters presented to them (Stein 1996).

As if these results were not bad enough, matters seem to be even worse. In these sorts of experiments many subjects not only give the wrong answer concerning which cards one needs to see in order to test the rule, but they also continue to give wrong answers in repeated trials (Wason and Johnson-Laird 1972). Even more worrisome is the fact that many subjects will insist on their original (erroneous) answers for testing the rule even after it has been explained to them that the general principles their answers rely upon are mistaken and fail to provide genuine tests of the rule (Wason 1968; Wason and Johnson-Laird 1972). For example, some subjects who said that only the covered side of card 1 needs to be seen in order to evaluate the rule will continue to insist that only card 1 needs to be examined. They will continue to insist this even after it has been explained to them why only examining card 1 fails to implement the correct principle, i.e. it fails to provide a genuine test of the rule in question. These results seem to suggest, at least prima facie, that we exhibit fairly robust and widespread irrationality when it comes to the application of some basic rules of deductive reasoning.

13.1.2 Errors in Inductive Reasoning: The Conjunction Fallacy

Amos Tversky and Daniel Kahneman's (1983) experiment examining our ability to make judgments concerning how comparatively likely various events are is perhaps the most widely discussed experiment suggesting that we make serious errors when reasoning inductively. In this experiment subjects read the following description of a woman:

> Linda is 31 years old, single, outspoken, and very bright. She majored in philosophy. As a student, she was deeply concerned with issues of discrimination and social justice, and also participated in anti-nuclear demonstrations. (Tversky and Kahneman 1983, p. 297)

After reading the description of Linda subjects are asked to rank a series of statements from the most probable to the least probable based on the information they have been given about her:

1. Linda is a teacher in an elementary school
2. Linda works in a bookstore and takes Yoga classes
3. Linda is active in the feminist movement
4. Linda is a psychiatric social worker
5. Linda is a member of the League of Women Voters
6. Linda is a bank teller
7. Linda is an insurance salesperson
8. Linda is a bank teller and is active in the feminist movement (Tversky and Kahneman 1983, p. 297)

Interestingly, 85 % of subjects ranked (3) as more likely than (8) and (8) as more likely than (6). This is of particular interest because in ranking (8) as more likely than (6) subjects are violating a basic rule of probability theory. When two events, such as being a bank teller (B) and being active in the feminist movement (F), are independent the probability that both events occur is equal to the product of the probability of each event occurring. In other words, the probability of (B&F) = the probability of (B) **x** the probability of (F). Hence, there is no way that the probability (B&F) can be higher than the probability of (B) (or the probability of (F) for that matter). The story is more complicated when the two events are not independent. However, even when the events are not independent there are no cases where the probability of (B&F) will be greater than the probability of (B) (Stich 1990). So, when subjects rank (8) as more likely than (6) they are violating a basic principle of probability theory—they are committing the *conjunction fallacy* (the fallacy of thinking that a conjunction is more probable than one of its component conjuncts).[3]

[3]The reason that committing the conjunction fallacy is irrational is that it makes one susceptible to a "Dutch book". A Dutch book is a series of bets which are such that no matter what the outcome of the series of events betted upon the person accepting the series of bets will lose all of her money. When someone commits the conjunction fallacy she is accepting a distribution of probabilities which could be used to construct a Dutch book against her. For more on how

Again, these results seem to be fairly robust. In order to make sure that subjects did not interpret (6) as meaning "Linda is a bank teller and is not active in the feminist movement" the experiment was run with (6) replaced with "Linda is a bank teller whether or not she is active in the feminist movement". Similar results were found—subjects still tended to rank (8) as more probable than the revised version of (6) (Stich 1990). Further, the subjects of these experiments were not just "statistically naïve undergraduates", but also "presumably *not* statistically naïve psychology graduate students"—suggesting that committing the conjunction fallacy is a fairly widespread error (Stein 1996, p. 94). As with the results of the selection task, the results of these experiments seem to suggest, at least prima facie, that we exhibit a fairly robust and widespread irrationality when it comes to a particular kind of reasoning. In this case the empirical evidence appears to suggest that we make systematic mistakes with respect to inductive reasoning (at least when that reasoning deals with probabilities).

13.2 Responses to the Threat of Irrationality

On the basis of experimental findings like those discussed in the previous section many suppose that we exhibit systematic irrationality. Given the systematic nature of this irrationality and the fact that it seems to be widespread, one might worry that in many of the cases where we do reason correctly it is largely a matter of luck. In other words, one might worry that we cannot sufficiently trust our reasoning abilities in a wide range of cases in order to have knowledge. Thus, it may seem that we know much less than we ordinarily take ourselves to know—this would seriously affect our knowledge in ordinary situations, and possibly our scientific knowledge as well.

Fortunately, as we will see, there are a number of ways to respond to empirical evidence of our irrationality. We will not examine any of these responses in exhaustive detail; nevertheless, we will explore them in sufficient detail to get a good idea of their effectiveness. While it may be that none of these responses is ultimately decisive, they do give us reason to be optimistic about our rationality. More importantly for our purposes, we will see that there is good reason to think that even if we are irrational in many of the ways suggested by these experiments, this does not undermine our scientific knowledge. Let us turn to some of the responses on behalf of our being rational.

committing the conjunction fallacy makes one Dutch bookable see Stein (1996). For an accessible general discussion of Dutch books see Skyrms (1986).

13.2.1 Impossibility Responses

One general category of responses to the threat these empirical studies seem to pose for our rationality is impossibility responses. These responses argue that it is impossible for humans to be systematically irrational. If this is correct, then while the empirical research suggests that we make a variety of mistakes in certain situations it does not (and could not) establish that irrationality is widespread. It will be worth briefly looking at a few attempts at impossibility responses.

The first impossibility response comes from consideration of the principle of charity. The principle of charity was originally developed as a principle of how language translations should be conducted (Quine 1960). According to this principle, when translating the assertions of someone speaking an unknown language we should translate what she says in such a way that we do not understand her to be speaking absurdities. As Edward Stein (1996, p. 112) puts the point, "if I translate someone speaking an unknown language as saying something absurd, I should scrutinize my translation, not the rationality of the person." Of course, this is not to say that it is never appropriate to translate someone as saying something absurd. There are clearly times when the speaker intends to say absurd things. There are also times when a speaker inadvertently says absurd things. Nonetheless, the idea is that in the vast majority of cases when a speaker is making sincere assertions we should interpret her in such a way that what she says is rational.

One way to argue that people must be rational is to argue that the principle of charity applies to how we should understand the principles of reasoning which people employ.[4] Essentially, the idea is that "in order to characterize a person's reasoning competence, you have to assume that the principles she uses are basically rational" (Stein 1996, p. 116). Given this, one might think that it is simply impossible for us to be systematically irrational despite what the empirical research initially seems to suggest.

At first glance this appears to be a promising response to the suggestion that studies like those we explored above show we are irrational. Before this can be settled though, we have to first recognize that there are two version of the principle of charity which might be applied to our reasoning. The *strong version* of this principle is that "people should *never* be interpreted as irrational" (Stein 1996, p. 116). The *weak version* of this principle is that "*unless* there is strong empirical evidence to the contrary, people should be interpreted as rational" (Stein 1996, p. 116). The strong version of the principle of charity is clearly too strong. Surely, there are times when people are irrational. So, the strength of this response rests on the weak version of the charity principle. While the weak version seems to be a plausible principle, it does not seem to show that it is *impossible* for us to be systematically irrational. After all, this principle allows that when there is sufficient empirical evidence it is correct to interpret people as irrational. Thus, it seems that

[4]Davidson (1985) and Dennett (1978, 1987) both argue that the principle of charity applies to our reasoning.

the principle of charity does not provide good grounds for thinking it is impossible that we are systematically irrational.

The second impossibility response also comes by way of a comparison with language. One plausible way of understanding the grammatical rules of a particular language is that they are generalizations, or idealizations, of how competent speakers actually use the language. Hence, when it comes to the English language, for example, the grammatical rules which govern that language are generalizations from how people who speak English use the language. If this is correct, then while competent speakers of a language might at times break the grammatical rules, they cannot be totally off base in their application of the grammatical rules since the rules themselves come from how people actually speak.

Some argue that rules of good reasoning are analogous to grammatical rules for a language (Cohen 1983). The idea is that the rules for good reasoning, like the grammatical rules of a language, are generalizations of how people actually reason. If this is correct, then it simply cannot be that people tend to reason incorrectly—we cannot be systematically irrational.

One might worry about this sort of response for at least two reasons.[5] First of all, one might simply doubt the analogy between reasoning and language use (Feldman 2003). After all, it seems strange to say that people routinely misuse the language that they speak; it does not seem all that strange to think that people routinely make errors of reasoning. The first would make it impossible to understand speakers of a language, but the latter does not appear to yield any analogous results. Second, one might worry that even if we grant the analogy between language and reasoning, it will not be enough to show that people are not systematically irrational. It could be that the rules of good reasoning—what it is rational to infer on the basis of a given body of evidence—are generalizations from when we think very carefully about our evidence and what it supports, i.e. when we are reasoning at our very best. Of course, "our common and unconsidered judgments could be frequently irrational, even if rationality is determined by what we do when we are more careful" (Feldman 2003, p. 162). So, it may be that the claim that we are systematically irrational is consistent with the rules of good reasoning being like those of correct grammar. Thus, it is not clear if this response can provide an adequate defense of human rationality.

Perhaps a different approach to arguing that systematic irrationality is impossible will succeed. One might think that whereas the first two responses seek to show that it is impossible for us to be systematically irrational on the basis of conceptual grounds a third impossibility response which seeks to establish this point on empirical grounds would be more effective. Specifically, this third response involves

[5]Cohen's (1983) defense of this response is quite sophisticated, and it is worthy of careful consideration. However, for our purposes it is enough to grasp the general nature of this sort of response and some reasons that one might be skeptical of its effectiveness. For detailed critical discussion of Cohen's arguments concerning human irrationality see Stein (1996) and Stich (1990).

arguing on the basis of evolution that humans must be rational.[6] This response has two basic steps (Stein 1996). The first step relies on the claim that natural selection will select for mechanisms which yield true beliefs. The second step involves arguing that there is a close connection between our being rational and our having mechanisms which yield true beliefs. The idea is that natural selection "guarantees that all normal cognitive systems will be rational" (Stich 1990, p. 16). Simply put, this response is based on the idea that since we have evolved—as a species we have withstood the battle of survival of the fittest—we must be rational.

While this evolutionary response does have some appeal, it faces some serious objections.[7] One problem with this response is that many of the beliefs and reasoning tasks that the empirical studies we have discussed examine concern fairly abstract matters. It is far from clear that beliefs about abstract and theoretical matters have much connection with survival (Feldman 2003). Another problem is that it seems clear that it is not necessary for beliefs to be true or even rational in order to be conducive to survival. As Stephen Stich (1990, p. 62) explains "a very cautious, risk-aversive inferential strategy—one that leaps to the conclusion that danger is present on very slight evidence—will typically lead to false beliefs more often, and true ones less often, than a less hair-trigger one that waits for more evidence before rendering judgment." Yet, as Stich points out, it is possible that the first sort of cognitive system is selected for and the second is not. The reason for this is that "natural selection does not care about truth; it cares only about reproductive success. And from the point of view of reproductive success, it is often better to be safe (and wrong) than sorry" (Stich 1990, p. 62). Obviously, these considerations do not show that natural selection has failed to lead to our having cognitive processes which guarantee our rationality. They do, however, render the claim that natural selection *must* have so equipped us doubtful.

13.2.2 Questioning the Evidence

Given the challenges facing impossibility responses, one might be tempted to take a different approach to defending our rationality from the sorts of studies discussed above. A plausible way of doing this is to argue that although it is possible we are irrational, there is good reason to doubt that the empirical research has shown that we are. In other words, one might question the evidence of our irrationality purportedly provided by these studies.

Researchers have examined both the selection task and the conjunction fallacy studies by running variations of the studies. The results of these additional studies might be taken to cast doubt on how much evidence the original studies really give

[6]Several philosophers have expressed sympathy for this sort of argument including, among others, Dennett (1987), Goldman (1986), Lycan (1988), Millikan (1984), and Sober (1981).

[7]For detailed discussion and criticism of this argument see Stein (1996) and Stich (1990).

us for thinking that we are systematically irrational. Studies have found that when it comes to selection tasks subjects do considerably better when the selection task is less abstract than the task of the original experiment.[8] In fact when given a selection task involving the evaluation of a rule like "Every time I go to New York, I travel by train" subjects correctly determine which cards need to be uncovered nearly 66 % of the time (Stein 1996, p. 83). This is considerably better than subjects' performance in the original, rather abstract, selection task. In light of results like this, one might think that in the original selection task subjects are simply misinterpreting the task, and so applying a different rule of reasoning than they normally would (Cohen 1983).

Similar results were found for experiments concerning the conjunction fallacy. Studies have shown that when "the problem is phrased in such a way that subjects are being asked to indicate *frequency* rather than *probability*, their responses are in accord with the conjunction principle" (Stein 1996, p. 100).[9] For instance, when subjects are given the description of Linda from above and told that a certain number of people fit it, they answer the question as to how many of those people are (1) bank tellers and how many are (2) bank tellers and feminists in ways that are consistent with the conjunction principle (Gigerenzer 1991). That is to say, when the experiment is formed this way subjects do not tend to commit the conjunction fallacy. Again, this may lead one to think that subjects are merely misunderstanding what is asked of them in the original experiment (Feldman 2003).

The results of these further studies seem to weaken the empirical evidence for our irrationality provided by the original experiments. Nevertheless, it is not clear that this is enough to defend us from the charge of systematic irrationality. One reason this may not be enough is that at least in some cases, such as subjects' misinterpretation in the original selection task experiments and their correct interpretation in the concrete cases, it seems subjects are relying on content-specific rules. That is, they are using rules of reasoning which depend upon the specific subject matter to which the rule is being applied. Their reliance on content-specific rules explains why they use the correct rule when reasoning about the concrete cases, but they get things wrong in the original case with its more abstract content. The reliance on content-specific rules is worrisome because the correct rules of reasoning, such as the Conditional-Testing Principle, are content-neutral (Stein 1996). The correct rules of reasoning apply regardless of the particular subject being discussed. In light of this, one might worry that even if we tend to use content-specific rules which yield the appropriate results in some cases, we are still using the incorrect reasoning rules. Hence, one might worry that we are still exhibiting systematic irrationality—even when we are getting the correct answers to selection tasks!

Another reason that this may not be enough to defend our rationality is that there are a large number of additional studies which purport to show that we are

[8] See Wason and Shapiro (1971) and Johnson-Laird et al. (1972).

[9] See Fielder (1988).

systematically irrational. For example, studies seem to show that belief polarization is quite common. Belief polarization is the phenomenon which occurs when someone begins with a particular view on a topic and then receives mixed evidence on the topic (evidence that is balanced between the two sides of the debate). The person's belief tends to become more polarized; she tends to believe the position that she started with even more strongly than she initially did after receiving the mixed evidence (Lord et al. 1979).[10] It is not clear how subjects could be misinterpreting the task here. So, one might worry that this provides evidence for systematic irrationality. Belief polarization is particularly interesting for our purposes because it may make one worry that while it is widely acknowledged that scientific knowledge is theory-laden to some extent, perhaps these results show that it is problematically so. In addition to this sort of phenomenon, there are numerous studies suggesting that we exhibit a wide range of cognitive biases in our reasoning.[11] One might worry that it will be difficult to show that in all of these cases in which widespread irrationality is suggested by the empirical evidence there is a misinterpretation going on which allows for our being rational despite making mistakes. Consequently, it at least seems to remain an open, empirical question whether we exhibit widespread irrationality (Stein 1996).

13.2.3 A Modest Response

It appears that the impossibility responses and questioning the empirical evidence do not provide clear defenses of our rationality. Given this, one might conclude that the empirical evidence shows we are systematically irrational in various ways—we fall prey to cognitive biases, make mistakes when it comes to deductive and inductive reasoning, and so on. In light of this systematic irrationality one might think that we simply do not, and cannot, have much of the knowledge that we take ourselves to have. Further, one might conclude that we do not have much, if any scientific knowledge. This is a very dismal assessment indeed.

Although one might conclude that we do not have much scientific or ordinary knowledge on the basis of these empirical results, doing so would be a mistake. What is more, this would be a mistake *even if* we grant that the empirical evidence does demonstrate we are systematically irrational (something that we have already noted is not clearly the case). The inferences from our being systematically irrational in various ways to our lacking ordinary knowledge and to our lacking scientific knowledge are both dubious. Here is where a modest response to the empirical evidence can provide us with succor from the skeptical conclusions one might attempt to draw from these studies.

[10]See Gilovich (1991) for discussion.

[11]See Fine (2006), Gilovich (1991), Kahneman (2011), Kahneman et al. (1982), and Nisbett and Ross (1980).

Rather than leading us to skepticism, recognizing that we are systematically irrational (if we really are) and prone to make certain common errors of reasoning should leave us optimistic about our prospects for both ordinary and scientific knowledge. The mere fact that we are in a position to evaluate our reasoning—we can determine that our reasoning in these experimental situations is poor—suggests that we tend to reason correctly a lot of the time. Plausibly, in order to recognize the mistakes that we are making in these sorts of cases *as mistakes* requires reasoning in the correct way. Further, these deviations from correct reasoning would not seem so jarring if we did not accept, and presumably take ourselves to reason in accordance with, the correct ways of reasoning. Consequently, it is reasonable to think that the fact that we recognize the results of these experiments as evidence for thinking we make errors in reasoning suggests that we often do reason correctly. In other words, "the focus on error does not denigrate human intelligence, any more than the attention to diseases in medical texts denies good health. Most of us are healthy most of the time, and most of our judgments and actions are appropriate most of the time" (Kahneman 2011, p. 4).

Another reason that we should be optimistic is that because of the fact that we recognize our susceptibility to various mistakes we can correct for those mistakes. Once we are aware of the effects of belief polarization, for example, we can take steps to temper our assessments of our viewpoints on controversial issues (Kelly 2008). The same is true of other cognitive biases which we are prone toward (Kahneman 2011). In fact, we often take steps to help catch biases and mistakes already. When we recognize that we might be prone to making a particular mistake we will double-check our calculations, ask others for their opinions, and so on. The fact that we are aware that we tend to fall prey to certain mistakes helps us to seek methods for avoiding those mistakes. It is much easier to avoid a pit that you see in the path ahead than it is to avoid one of which you are totally unaware. Recognizing the sorts of errors of reasoning we are likely to commit gives us hope that we can correct for our mistaken natural tendencies.

This sort of correcting or vigilance for avoiding cognitive errors is something that we seem to be particularly good at when we work with others. For instance, when David Moshman and Molly Geil (1998) conducted a version of the selection task experiment in which some participants were given the task individually and others worked in groups they found that groups did dramatically better. Similarly to the original selection task experiment, individuals working on the task only produced the correct answer about 9 % of the time. However, groups had a 70 % success rate with respect to the selection task. Groups that consisted of participants who had already tried the task as individuals did even better—they had an 80 % success rate. In general groups do much better at tasks for which individuals fail to give correct answers.[12] Discussion and interactions with others can help us to avoid errors and greatly increase our performance. Unsurprisingly, peer discussion methods have been found to be very effective for learning (Slavin 1995). Some go so far as to

[12]See Bonner et al. (2002), Laughlin and Ellis (1986), and Moshman and Geil (1998).

claim that various cognitive biases we commit as individuals are not only to be expected, but perhaps are conducive to greater performance on the whole, given the social nature of our reasoning practices (Mercier and Sperber 2011; Sperber and Mercier 2014). Given these findings, it is not surprising that in science we employ particular methods when conducting research—we rely on things like peer review, participate in discussions, and so on.[13] By doing so, we can compensate for, and maybe even take advantage of, various cognitive mistakes that we are prone to make as individuals and attain scientific knowledge in spite of our shortcomings.

13.3 Conclusion

We have seen that there are numerous experiments which seem to suggest that humans are systematically irrational. Although some might take this empirical evidence to show that we lack much of the ordinary and scientific knowledge which we take ourselves to have, this conclusion is premature. While there are difficulties for the responses which seek to establish that widespread irrationality of the sort that might be thought to undermine our knowledge is impossible, it is not obvious that such a response cannot succeed. Additionally, we have seen that there are some grounds for doubting the empirical evidence that is purportedly yielded by experiments on our reasoning. However, we have also seen that there is reason for thinking that it will be difficult to provide sufficient evidence for thinking that we should doubt all of the myriad evidence of our irrationality in the numerous experimental situations that have been tested.

Despite these somewhat dismal prospects for showing that we do not make systematic mistakes of reasoning, we have seen that there is cause for optimism about our prospects for knowledge. By recognizing our tendency to make certain systematic mistakes we can take steps to correct for these effects both as individuals and as groups. Thus, we have seen that while there is empirical evidence that we make various mistakes of reasoning, there is no reason to think that this means we cannot have scientific knowledge or that we lack the knowledge we ordinarily take ourselves to have. At most what we seem to have is further reason for thinking that scientific knowledge should be held tentatively. But, this is something to which we were already committed.

[13]We will discuss these issues further in the final part of the book when we examine some of the social aspects of scientific knowledge.

References

Bonner, B. L., Baumann, M. R., & Dalal, R. S. (2002). The effects of member expertise on group decision making and performance. *Organizational Behavior and Human Decision Processes, 88*, 719–736.

Cohen, L. J. (1983). Can human irrationality be experimentally demonstrated? *Behavioral and Brain Sciences, 6*, 317–370.

Davidson, D. (1985). Incoherence and irrationality. *Dialectica, 39*, 345–354.

Dennett, D. C. (1978). *Brainstorms*. Cambridge, MA: MIT Press.

Dennett, D. C. (1987). *The intentional stance*. Cambridge, MA: MIT Press.

Feldman, R. (2003). *Epistemology*. Upper Saddle River: Prentice Hall.

Fielder, K. (1988). The dependence of the conjunction fallacy on subtle linguistic factors. *Psychological Research, 50*, 123–129.

Fine, C. (2006). *A mind of its own: How your brain distorts and deceives*. Cambridge, UK: Icon books.

Gigerenzer, G. (1991). How to make cognitive illusions disappear: Beyond "heuristics and biases". *European Review of Social Psychology, 2*, 83–115.

Gilovich, T. (1991). *How we know what isn't so: The fallibility of human reason in everyday life*. New York: Free Press.

Goldman, A. I. (1986). *Epistemology and cognition*. Cambridge, MA: Harvard University Press.

Johnson-Laird, P., Legrenzi, P., & Legrenzi, M. (1972). Reasoning and a sense of reality. *British Journal of Pyschology, 63*, 395–400.

Kahneman, D. (2011). *Thinking, fast and slow*. New York: Farrar, Straus, and Giroux.

Kahneman, D., Slovic, P., & Tversky, A. (Eds.). (1982). *Judgment under uncertainty: Heuristics and biases*. Cambridge, UK: Cambridge University Press.

Kelly, T. (2008). Disagreement, dogmatism, and belief polarization. *Journal of Philosophy, 105*, 611–633.

Laughlin, P. R., & Ellis, A. L. (1986). Demonstrability and social combination processes on mathematical intellective tasks. *Journal of Experimental Social Psychology, 22*, 177–189.

Lord, C., Ross, L., & Lepper, M. (1979). Biased assimilation and attitude polarization: The effects of prior theories on subsequently considered evidence. *Journal of Personality and Social Psychology, 37*, 2098–2109.

Lycan, W. G. (1988). *Judgement and justification*. Cambridge, UK: Cambridge University Press.

Mercier, H., & Sperber, D. (2011). Why do humans reason? Arguments for an argumentative theory. *Behavioral and Brain Sciences, 34*, 57–111.

Millikan, R. G. (1984). *Language, thought, and other biological categories*. Cambridge, MA: MIT Press.

Moshman, D., & Geil, M. (1998). Collaborative reasoning: Evidence for collective rationality. *Thinking and Reasoning, 4*, 231–248.

Nisbett, R., & Borgida, E. (1975). Attribution and the psychology of prediction. *Journal of Personality and Social Psychology, 32*, 932–943.

Nisbett, R., & Ross, L. (1980). *Human inference: Strategies and shortcomings of social judgment*. Englewood Cliffs: Prentice Hall.

Plous, S. (1993). *The psychology of judgment and decision making*. New York: McGraw-Hill.

Putnam, H. (1981). *Reason, truth and history*. Cambridge, UK: Cambridge University Press.

Quine, W. V. O. (1960). *Word and object*. Cambridge, MA: MIT Press.

Skyrms, B. (1986). *Choice and chance* (3rd ed.). Belmont: Wadsworth.

Slavin, R. E. (1995). *Cooperative learning: Theory, research, and practice*. London: Allyn and Bacon.

Sober, E. (1981). The evolution of rationality. *Synthese, 46*, 95–120.

Sperber, D., & Mercier, H. (2014). Reasoning as social competence. In H. Landemore & J. Elster (Eds.), *Collective wisdom: Principles and mechanisms* (pp. 368–392). Cambridge, UK: Cambridge University Press.

Stein, E. (1996). *Without good reason: The rationality debate in philosophy and cognitive science*. Oxford: Oxford University Press.

Stich, S. P. (1990). *The fragmentation of reason: Preface to a pragmatic theory of cognitive evaluation*. Cambridge, MA: MIT Press.

Tversky, A., & Kahneman, D. (1983). Extensional versus intuitive reasoning: The conjunction fallacy in probability judgment. *Psychological Review, 90*, 293–315.

Wason, P. (1966). Reasoning. In B. Foss (Ed.), *New horizons in psychology* (pp. 135–151). Harmondsworth: Penguin.

Wason, P. (1968). Reasoning about a rule. *Quarterly Journal of Experimental Psychology, 20*, 273–281.

Wason, P., & Johnson-Laird, P. (1972). *Psychology of reasoning: Structure and content*. Cambridge, MA: Harvard University Press.

Wason, P., & Shapiro, D. (1971). Natural and contrived experience in a reasoning problem. *Quarterly Journal of Experimental Psychology, 23*, 63–71.

Chapter 14
Anti-realism About Science

Abstract This chapter discusses one of the major debates in philosophy of science related to scientific knowledge, the debate between realists and anti-realists. Realists maintain that our best-confirmed scientific theories are true (or at least approximately true), but anti-realists think we should only accept that our best-confirmed scientific theories are useful in some sense without committing to their truth, approximate or otherwise. Some of the major arguments on both sides of this debate are evaluated in this chapter, though special attention is paid to the so-called "miracle argument" for scientific realism. Throughout the chapter a realist stance, which allows for genuine scientific knowledge, is defended. Ultimately, the chapter concludes that while anti-realist arguments are important and worth taking seriously, they do not pose an insurmountable threat to a realist conception of science. Such a conception holds that in the case of our best-confirmed theories the truth of those theories best explains their success, which gives us justification for believing that they are true.

In the three preceding chapters in this part of the book we examined various challenges to our scientific knowledge. In each of those cases the threat to our scientific knowledge arose as a consequence of a more widespread attack on our knowledge of the world. In this chapter we will consider a final challenge that specifically targets our scientific knowledge.[1] The challenge we will consider in this chapter comes by way of anti-realist threats to scientific realism, which is, roughly, the view that our best scientific theories yield genuine knowledge of the world around us (Boyd 1983; Chakravartty 2007, 2014; Gauch 2012; Psillos 1999; Smart 1963).

The debate concerning scientific realism is very important. As Anjan Chakravartty (2014, p. 1) notes, "debates about scientific realism are centrally connected to almost everything else in the philosophy of science, for they concern the very nature of scientific knowledge." So, increasing our understanding of scientific realism, and some of the challenge brought against it, will deepen our understanding of NOS.

[1]Giere (2006) also construes anti-realism as a species of skepticism.

K. McCain, *The Nature of Scientific Knowledge*, Springer Undergraduate Texts in Philosophy, DOI 10.1007/978-3-319-33405-9_14

14.1 Scientific Realism

In order to better understand the difficulties posed by the main anti-realist challenges to our scientific knowledge it will be useful to first get clear on the nature of scientific realism. There is some difficulty in this task because "it is only a slight exaggeration to say that scientific realism is characterized differently by every author who discusses it" (Chakravartty 2014, p. 2). Despite the difficulty in precisely specifying the best way to understand scientific realism, we should not despair. It is not too daunting of a task to come to grips with a general conception of scientific realism—a conception that will be sufficient for our purposes here.[2]

Perhaps the best way to proceed is to first briefly consider what *realism* is in general. "Realism" as a general term applies to "any position that endorses belief in the reality of something" (Chakravartty 2014, p. 4). So, the view that there is an external world around us—a world of trees, animals, rocks, and so on—is a kind of realism. The view that abstract objects like classes, concepts, and sets are real is another kind of realism. Like these other kinds of realism, scientific realism is the view that some particular things are real. Namely, scientific realism is the view that the sorts of things described by our best scientific theories, whether observable or unobservable, are real.[3] Of course, scientific realism goes beyond simply claiming these entities are real; it also claims that our best scientific theories provide us with knowledge of these entities and the laws governing their behavior.

With this understanding of the basic commitments of realist views, we can now better grasp the particular commitments of scientific realism. Scientific realism is actually committed to three positive theses about science (Chakravartty 2014; Psillos 1999). The first thesis scientific realism is committed to is the metaphysical claim that there is a mind-independent external world. In other words, there is a world external to the operations of our own minds, and there are objective facts

[2]The literature on scientific realism and anti-realism is vast. Our task would be made nigh impossible if we were to not only discuss realism and anti-realism in general, but also to discuss the myriad idiosyncratic formulations of these views. Additionally, there are many thought-provoking theories which fall somewhere in between realism and anti-realism such as "semirealism" (Chakravartty 1998, 2007), "scientific perspectivism" (Giere 1988, 2006), and "structural realism" (Ladyman 1998, 2011; Worrall 1989) among others. Providing a useful discussion of these theories would make this chapter unwieldy at best, and it is unnecessary for our current purpose.

[3]Roughly, things are thought to count as observable when it is possible for humans to observe them with our unaided senses. Anything that is not observable in this sense is considered unobservable. Interestingly, there is controversy concerning whether this distinction between observable and unobservable can actually be drawn in an unproblematic way. Churchland (1985) and Musgrave (1985) argue that this distinction is problematic. See Dicken and Lipton (2006) for insightful discussion of Musgrave's argument. This is worth pointing out because if there are problems concerning how to draw this distinction, these are problems for anti-realist views (such as that of van Fraassen 1980) which claim we can know about the observable, but not the unobservable—problems that are not shared by scientific realists (Chakravartty 2014).

about the features of this world.[4] The second is a semantic thesis. According to this commitment of scientific realism, scientific theories and their theoretical statements should be understood in a literal sense. That is to say, when a scientific theory makes a claim about a particular entity, such as a quark, scientific realism is committed to taking that claim to be making a literal assertion—the scientific realist holds that the claim is something which is either true or false. The third thesis scientific realism is committed to is that at least some theoretical claims of scientific theories constitute knowledge of the mind-independent external world. Putting this together gives us a good general description of scientific realism. It is the view that "our best scientific theories give true or approximately true descriptions of observable and unobservable aspects of a mind-independent world" and when we believe those descriptions are true on the basis of sufficiently strong evidence; we have scientific knowledge of the mind-independent external world (Chakravartty 2014, p. 6).

Before considering the primary reasons for accepting scientific realism and some of the various anti-realist challenges to scientific realism, it is worth briefly pausing to clarify two points about our general description of scientific realism. First, the scientific realist is not committed to claiming that all components of our best scientific theories are true. Instead, they are committed to the view that *some* claims of our best scientific theories are true, and that our best scientific theories are themselves at least approximately true (recall our discussion of approximate truth in Chap. 4). Second, it is important to be clear about what is meant by our "best scientific theories". Scientific realists tend to have in mind theories that are *mature* and *non*-ad hoc when they speak of our "best" theories. A mature scientific theory is one which has been thoroughly tested and has become established in its particular science. There are various ways to understand what it means to be non-ad hoc.[5] In simplest terms, a theory's being non-ad hoc means that the theory does more than simply accommodate the known data; the theory makes novel predictions.[6] A theory provides a novel prediction of some data if and only if the theory entails the data, the data is true, and the theory was not designed to entail that data. By contrast, a theory merely accommodates some data if and only if the theory entails that data, the data is true, but the theory was designed to entail that data (White 2003).[7] Now that we have clarified the nature of scientific realism it is time to see what can be said in support of it.

---

[4]Another way to understand this is that scientific realism is committed to the truth (at least for the most part) of the Real World Hypothesis defended in Chap. 11.

[5]See Psillos (1999) for a nice discussion of what it means for a theory to be ad hoc as well as for a discussion of what is required for a theory to be mature.

[6] Barnes (2008), Hitchcock and Sober (2004), Leplin (1997), Psillos (1999), and White (2003) all maintain that making novel predictions is important for being non-ad hoc.

[7]The use of entailment here, and in White's own discussion, is to simplify matters. Of course, it is plausible that a theory, the truth of which makes some data highly probable without entailing that it, predicts or accommodates that data too.

14.2 Support for Scientific Realism

Scientific realism is in many ways a commonsense position to hold. For instance, if we were to ask people in ordinary situations whether our best scientific theories provide us with knowledge of the world, most would answer "yes". Further, if we were to ask people whether they believe in various unobservable entities posited by science, such as atoms, DNA, and so on, they would affirm that they do. Perhaps even more importantly, scientific realism seems to be assumed by many scientists, and maybe even by scientific practice itself. This assumption is so widespread among scientists that Hugh G. Gauch, Jr. (2012, p. 29) claims "ordinary science is so thoroughly tied to realism that realism's competitors seem to scientists to be somewhat like [a] philosophical joke."

Although being plausible and widely accepted are positive marks in favor of scientific realism, this alone is not overwhelming support for the view. It is for this reason that philosophers have produced arguments in support of scientific realism. The most prominent of those arguments is the "Miracle Argument".[8] Perhaps the most famous expression of the key idea of this argument is that given by Hilary Putnam (1975, p. 73), "the positive argument for realism is that it is the only philosophy that does not make the success of science a miracle." Hilary Kornblith (2013, p. 260) nicely articulates the thinking which underlies the Miracle Argument in this way:

> The achievements of science over the past few hundred years show progress of three very striking sorts: we are able to make predictions which are both more accurate and more wide-ranging over time; we have better and better explanations for a wider and wider range of phenomena; and we have ever more elaborate technological innovations. There is a single underlying explanation for these progressive features of science: our theories tend to be at least approximately true, and the enterprise of science, the greatest intellectual achievement of the human species, thus provides us with an ever-expanding body of knowledge of the world around us.

Essentially, the Miracle Argument for scientific realism is an inference to the best explanation. Recall from Chap. 10 that inferences to the best explanation have this general structure:

$F_1, F_2, \ldots, F_n$ are facts in need of explanation.
Hypothesis H explains the F_i.
No available competing hypotheses would explain the F_i as well as H does.

Therefore, H is true.

The facts in need of an explanation in this case are the empirical successes of our best scientific theories. That is to say, among other things, our best scientific theories

---

[8]This argument is also known as the "No Miracles Argument" and the "Ultimate Argument". For more detailed examination of the nature of this argument than we will be able to engage in here see Musgrave (1988) and Psillos (1999).

unify a wide range of data while making successful novel predictions (predictions that are correct when tested). The truth (or at least approximate truth) of our best scientific theories explains their successes, and it explains these facts better than any other hypothesis does.[9] In fact, some scientific realists accept Putnam's view that the truth of our best scientific theories is the *only* explanation of their success that is worth taking seriously. Although many scientific realists are sympathetic to this position, it is worth noting that they do not need to claim anything this strong in order for the Miracle Argument to work. All that is needed for this inference to the best explanation is that the truth of our best scientific theories is the *best* explanation of their successes. From this we can reasonably infer that our best scientific theories are true, i.e. we can reasonably conclude that scientific realism is correct.

14.3 Anti-realist Challenges

Although there is a variety of versions of anti-realism, we will simply focus on the most powerful anti-realist attacks on scientific realism and what can be said in response to them. This is a much more manageable task than attempting to consider the nuances of the various positive accounts of anti-realism that are put forward as competitors to scientific realism. Each of these anti-realist attacks on scientific realism targets some aspect of the Miracle Argument. So, responding to the challenge of anti-realism for our scientific knowledge consists of defending the Miracle Argument from the strongest anti-realist attacks.

14.3.1 Attacking Inference to the Best Explanation

The first way that some anti-realists attack the Miracle Argument is by attacking its very structure. That is, some anti-realists seek to undermine the support for scientific realism provided by the Miracle Argument by arguing that inference to the best explanation (IBE) is not a good way to reason in general. If correct, this would undercut the support the Miracle Argument provides for scientific realism because the fact that the truth of our best scientific theories best explains their successes would fail to provide good reason to accept that they are in fact true. Here we will consider two of the strongest objections to IBE put forward by anti-realists.[10]

The first objection to IBE is Bas van Fraassen's (1989) well-known "Best of a Bad Lot" objection. According to van Fraassen (1989, p. 143) when we choose the

[9]For simplicity I will often drop the qualification "approximate truth" and speak just of the truth of our best scientific theories. However, the discussion should be understood in terms of our best scientific theories being true or approximately true.

[10]See McCain (2012) for further discussion of both of these objections.

best available explanation from a set of competing explanations "our selection may well be the best of a bad lot." That is, van Fraassen argues that recognizing that H is the best available explanation of F is not enough to justify belief in H. As he says:

> To believe is *at least* to consider more likely to be true, than not. So to believe the best explanation requires more than an evaluation of the given hypothesis. It requires a step beyond the comparative judgment that this hypothesis is better than its actual rivals... For me to take it that the best of set *X* will be more likely to be true than not, requires a prior belief that the truth is already more likely to be found in *X*, than not. (1989, p. 143)

There are two ways of understanding van Fraassen's Best of a Bad Lot objection. One way is to understand it as a version of the "Truth Demand", which we discussed in Chap. 12. Though it is plausible that understanding van Fraassen's objection as a version of the Truth Demand is most in line with what he intends and is the strongest form of this objection because it challenges the idea that being a good explanation is correlated with being likely to be true, we can set this way of understanding van Fraassen's objection aside here. The reason we can set aside this way of understanding the Best of a Bad Lot objection is because we have already examined the Truth Demand and considered plausible responses to it in Chap. 12. The second way of understanding van Fraassen's objection is as challenging the thought that one can legitimately infer that the best explanation is true on the grounds that the set of available explanations might all be bad explanations. Although this way of construing the objection has a straightforward response, it is worth considering because one might think of the Best of a Bad Lot in this way and because it does help clarify the need for an important qualification for IBE in general—a qualification that we noted in Chap. 10.

The second way of construing the Best of a Bad Lot objection admits of a straightforward response. In general, IBE can be defended from this objection by placing a minimum threshold on how good an explanation has to be in order to be inferred. That is to say, van Fraassen's objection illuminates the fact that IBE should be understood as inference to the best available explanation that is "good enough". As we noted in Chap. 10, an important caveat for our understanding of IBE is that it is not enough to be the best available explanation, the inferred explanation must be the best available explanation, *and* it must be a sufficiently good explanation of the relevant data. The Best of a Bad Lot objection helps to make the importance of this caveat clear, but it does not provide any reason to doubt that IBE is a good way to reason once we have properly understood the ways in which IBE should be qualified.

The second well-known objection to IBE is what is often called the "Argument from Indifference". Like the previous objection, van Fraassen (1989) is the originator of this objection. Here is van Fraassen's expression of the objection:

> I believe, and so do you, that there are many theories, perhaps never yet formulated but in accordance with all evidence so far, which explain at least as well as the best we have now. Since these theories can disagree in so many ways about statements that go beyond our evidence to date, it is clear that most of them by far must be false. I know nothing about our best explanation, relevant to its truth-value, except that it belongs to this class. So I must

> treat it as a random member of this class, most of which is false. Hence it must seem very improbable to me that it is true. (1989, p. 146)

Thus, van Fraassen claims the fact that H is the best available explanation of F does not justify us in thinking that H is true because H is simply one of a very large number of mostly false theories (most of which we have not thought of yet) which explain F equally well.

One plausible way to respond to van Fraassen on this point is by maintaining that the way in which we come to form theories, at least in the empirical sciences, precludes our best explanations from being merely members of a large set of mostly false theories (Lipton 2004; Psillos 1996, 1999). We have a large amount of background knowledge that goes into our theory formation practices, and this background knowledge increases the likelihood that our best explanations are correct. Hence, it is far from clear that we should think that our best scientific theories are simply members of large sets of mostly false theories.

There is more that can be said to defend IBE from this objection, however. Consider van Fraassen's claim that "I believe, and so do you, that there are many theories, perhaps never yet formulated but in accordance with all evidence so far, which explain at least as well as the best we have now." Setting aside the fact that his rhetorical flourish "and so do you" makes this sentence straightforwardly false, at least if by "you" he means most supporters of scientific realism, there is an obvious problem with van Fraassen's claim. The problem is that van Fraassen uses this claim about what is *believed* about theories to establish a claim about what is *true* of theories. van Fraassen uses the claim that it is believed that our best theories are members of sets of theories which all explain the relevant data equally well to support the claim that our best explanations are in fact members of such sets of theories. What van Fraassen needs instead of this claim about our beliefs concerning theories is the claim that there *are* many theories which are equally good or better explanations than the explanations provided by our current best explanations. Of course, this is a claim that a supporter of scientific realism will deny with respect to our best scientific theories. In order to provide reasons for accepting the claim that there are many theories which are just as good as our current best scientific theories van Fraassen needs to provide an argument for thinking this is true. The best bets for such an argument involve appealing to the Underdetermination of Theories by Evidence or the Pessimistic Induction; however, as we will see in the sections that follow neither of these arguments seem to be successful. So, both of van Fraassen's objections to IBE fail to pose a problem for IBE in general or for the Miracle argument in particular.[11]

[11]As we noted in Chap. 10, another significant criticism of IBE is that it is inconsistent with accepted theories of probabilistic reasoning such as Bayesianism. Although the literature on this line of criticism is quite interesting, we will not explore it here. For discussion of, and responses to, this criticism of IBE see Lipton (2004), McCain & Poston (2014), McGrew (2003), Okasha (2000), Psillos (1999), and Weisberg (2009).

14.3.2 Underdetermination of Theories by Evidence

The next challenge to scientific realism is the Underdetermination of Theories by Evidence (UD). UD seeks to show that for any scientific theory there is a rival theory which fits the evidence just as well as that theory. Similar to the underdetermination argument for external world skepticism we considered in Chap. 11, the idea here is that if the evidence does not better support a given theory than some rival theory we are not justified in believing either theory to be true. How does this pose a problem for the Miracle Argument? It does so by casting doubt on the claim that the best explanation for the success of our best theories is their truth. After all, if we should think that the evidence underdetermines between our best theories and rival theories, we should not think that our best theories are true. Plausibly, if we should not accept that our best theories are true, their truth is not going to be the best explanation of their empirical success.

Let us make the argument for UD more precise.

UD-1) If the evidence is underdetermined between T (one of our best scientific theories) and R (some rival to T), then the evidence does not support thinking that T is true.
UD-2) T and R are empirically equivalent.
UD-3) If T and R are empirically equivalent, then the evidence is underdetermined between T and R.
UD-4) Therefore, the evidence does not support thinking that T is true.
UD-5) If the evidence does not support thinking that T is true, then T's truth is not the best explanation of T's empirical success.
UD-6) Therefore, T's truth is not the best explanation of T's empirical success.[12]

Supporters of UD maintain that each of our best scientific theories has empirically equivalent rivals—rival theories which "have the same class of empirical, viz., observational consequences" (Laudan and Leplin 1991, p. 451). The thought is that since each of our best scientific theories has rivals which make all of the same empirical claims—the same claims about what we will observe in various circumstances—then the evidence cannot determine between our best scientific theories and their rivals. But, if the evidence cannot determine between them, then it seems we cannot reasonably believe of either our best scientific theories or of their rivals that they are true. If we cannot reasonably believe of our best scientific theories that they are true, then it seems we have to look elsewhere for an explanation of their empirical success. Thus, if the argument for UD is sound, then there is a serious problem for the Miracle Argument because the truth of our best scientific theories does not best explain their empirical success.

[12]There are, of course, other ways of raising underdetermination problems for scientific realism. Nonetheless, the argument presented here represents the most common way of making this sort of objection. For an interesting variation on the original underdetermination argument see Stanford (2006), and for responses to Stanford see Chakravartty (2008) and Godfrey-Smith (2008).

Fortunately, there is good reason to think that the argument for UD is unsound. Larry Laudan and Jarrett Leplin (1991) convincingly argue that underdetermination is not entailed by empirical equivalence. That is, they provide a convincing argument that UD-3 is false. Laudan and Leplin's argument is roughly the following:

1. A theory can be confirmed by observations that are not part of its empirical consequences.
2. If (1), then it is not the case that if two theories are empirically equivalent, then they are underdetermined.
3. Therefore, it is not the case that if two theories are empirically equivalent, then they are underdetermined.[13]

Let us start our examination of this argument by looking at (2). It is pretty clear that this premise is true. It seems clear that the truth of (1) entails that an observation's being an empirical consequence of a theory is not necessary for that observation to provide evidential support for the hypothesis. Since the empirical consequences of a hypothesis are not the only things that are relevant to the evidential support of a theory, we have no reason to infer that two hypotheses are underdetermined simply because they are empirically equivalent. After all, if (1) is true, then it is possible that the evidence can favor one empirically equivalent theory over another because the first, but not the second, is supported by observations which are not empirical consequences of either theory. Consequently, the real action with this argument concerns the case for the truth of (1).

In support of (1), Laudan and Leplin describe the sort of situation where a theory is confirmed by an observation which is not part of its set of empirical consequences. Further, the situation they describe is also one in which an empirically equivalent theory is not confirmed by the observation which confirms a rival theory. That is, they present an example in which two empirically equivalent theories are supported to different degrees by a given piece of evidence.[14] The sort of example Laudan and Leplin present involves two distinct, empirically equivalent theories, T and R.

[13]This is actually a simplified version of Laudan and Leplin's argument. They argue for the additional claim that observing empirical consequences of a hypothesis does not always provide evidence for the hypothesis. I do not include this part of their discussion here for two reasons. First, there are problems with the argument that Laudan and Leplin provide in support of this additional claim. Second, this additional claim is not necessary for showing that the argument in support of UD is unsound. For our purposes this is all we need to concern ourselves with, so we do not need to explore Laudan and Leplin's additional claim. For further discussion of Laudan and Leplin's arguments and the relation of empirical equivalence to underdetermination see Kukla (1993, 1996), Leplin (1997), and Leplin and Laudan (1993).

[14]It should be noted that if the example presented by Laudan and Leplin actually demonstrates that two empirically equivalent theories can gain different degrees of evidential support from an observation, the example alone is sufficient to demonstrate that empirical equivalence does not entail underdetermination. This is a fact that Laudan and Leplin recognize, however, they present the rest of their argument to drive home their point. It is worth following their lead and presenting the entire argument instead of just this example because understanding the full argument helps to better illuminate the nature of the relation between empirical equivalence and underdetermination.

T, but not R, is a consequence of a larger theory, **Theory**. Additionally, a further hypothesis, H, which is not a consequence of either T or R, is entailed by **Theory**. Now, in this sort of situation Laudan and Leplin argue that when E, an empirical consequence of H, obtains it provides evidential support for H. E also provides some evidential support for **Theory** because H is entailed by **Theory**. Since E provides evidential support for **Theory**, it provides some indirect support for theories which are entailed by **Theory** including T. As a result, it is possible for a theory to be confirmed by observations which are not empirical consequences of that theory. The theory can be supported by empirical consequences of some other larger theory which entails both the original theory and a hypothesis that has those empirical consequences. What is more, Laudan and Leplin's example also demonstrates that two empirically equivalent theories can gain different degrees of evidential support from an observation. It is quite reasonable to assume in this case that E does not provide evidential support for R via supporting some other theory which entails R. Thus, E provides evidence for T but not R. Therefore, there can be evidence which supports empirically equivalent theories to different degrees. Thus, the fact that two theories are empirically equivalent does not entail that they are underdetermined.

The fact that the sort of situation that Laudan and Leplin describe is possible implies that it is not the case that empirical equivalence entails underdetermination. This provides good reason to think that UD-3 is false. So, this example gives us good reason to think that the argument for UD is unsound. It also gives us good reason to deny UD, the anti-realist's claim that each of our best scientific theories has rivals which are equally well supported by our evidence.

Now, anti-realists might attempt to defend the claim that our best scientific theories are underdetermined by noting that the example Laudan and Leplin employ is dependent upon the fact that the empirically equivalent rivals under consideration are partial theories (Hoefer and Rosenberg 1994). In the example that Laudan and Leplin present in support of (1), what allows for one theory to gain evidential support which its empirically equivalent rival does not is the fact that only the former is entailed by a broader theory, which gains evidential support from a particular observation. Yet, in the case of total theories, theories that cover all observations, there are no broader theories in which they can be embedded. Thus, Carl Hoefer and Alexander Rosenberg (1994, p. 600) rightly claim that the example, which Laudan and Leplin offer, "cut[s] no ice in the case of empirically equivalent total theories." Hence, the anti-realist might try to save UD by arguing that we should understand the underdetermination claim to apply to total theories rather than partial ones.

Although the restricted principle—that empirically equivalent total theories are underdetermined by the evidence—is not shown to be false by Laudan and Leplin's argument, there is still good reason to doubt its truth. First, one might argue that individual observations can still provide more evidence in support of a particular total theory than they do for its empirically equivalent rivals. According to Jarrett Leplin (1997), the reason this is the case is that one total theory may *predict* the relevant observation, whereas the other total theory may only *accommodate* the observation.[15]

Second, as we saw in Chap. 11, there is good reason to think that empirically equivalent total theories are not necessarily underdetermined. As we saw when discussing external world skepticism, the Real World Hypothesis (RWH) and its skeptical rivals are total theories. They cannot be embedded in larger theories about the external world, and they account for all observations. Nevertheless, we also saw that the RWH and its skeptical rivals are not underdetermined despite being empirically equivalent. Explanatory considerations show that the evidence better supports the RWH than it does skeptical rivals to the RWH. The general lesson here is clear: even when a group of total theories are empirically equivalent it is possible that they are not underdetermined by the evidence because one of the theories might provide the best explanation of the observations they each entail. Given this and the point about prediction versus accommodation, it appears that the anti-realist's retreat to total theories does not save UD. Thus, the threat of the Underdetermination of Theories by Evidence does not seem to pose a genuine threat to the Miracle Argument or scientific realism.

14.3.3 *The Pessimistic Induction*

The final anti-realist challenge to scientific realism we will consider is the Pessimistic Induction (PI).[16] The worry at the heart of the PI has been around for quite some time (see Poincaré 1905/1952; Putnam 1978). However, it was not pressed as an argument against scientific realism until the 1980s (Laudan 1981). The general idea behind the PI is that when we look at the history of science we discover it consists of a vast number of empirically successful scientific theories that were later rejected. Since all of our previous empirically successful scientific theories have turned out to be false, by inductive reasoning it is likely that our current successful scientific theories are also false. So, the PI provides an argument which seeks to undercut the claim in the Miracle Argument that the best explanation of our current best scientific theories' empirical successes is that they are true. We will first take a closer look at the PI before considering how scientific realists can respond to the challenge it presents.

Stathis Psillos (1999, p. 101) nicely summarizes the heart of the PI in this way:

> The history of science is full of theories which at different times and for long periods had been empirically successful, and yet were shown to be false in the deep-structure claims they made about the world . . . Therefore, by a simple (meta-) induction on scientific theories, our

[15]For discussion of the prediction versus accommodation debate see Barnes (2008), Hitchcock and Sober (2004), Harker (2006, 2008), Horwich (1982), Leplin (1997), Lipton (2004), Psillos (1999), Schlesinger (1987), and White (2003).

[16]This is also sometimes referred to as the "Pessimistic Meta-Induction".

> current successful theories are likely to be false (or, at any rate, are more likely to be false than true), and many or most of the theoretical terms featuring them will turn out to be non-referential... Therefore, the empirical success of a theory provides no warrant for the claim that the theory is approximately true.

There are a number of ways to turn the driving concern of the PI into a more formal argument of either a deductive or an inductive variety (Mizrahi 2013). Perhaps the simplest way to understand the PI as an argument, and as a challenge to the Miracle Argument, is as a *reductio ad absurdum* argument. This way of formulating the PI starts with the assumption that the empirical success of a scientific theory is a reliable indication of the theory's truth. It is then shown that this assumption coupled with other (purportedly) true premises leads to a contradiction. Since the assumption that empirical success is a reliable guide to truth leads to a contradiction, it is concluded that this assumption is false. As we noted above, if it is false that a scientific theory's being empirically successful is a reliable way of determining whether it is true, then there is a serious problem for the Miracle Argument. A key claim of that argument, that the best explanation of the empirical success of our best scientific theories is their truth, would be false.

Here is a more formal presentation of the PI:

PI-1) Assume (for *reductio*) that the empirical success of a scientific theory is a reliable indication of its truth.
PI-2) If (PI-1), then most of our current empirically successful theories are true.
PI-3) Therefore, most of our current empirically successful theories are true.
PI-4) If (PI-3), then all past scientific theories are false because they are different from our current empirically successful theories in important ways.
PI-5) Therefore, all past scientific theories are false because they are different from our current empirically successful theories in important ways.
PI-6) Many past scientific theories were empirically successful.
PI-7) If (PI-1) and (PI-6), then many past scientific theories are true.
PI-8) Therefore, many past scientific theories are true.
PI-9) Therefore, the empirical success of a scientific theory is not a reliable indication of its truth. (This is because (PI-5) and (PI-8) are contradictory)[17]

With this more formal presentation of the PI in hand it will be easier to understand how scientific realists might best respond to the PI.

The most common strategy employed by scientific realists for responding to the PI is to attack (PI-6).[18] According to Larry Laudan (1984, p. 110), a past scientific

[17] See Lewis (2001) and Saatsi (2005) for similar formulations of the PI.

[18] Another strategy for responding to the PI which has been put forward in recent years is to argue that the PI is actually a fallacious argument. Lange (2002) argues that the PI commits what he calls the "turn over fallacy". Lewis (2001) argues that the PI commits the fallacy of false positives. Saatsi (2005) argues that both Lange and Lewis are mistaken—the PI does not commit either fallacy. Mizrahi (2013) argues that Saatsi is mistaken because the PI can be understood in three different ways, and on each understanding it is fallacious. We will not explore the merits of this strategy for responding to the PI here.

theory counts as empirically successful "so long as it has worked reasonably well, that is, so long as it has functioned in a variety of explanatory contexts, has led to several confirmed predictions, and has been of broad explanatory scope." He claims that the scientific theories he appeals to in the PI meet these requirements. As a result, he claims they are empirically successful.

Scientific realists can plausibly deny that the theories Laudan appeals to are genuinely successful though. As Stathis Psillos (1999, p. 105) aptly notes, "the notion of empirical success should be more rigorous than simply getting the facts right, or telling a story that fits the facts." This is why, as we noted above, scientific realists limit the focus of the Miracle Argument to our *best* scientific theories—those that are both mature and non-ad hoc. The problem for the PI here is that anti-realists pressing this argument require "the treatment of all theories' empirical success as a unitary phenomenon" (Doppelt 2007, p. 112). However, this seems to be a mistaken understanding of the empirical success of our current best scientific theories compared to the empirical success of past scientific theories. Only our current best scientific theories "meet the standards of confirmation that are the most reasonable criteria of theoretical knowledge in science" (Doppelt 2007, p. 112). According to Gerald Doppelt (2007, p. 112), the fact that our current best scientific theories meet these criteria "differentiates them, their success, from that of superseded theories." Consequently, realists can reasonably claim that (PI-6) is false because the past scientific theories which are used in the PI are not genuinely successful in the way that our current best scientific theories are. Simply put, our current best scientific theories meet the appropriate standards for counting as empirically successful (the rigorous testing and explanatory requirements that we demand of our scientific theories before they are accepted today), but the past scientific theories appealed to in the PI do not.[19]

A further point about the PI and the success of our current best scientific theories is worth mentioning. It is not all that surprising how we might come to have scientific theories that are true (or at least approximately true) now on the basis of a history of false scientific theories. Given the nature of scientific progress, it is plausible that our current best scientific theories are such that we cannot reasonably think they are likely to be false on the grounds that empirically successful scientific theories in the past turned out to be false. As Alexander Bird (2007, p. 80) explains, "The falsity of earlier theories is the very reason for developing new ones—with a view to avoiding that falsity . . . Later scientific theories are not invented independently of the successes and failures of their predecessors. New theories avoid the pitfalls of their falsified predecessors and seek to incorporate their successes."[20] So, not only can realists maintain that the PI does not provide reasons

[19]See Psillos (1999) for an excellent discussion of why many of the particular past scientific theories that Laudan refers to in pressing the PI are not actually successful.

[20]The idea that the elements responsible for past scientific theories being empirically successful are retained in our current best theories is another point that is often claimed in response to the PI. This approach allows that past scientific theories were successful, but it casts doubt on the claim that past scientific theories were totally false. What this approach seeks to do is show that the parts

to doubt that the best explanation of our current best scientific theories' empirical successes is their truth, but they can maintain that we should expect our current best scientific theories to be built upon our past scientific theories in such a way that past failures do not provide grounds for doubting our current and future successes.

14.4 Conclusion

In this chapter we have considered the Miracle Argument in support of scientific realism as well as various anti-realist challenges to that argument (and by extension to our scientific knowledge). In each case we have seen that scientific realists have the resources to defend the Miracle Argument, and our scientific knowledge, from these anti-realist threats. Thus, we have seen that anti-realist challenges, while important and worth taking seriously, are not a cause for genuine concern with respect to our scientific knowledge.

We have now come to the end of the third part of this book. Up to this point we have been primarily concerned with scientific knowledge as had by individuals. We have considered what it takes for someone to know a claim in general. We have also considered how individuals might come to have scientific knowledge and how challenges to that knowledge might be overcome. In each case, though, we have primarily focused on the epistemic situation of individual knowers. In the final part of the book (part IV), we will expand our focus from looking at the scientific knowledge of individuals to how that knowledge arises and is spread throughout a scientific community.

References

Barnes, E. C. (2008). *The paradox of predictivism*. Cambridge, UK: Cambridge University Press.

Bird, A. (2007). *Nature's metaphysics: Laws and properties*. Oxford: Oxford University Press.

Boyd, R. (1983). On the current status of the issue of scientific realism. *Erkenntnis, 19*, 45–90.

Chakravartty, A. (1998). Semirealism. *Studies in the History and Philosophy of Science, 29*, 391–408.

Chakravartty, A. (2007). *A metaphysics for scientific realism: Knowing the unobservable*. Cambridge, UK: Cambridge University Press.

Chakravartty, A. (2008). What you don't know can't hurt you: Realism and the unconceived. *Philosophical Studies, 137*, 149–158.

of past scientific theories responsible for their empirical successes are the true parts which have been retained in our current best scientific theories. If this is correct, then the PI fails to provide a reason to doubt that empirical success is a reliable guide to the truth of a scientific theory. For this sort of approach see Kitcher (1993), Psillos (1999), and Worrall (1989).

Chakravartty, A. (2014). Scientific realism. In E. N. Zalta (Ed.), *The Stanford encyclopedia of philosophy* (Spring 2014 Edition). http://plato.stanford.edu/archives/spr2014/entries/scientific-realism/

Churchland, P. M. (1985). The ontological status of observables: In praise of the superempirical virtues. In P. Churchland & C. Hooker (Eds.), *Images of science: Essays on realism and empiricism* (pp. 35–47). Chicago: University of Chicago Press.

Dicken, P., & Lipton, P. (2006). What can Bas believe? Musgrave and van Fraassen on observability. *Analysis, 66*, 226–233.

Doppelt, G. (2007). Reconstructing scientific realism to rebut the pessimistic meta-induction. *Philosophy of Science, 74*, 96–118.

Gauch, H. G., Jr. (2012). *Scientific method in brief*. Cambridge, UK: Cambridge University Press.

Giere, R. N. (1988). *Explaining science: A cognitive approach*. Chicago: University of Chicago Press.

Giere, R. N. (2006). *Scientific perspectivism*. Chicago: University of Chicago Press.

Godfrey-Smith, P. (2008). Recurrent transient underdetermination and the glass half full. *Philosophical Studies, 137*, 141–148.

Harker, D. (2006). Accommodation and prediction: The case of the persistent head. *British Journal for the Philosophy of Science, 57*, 309–321.

Harker, D. (2008). On the predilections for predictions. *British Journal for the Philosophy of Science, 59*, 429–453.

Hitchcock, C., & Sober, E. (2004). Prediction versus accommodation and the risk of overfitting. *British Journal for the Philosophy of Science, 55*, 1–34.

Hoefer, C., & Rosenberg, A. (1994). Empirical equivalence, underdetermination, and systems of the world. *Philosophy of Science, 61*, 592–607.

Horwich, P. (1982). *Probability and evidence*. Cambridge, UK: Cambridge University Press.

Kitcher, P. (1993). *The advancement of science: Science without legend, objectivity without illusions*. New York: Oxford University Press.

Kornblith, H. (2013). Is philosophical knowledge possible? In D. E. Machuca (Ed.), *Disagreement and skepticism* (pp. 260–276). New York: Routledge.

Kukla, A. (1993). Laudan, Leplin, empirical equivalence and underdetermination. *Analysis, 53*, 1–7.

Kukla, A. (1996). Does every theory have empirically equivalent rivals? *Erkenntnis, 44*, 137–166.

Ladyman, J. (1998). What is structural realism? *Studies in the History and Philosophy of Science, 29*, 409–424.

Ladyman, J. (2011). Structural realism versus standard scientific realism: The case of phlogiston and dephlogisticated air. *Synthese, 180*, 87–101.

Lange, M. (2002). Baseball, pessimistic inductions and the turnover fallacy. *Analysis, 62*, 281–285.

Laudan, L. (1981). A confutation of convergent realism. *Philosophy of Science, 48*, 19–48.

Laudan, L. (1984). *Science and values*. Berkeley: University of California Press.

Laudan, L., & Leplin, J. (1991). Empirical equivalence and underdetermination. *Journal of Philosophy, 88*, 449–472.

Leplin, J. (1997). *A novel defense of scientific realism*. Oxford: Oxford University Press.

Leplin, J., & Laudan, L. (1993). Determination underdeterred: Reply to Kukla. *Analysis, 53*, 8–16.

Lewis, P. (2001). Why the pessimistic induction is a fallacy. *Synthese, 129*, 371–380.

Lipton, P. (2004). *Inference to the best explanation* (2nd ed.). New York: Routledge.

McCain, K. (2012). *Inference to the best explanation and the external world: A defense of the explanationist response to skepticism*. PhD dissertation, University of Rochester.

McCain, K., & Poston, T. (2014). Why explanatoriness is evidentially relevant. *Thought, 3*, 145–153.

McGrew, T. (2003). Confirmation, heuristics, and explanatory reasoning. *British Journal for the Philosophy of Science, 54*, 553–567.

Mizrahi, M. (2013). The pessimistic induction: A bad argument gone too far. *Synthese, 190*, 3209–3226.

Musgrave, A. (1985). Constructive empiricism and realism. In P. Churchland & C. Hooker (Eds.), *Images of science: Essays on realism and empiricism* (pp. 197–221). Chicago: University of Chicago Press.
Musgrave, A. (1988). The ultimate argument for scientific realism. In R. Nola (Ed.), *Relativism and realism in science* (pp. 229–252). Dordrecht: Kluwer.
Okasha, S. (2000). Van Fraassen's critique of inference to the best explanation. *Studies in History and Philosophy of Science, 31*, 691–710.
Poincaré, H. (1905/1952). *Science and hypothesis*. New York: Dover.
Psillos, S. (1996). On van Fraassen's critique of abductive reasoning. *Philosophical Quarterly, 46*, 31–47.
Psillos, S. (1999). *Scientific realism: How science tracks truth*. London: Routledge.
Putnam, H. (1975). *Philosophical papers, vol. 1: Mathematics, matter and method*. Cambridge, UK: Cambridge University Press.
Putnam, H. (1978). *Meaning and the moral sciences*. London: Routledge.
Saatsi, J. T. (2005). On the pessimistic induction and two fallacies. *Philosophy of Science, 72*, 1088–1098.
Schlesinger, G. (1987). Accommodation and prediction. *Australasian Journal of Philosophy, 65*, 33–42.
Smart, J. J. C. (1963). *Philosophy and scientific realism*. London: Routledge & Kegan Paul.
Stanford, P. K. (2006). *Exceeding our grasp: Science, history, and the problem of unconceived alternatives*. Oxford: Oxford University Press.
van Fraassen, B. C. (1980). *The scientific image*. Oxford: Oxford University Press.
van Fraassen, B. C. (1989). *Laws and symmetry*. Oxford: Oxford University Press.
Weisberg, J. (2009). Locating IBE in the Bayesian framework. *Synthese, 167*, 125–143.
White, R. (2003). The epistemic advantage of prediction over accommodation. *Mind, 112*, 653–683.
Worrall, J. (1989). Structural realism: The best of both worlds? *Dialectica, 43*, 99–124.

Part IV
Social Dimensions of Scientific Knowledge

Chapter 15
Gaining Scientific Knowledge from Others

Abstract This chapter explores how we gain knowledge from other people. In particular two of the primary ways that we come to have social evidence are explored. The first is the most prevalent form of social evidence: testimony. This chapter examines some of the best explanations of how it is that we come to have knowledge via the testimony of others. The second way that we come to have social evidence is by learning of disagreements. This chapter explores the epistemic significance of disagreement as it occurs in science. In particular it discusses how we should respond when we discover that someone disagrees with us about a scientific claim. When we discover that equally informed experts in an area of science disagree we should refrain from believing one side or the other is correct until we have further deciding evidence. When novices discover that experts disagree with them concerning a scientific claim, the novices should defer to the experts.

Our discussion up to this point has been predominately focused on individualistic aspects of scientific knowledge and of knowledge in general. The reason for this is twofold. First, epistemology has historically been, and largely continues to be, focused on the epistemic appraisal of individual agents and how such agents can come to have knowledge by way of their own individual cognitive faculties. Second, this investigation of individualistic aspects of knowledge provides a critical foundation for understanding key features of knowledge in general and scientific knowledge in particular. Despite the importance of grasping the primarily individualistic aspects of knowledge which we have hitherto discussed, it would be a mistake to conclude our exploration of scientific knowledge without delving into some of the social dimensions of epistemology.

The idea of a scientist working in isolation discovering the great secrets of the universe, such as when Isaac Newton came up with many of his ideas related to calculus, gravitation, and optics while away from Cambridge at his family's farm, is no longer the norm of science (Gleick 2004). Perhaps it never was. As Newton himself famously claimed, "if I have seen farther, it is by standing on the shoulders of giants" (Hawking 2002, p. 725). Although Newton may have done a lot of work on his own, it seems that even he depended on the work of others in important ways in coming to his own conclusions. Whether it was ever the case that science proceeded by way of lone researchers working in total isolation, it no

K. McCain, *The Nature of Scientific Knowledge*, Springer Undergraduate Texts in Philosophy, DOI 10.1007/978-3-319-33405-9_15

longer proceeds in this fashion. In fact, at least since the 1800s with the work of philosophers like John Stuart Mill and Charles Sanders Peirce, a thoroughgoing social understanding of science has been on the rise (Longino 2013).[1] The social nature of science is even more readily apparent since the emergence of "Big Science" in the latter half of the twentieth century. Big Science projects like the Manhattan Project and the Humane Genome Project involve large groups of scientists with differing areas of expertise working together on common research projects (Longino 2013).

Even setting aside Big Science, it is widely recognized that science is a distinctly social enterprise. In practice only some scientific results are checked by independent studies. "Many are simply accepted on trust . . . in science, knowledge grows by depending on the testimony of others" (Longino 2013, p. 7). Further, there are several features of science which make it particularly well suited to yielding knowledge of the world. As Alvin Goldman (1999, pp. 250–251) notes, many of these features are social in nature:

- The marshaling and distribution of resources to facilitate scientific investigation and observation.
- A system of credit and reward that provides incentives for workers to engage in scientific research and to distribute their efforts in chosen directions.
- A system of disseminating scientific findings and theories as well as critical assessments of such findings and theories.
- The use of domain-specific expertise in making decisions about dissemination, resource allocation, and rewards.

Undeniably, science is a social practice and scientific knowledge is social in important ways. Given this, a better grasp of some of the social aspects of scientific knowledge can facilitate a deeper understanding of NOS. Hence, the focus of this final part of the current book is on important social dimensions of scientific knowledge.

Before diving into the specific social features of scientific knowledge it will be helpful to first get a broad picture of the focus of social epistemology. That is, it will be helpful to very briefly consider what kinds of epistemic issues are considered particularly social. A helpful way of differentiating social epistemological issues is in terms of the agents involved (Goldman 2011). Our focus in the earlier parts of this book was predominately on the knowledge (and justification) had by individual agents. Of course, such knowledge often falls under traditional, individualistic epistemology, however, some key social epistemological issues can arise for individual agents. Typically, the knowledge of individual agents is social when it is based on social evidence—evidence that "concerns acts of communication by others, or traces of such acts such as pages of print or messages on computer screens" (Goldman 2011, pp. 14–15). Additionally, "social evidence can consist in other people's doxastic states that become known to the agent" (Goldman 2011,

[1] See in particular Mill (1859/2008) and Peirce (1868/1982, 1878).

p. 15). Very roughly, one way to have social knowledge is for S to come to know that *p* on the basis of information provided to her by others. This information from others can come through the acts of communication whereby various people engage with S or by way of S learning of the beliefs of others in some way other than their intentional communication of the information to her.

Another way that knowledge can be social is for the knowing agent to be a group agent (a collective entity), rather than an individual. For instance, if a jury knows that the defendant is innocent or if a corporation knew that dumping would cause harm, such knowledge is social knowledge.[2]

A final way for knowledge to be social is for that knowledge to be generated by an epistemic system. As Goldman (2011, p. 13) explains, "an epistemic system is a social system that houses a variety of procedures, institutions, and patterns of interpersonal influence that affect the epistemic outcomes of members." The clearest examples of epistemic systems are "formal institutions with publicly specified aims, rules, and procedures" such as science and education (Goldman 2011, p. 19).

Although all three ways of generating social knowledge are important and worth exploring, we will limit our focus here to individuals' use of social evidence to gain knowledge (the focus of the remainder of this chapter) and science as an epistemic system (the focus of the next chapter). The reasons for this are that the idea of group agents having doxastic attitudes over and above those held by the individual members of the group is controversial, and the knowledge had by group entities, if such a thing occurs, is not central to understanding NOS.[3] With these preliminaries out of the way, let us turn our attention to social evidence.

15.1 Testimony

There are a number of issues to get clear on when it comes to the first sort of social evidence that we will discuss, testimony, and the role that it plays in our knowledge in general as well as scientific knowledge in particular. Let us begin by getter clearer on what testimony is.

15.1.1 What Testimony Is

Often when we hear the word "testimony" we think of formal settings such as when a witness in a courtroom has sworn to give honest testimony concerning events she

[2]See Bratman (1993), Gilbert (1989), Pettit (2003), and Searle (1995) for reasons to think that groups can have doxastic attitudes.

[3]See Quinton (1975), Rupert (2011), and Wilson (2004) for considerations against thinking groups can have doxastic attitudes that are not reducible to the attitudes of the members of the group. See List (2014) for helpful discussion of different kinds of group agents and the degree to which their attitudes may be reducible to those of their members.

observed (Audi 2011). Obviously, this sort of formal testimony is very important, and it can provide us with key social evidence in some settings. However, much of our social evidence comes from more informal testimony—testimony which occurs outside of a court of law. In fact, informal testimony is so pervasive and important to our epistemic lives that it seems true to say that "virtually everything we know depends in some way or other on the testimony of others—what we eat, how things work, where we go, even who we are" (Lackey 2011, p. 71).[4] Consider some of the things we know, or come to know, on the basis of testimony: our birth dates (after all, we do not remember being born), the average summer temperature of a country we have never visited, how to get to a restaurant in an unfamiliar city, and the list goes on and on. In each case we come to know these facts by way of the testimony of others. Perhaps we are told our birth dates by our families, we might read on a weather app that the average summer temperature of a particular country is *n* degrees, and a passerby might tell us how to get to the restaurant. What exactly is it that all of these instances of testimony have in common?

A tempting answer to the above question would be that they are all instances where someone else engages in an act of communication with us. Consequently, one might think that any act of communication from one person to another is an instance of testimony. This, however, is much too broad. There are times when we communicate with each other without the intention of sharing information. For example, if Laura is running a race and Bruce shouts at her "Yay Laura!", Bruce is not testifying. In this case Bruce is merely expressing support for Laura. Now, Laura might come to know things by hearing Bruce's shout, such as that Bruce wants her to win, but the knowledge that Laura gains from Bruce's shout is not based on his "testimony" because he provided none. This would be an instance where someone gets knowledge from another's speech act, but that knowledge is not itself testimonial. Another example of gaining non-testimonial knowledge from another's speech act would be when Charlotte hears Chuck speak for the first time and comes to know that Chuck has a deep voice. Charlotte's knowledge of the tone of Chuck's voice comes from her hearing him speak, but it is not testimonial knowledge. Admittedly, in both of these cases we might think that the knower is relying on social evidence in the broad sense of social evidence—evidence gained from another person—however, for our purposes here we will focus on the most prevalent form of social evidence: testimony. So, we will need a more precise definition of testimony before we continue, such as: "a speaker's making an act of communication—which includes statements, nods, pointing, and so on—that is intended to convey the information that *p* or is taken as conveying the information that *p*" (Lackey 2011, p. 72).[5] Undeniably, this definition of testimony could be yet

[4]Also see Adler (2014), Audi (2011), and Goldman (1999) on the prevalence of testimony in our epistemic lives. Goldman points out that this sort of epistemic practice is not only ubiquitous throughout human cultures, it also occurs frequently among many non-human animals to some extent.

[5]See Lackey (2008) for a more complete discussion of how to best define testimony.

more precise, but for our purposes it will suffice. Now that we have a grasp of what testimony is, let us consider how it works.

15.1.2 How Testimony Works

Alvin Goldman (1999, p. 104) aptly notes that there seem to be four stages which occur when acts of testimony add to the overall knowledge of a social group: "(1) discovery, (2) production and transmission of messages, (3) message reception, and (4) message acceptance." Let us consider a simple example in order to illustrate this. Rosa and Evelyn have shipwrecked on a deserted island. They have decided that their first priority is to find water. They agree that it will be best to try different routes and reconvene on the beach in an hour. Rosa discovers a clear spring close to the path that she has taken (stage 1). When Rosa and Evelyn meet up on the beach Rosa begins describing the spring and its location to Evelyn (stage 2). Evelyn listens and understands what Rosa is saying (stage 3). As a result, Evelyn believes that there is a spring where Rosa says (stage 4).

Here is one more example to help make the stages of testimony clearer. Clark is an investigative reporter for the *Daily Planet*. Late one night Clark discovers that Luther Industries is dumping barrels of waste into the harbor (stage 1). Clark writes up an article describing the actions of Luther industries, which his editor puts in the Sunday edition of the *Daily Planet* (stage 2). Lois and several other readers of the newspaper read Clark's article in the *Daily Planet* on Sunday morning (stage 3). After finishing the article, Lois and several others believe that Luther Industries has been dumping waste in the harbor (stage 4).

Although the four-stage model of testimony Goldman offers is perhaps a bit simplistic, it is helpful for getting clearer about how testimony works when it leads to increased social knowledge. It seems clear that when all of the stages of the process are completed well, testimonial knowledge is gained. And so, social knowledge is increased.

Goldman's four-stage model also helps to make it clear that there are various places where testimony can go wrong, i.e. fail to lead to knowledge. As we can see from the model, or even from thinking about testimonial exchanges in a less formal fashion, there are two primary ways that testimony can go wrong. First, the testifier can be incompetent in some way. For instance, Rosa might mistakenly take a mirage to be a spring full of fresh water or Clark might misread the names on the trucks dumping waste, and so mistakenly think that Luther Industries is dumping waste. Clearly, testimony can go wrong when the testifier is mistaken in what she reports. Second, the testifier can know the truth of the matter in question, but decide to lie or deceive. Unfortunately, it is not difficult to come up with cases where testifiers have a motivation to be dishonest in the things they say.

In light of some of the ways that testimony can go wrong one might worry that skepticism about relying on testimony will arise. After all, it seems that in any instance where someone testifies to us that p it could be that her report is false—either because she is merely mistaken or because she is trying to deceive us about *p*. If these are always possibilities that we face when we receive another's testimony, one might worry that we are never justified in accepting what we are told. Clearly, if we cannot justifiably accept the testimony of others, then the vast number of things we believe on the basis of testimony would fail to constitute knowledge. This is a very troubling conclusion indeed.

Fortunately, we have good reason to think that such a far-reaching skepticism of testimony is unfounded. For one thing, there are ways we can help determine that these skeptical possibilities do not obtain. For instance, the speaker might convey her testimony in a way which makes it clear that she has expertise on the subject—consider how your physician tells you about various options you have concerning medications. She does so using specific jargon which suggests her expertise, the setting in which she provides you the testimony is one that gives evidence that she is well-informed, and so on. Also, various observable characteristics of a speaker may help us to determine her sincerity with respect to what she asserts—we tend to pick up on things like "her tone of voice, her facial aspect, and her body language" (Goldman 1999, p. 108). All of these things can give us reason to think that she is testifying honestly. Of course, these are not perfect indications of accurate and honest testimony, but they do help us rule out egregious violations.

In addition to methods that we can use to help determine the quality of a speaker's testimony there are also reasons to think these skeptical possibilities do not generally obtain. First of all, there is simply the empirical fact that, overwhelmingly, testimony tends to be truthful (Adler 2014). Some argue that the reason for this is evolutionary. They claim that there are credible evolutionary reasons for thinking that humans have evolved with dispositions to be truthful most of the time (Goldman 1999). Another plausible explanation of this comes from consideration of the nature of communication itself. Plausibly, it is a presupposition of communication that participants are, at least for the most part, truthful (Lewis 1969; Schiffer 1972). While not all situations are ones in which the norm is to be truthful such as when a marketing campaign is designed to convince you of the superiority of a particular brand, we have information about which domains are the ones in which truthfulness is not the norm (Adler 2014). So, we can typically recognize whether we are in a setting where the norm is for testimony to be trustworthy or not. In addition to these presuppositions and our, perhaps natural, disposition to tell the truth there are also rewards and punishments in our social settings which encourage accurate, honest testimony (Adler 2014; Goldman 1999). This is particularly clear in the way that reputation plays a significant role in the scientific community (Adler 2014). The emphasis science places on things like the replication of experiments and the peer review process for publication also helps to encourage truthful and well-informed testimony when it comes to sharing scientific knowledge (Shatz 2004).

Given all of the methods we have in place in social settings for ensuring accurate testimony, we have good reason to think that generally trusting the testimony of others is likely to lead us to the truth. Nevertheless, we should "always engage in some assessment of the speaker for trustworthiness" because "to believe what is asserted without doing so is to believe blindly, uncritically" (Fricker 1994, p. 145).

By doing this we help to ensure that we are not led astray in the cases where the testimony is not genuine.

Although we have seen that there is good reason to think we should generally trust testimony because of the ways in which it is typically constrained and our natural dispositions, it is an open question whether one has to have this information about testimony before she can rely on testimony to gain knowledge. So, we are now faced with an important question: can we come to know that *p* simply because someone tells us that *p*, or must we have evidence from other sources like perception, memory, and reasoning to support our trusting the speaker's testimony when she tells us that *p*? In other words, "is testimony an autonomous source of epistemic authority" (Adler 2014, p. 1)? It is toward this, the central question of the epistemology of testimony, that we will now turn our attention.

15.1.3 How Testimony Provides Knowledge

Most views concerning testimonial justification/knowledge fall into two camps. Each side traces its origin to the writings of a major figure in the history of philosophy—reductionists trace their view to David Hume and non-reductionists (sometimes called "anti-reductionists") to Thomas Reid. Very roughly, reductionists answer the question concerning the autonomy of testimony as an epistemic source negatively—"testimony cannot provide justification or knowledge independently of other sources like perception, memory, and inductive inference". Non-reductionists answer it positively—"testimony can provide justification/knowledge on its own". Although for our purposes it does matter which camp has the truth of the matter, it will be worth briefly considering both views and some of the major challenges that each face because doing so will help to add depth to our understanding of testimony.

Let us start with reductionism. It will be helpful to first define reductionism more precisely. There are actually two versions of reductionism, which Jennifer Lackey (2011) refers to as "global reductionism" and "local reductionism". Global reductionism is the view that the justification of relying on testimony as a general source of belief "reduces to the justification of sense perception, memory, and inductive inference" (Lackey 2011, p. 74). Local reductionism, the more prominent of the two kinds of reductionism, is the view that in each particular case of testimonial belief the justification for that belief "reduces to the justification of instances of sense perception, memory, and inductive inference" (Lackey 2011, p. 75). More simply, the idea is this: global reductionism claims that you have to be justified in believing that testimony *in general* is reliable in order to have justified beliefs/knowledge on the basis of a particular instance of testimony whereas local reductionism claims that you only need to have justification for accepting the *particular* testimonial report in question in order to have a justified belief/knowledge on the basis of that testimony. To see the difference consider the following: Thelma believes that *p* on the basis of Tim's testimony that *p*. According to global reductionism, in order for Thelma's belief that *p* to be justified she must be justified in believing that testimony in general is reliable, i.e. she must be justified in believing that most instances of testimony are accurate. According to local reductionism, Thelma does not have to be justified in believing that testimony in general is reliable in order to justifiedly believe/know that *p* on the basis of Tim's testimony. Instead, local reductionism only requires that Thelma be justified in thinking that Tim's testimony in this instance is likely to be accurate.

The major objection that is often put forward against reductionist views is that ordinary people simply do not have enough information to have the positive reasons required to have justification for accepting either that testimony in general is reliable or that a particular instance of testimony is reliable.[6] Perhaps the most formidable form of this objection is what has been called the "infant/child objection" (Lackey 2005). This objection is simply the idea that children and infants seem to learn a lot about the world strictly via testimony. Furthermore, it seems that they can gain this knowledge without first having a sufficiently large non-testimonial basis for accepting the reliability of testimony. This seems to run counter to the claims of reductionism though. So, the objection claims that reductionism cannot be correct about testimony.

Now let us take a look at the other side of the debate. Non-reductionism is the view that testimony is like sense perception, memory, and reasoning—it is a basic source of justification/knowledge. As Lackey (2011, p. 73) explains, "nonreductionists maintain that, so long as there are no relevant undefeated defeaters, hearers can be justified in accepting what they are told *merely* on the basis of the testimony of speakers." With this more precise characterization of non-reductionism in hand we can characterize the general dispute between reductionists and non-reductionists quite simply. Reductionists think that in order for S's belief that is based on testimony to be justified/knowledge S has to have reasons for thinking that she can trust testimony (either in general or in this particular case), but non-reductionists think that S's testimonial belief is justified so long as she does not have reasons to distrust the testimony she has received. Hence, non-reductionists hold that one does not need positive reasons for thinking that testimony is trustworthy in order to gain testimonial justification/knowledge, one simply has to lack reasons for thinking that testimony is not trustworthy.

The primary objection to non-reductionism is that it makes testimonial justification/knowledge too easy to come by.[7] It is claimed that non-reductionists "sanction gullibility, epistemic irrationality, and intellectual irresponsibility" (Lackey 2011, p. 75). The problem with non-reductionism is that according to this view if one were to simply believe everything she saw on television, she would be justified in doing so as long as she did not possess positive reasons to be distrustful of the information she is given. Opponents of non-reductionism claim that this is ridiculous. It seems that one needs some positive reasons for trusting the information she gains from

[6] In addition to this objection there are others which are raised against each particular kind of reductionism. For more on these see Adler (2014) and Lackey (2011).

[7] For additional objections and discussion see Adler (2014) and Lackey (2011).

other people. If such positive reasons are not required, then one could justifiedly believe on the basis of the testimony of "randomly selected speakers, arbitrarily chosen postings on the Internet, and unidentified telemarketers" (Lackey 2011, p. 75). Unfortunately, this seems to be exactly the sort of thing that is licensed by non-reductionism.

Of course, supporters of both reductionism and non-reductionism attempt to respond to these objections as well as the various other challenges faced by their preferred views. Some argue that the objections facing reductionism and non-reductionism show that both views are unacceptable. They maintain that we should look to hybrids of reductionism and non-reductionism or to alternative approaches to both views in order to adequately capture the truth about testimony.[8] We will not concern ourselves with attempting to settle the debate between reductionists and non-reductionists (or supporters of hybrid views and alternative approaches) here. Simply understanding the issues that are of concern to these various theorists is sufficient for our purposes. However, before turning our attention to another key kind of social evidence and the important epistemological questions that consideration of it raises, we should briefly explore a particular account of testimonial justification/knowledge. This account of testimony fits quite nicely with the explanatory account of knowledge we have developed in earlier chapters.

A plausible account of testimony relies on inference to the best explanation (IBE). The idea behind this approach to testimony is quite simple. S is justified in believing that *p* on the basis of T's testimony when the best explanation S has for T's asserting *p* is that *p* is true.[9] Returning to our above examples, Evelyn is justified in believing that there is a spring near the path that Rosa took so long as the best available explanation she has for Rosa telling her this is that it is true that there is a spring near the path. Similarly, Lois is justified in believing that Luther Industries is dumping waste in the harbor just in case the best available explanation of the *Daily Planet* printing the article saying this is that it is true that Luther Industries is dumping waste in the harbor. It is not difficult to see how this would equally apply in scientific settings—we are justified in accepting the articles we read in *Nature* or *Science* because the best explanation of why the empirical findings reported in such articles appear in these journals is that they are accurate, we are justified in accepting what the professors in the physics departments at reputable universities tell us about General Relativity because the best explanation for why they are telling us this is that it is true, and so on. This IBE account seems to offer a straightforward account of testimonial justification/knowledge which fits very well with the view of scientific knowledge we have been exploring in earlier chapters.

[8] See, for example, Goldberg (2008), Faulkner (2000, 2007), Hinchman (2005), Lackey, (2008), Lehrer (2006), and Moran (2006).

[9] Fricker (1995), Harman (1965), Lipton (2007), Malmgren (2006), and Schiffer (2003) all provide considerations in support of this sort of view of testimony.

Now one might worry that the IBE account of testimony faces a serious problem because it appears to be a kind of local reductionist theory. As such, it would seem to face the infant/child objection described above. While it is true that the IBE account of testimony is plausibly considered a local reductionist theory because it requires S to have positive reasons for accepting any particular instance of testimony, it is not clear that the infant/child objection poses a serious problem for it. The IBE account of testimony does not seem to require nearly as much information as those pressing the infant/child objection typically presuppose. All that is required is that S be such that the truth of what she is told is the best explanation she has for why she is told this. It is not unreasonable to think that unsophisticated agents such as infants and children could have such reasons. After all, as we noted in Chap. 10, even very small children are fairly adept at evaluating explanations. Additionally, there is evidence that children keep track of both the reliability of informants and of informants' standing as testifiers in the community (Harris and Corriveau 2011). Finally, if the IBE view of testimony only requires that *p*'s truth be the best *available* explanation of why T asserts *p* in order for S to be justified in accepting T's testimony that *p*, it is likely that infants and children may be in position to have a lot of testimonial knowledge. At a very young age the truth of what others testify may be the only explanation that children have available for the testimony of others. As children develop the cognitive capacities to have alternative explanations available for why someone would assert that *p* it is likely that they also accumulate information concerning the reliability of informants. While these considerations are far from decisive, they do at least show that an IBE approach to testimony is plausible. They also show that such an account may be able to overcome the major challenge facing reductionist views. The IBE approach to testimony is not required by the explanatory account of scientific knowledge that we have developed throughout the previous chapters, but it does fit quite nicely with it.

15.2 Disagreement

Disagreement is a fact of our lives. People disagree with us about all sorts of things: religion, politics, philosophical theories, and many other issues. What is so important about the fact of disagreement for our purposes is that discovering disagreements can provide us with an important kind of social evidence. In many cases we gain evidence of disagreement by way of testimony. For example, Derek can gain evidence of disagreement when Hansel tells him that he disagrees with him about a particular issue. It is possible to gain evidence of disagreement in other ways too. Derek might learn of Hansel's disagreement with him not via Hansel's testimony, but instead by observing Hansel's behavior in various situations. How exactly we come to discover the disagreement is not all that important—what matters is our awareness of the disagreement and what it means for the justification we have for the things we believe.

There are some cases where the epistemic impact of discovering disagreement is pretty obvious. As Adam Elga (2007, p. 478) explains, "There are experts and gurus, people to whom we should defer entirely. There are fakes and fools, who should be ignored." The idea that Elga is expressing seems both plausible and widely shared. When we recognize that we disagree with someone who is an expert on a topic that we are not, we should generally defer to her opinion on the topic (Carey and Matheson 2013; Conee 2009; Kornblith 2013; Matheson 2015b). Alternatively, when we recognize that we disagree with someone who is a largely uninformed novice, much less a fake or fool, on a topic on which we are experts or at least much better informed, we should generally be unimpressed by her difference of opinion (Frances 2014).

Not all cases of disagreement are this easy to adjudicate though. The epistemological literature on the significance of disagreement has predominately focused on cases where the justificatory impact of disagreement is less clear—disagreements among epistemic peers (individuals who are in equally good epistemic positions to arrive at the truth of the disputed matter). There are a variety of theories concerning the impact on one's justification for believing that *p* when she discovers that an epistemic peer disagrees with her. These views fall along a spectrum from very "conciliatory"—discovering disagreement should make one much less certain of *p*'s truth—to very "steadfast"—discovering disagreement should have little to no effect on one's confidence in *p*'s truth (Christensen 2009, p. 756).[10]

Although there is much of value that can be gleaned from exploring the burgeoning epistemological literature on disagreement, examining it in detail will not be especially helpful for our purposes. One of the primary reasons for this is that the sorts of cases focused on in the epistemological literature tend to involve extremely idealized disagreements. The disagreements that are typically discussed are those which occur between only two individuals who are perfectly equal with respect to their epistemic positions concerning their dispute. As Jonathan Matheson (2015a, p. 5) aptly notes, "there is good reason to believe that peer disagreements [of the sort the epistemological literature tends to focus on] simply don't occur in the actual world." While there are numerous disagreements in the real world, "typically there is at least some epistemic advantage held by one of the parties: one of them has a little more evidence, one of them has thought about the issue a little more, one of them is a little more open-minded, and so forth" (Matheson 2015a, p. 5). So, the sort of idealized disagreements that the epistemological literature focuses on are not likely to be encountered.

A consideration of real world disagreements will be much more helpful for our purposes. The idealization that is perhaps farthest from how things actually are in the world is the restriction of disagreements to those occurring between

[10]For a sampling of the many theories of the epistemic significance of disagreement which have been put forward recently see Christensen and Lackey (2012), Feldman and Warfield (2010), and Machuca (2013). See Matheson (2015b) for in-depth critical discussion of many of the key points of contention among theories of the epistemic significance of disagreement.

just two individuals. In the disagreements we actually encounter though, "we are typically aware of a vast multitude of opinions" (Matheson 2015b, p. 125). Such disagreements are widespread in a variety of domains, even the sciences (Christensen 2009; Kornblith 2013; Matheson 2015b). Fittingly, our focus will be on the sort of evidence that is gained from disagreements in the sciences and how we should respond to that evidence.

When it comes to disagreements no intellectual endeavor is immune—even the sciences. Despite this fact, disagreement within the sciences is different than the disagreements which arise in politics, religion, philosophy, and countless other domains in a very important way. Disagreements in the sciences "tend to be resolved over time. Opinions in the sciences tend to converge, and . . . there is a great deal of reason to believe that these opinions tend to converge to the truth, or at least to closer and closer approximations to the truth" (Kornblith 2013, p. 261). As a result, in the sciences we tend to see "an ever-increasing body of accepted opinion among experts . . . there are large bodies of well-established results, and smaller areas of dispute"—areas of dispute which for the most part tend to be resolved as time progresses (Kornblith 2013, p. 266). This, at least upon a cursory examination, seems very different than what we find in other areas such as politics, religion, and philosophy. Disagreements in these domains appear very longstanding and perhaps intractable. Whether or not these other disciplines are worse off than the sciences with respect to disagreement is not all that important for our purposes. What is important is that the sciences tend to be *progressive*—resolving disagreements and coming closer to the truth over time. The progressive nature of science and the consensus of experts in the sciences over time give us some reason to think that the opinions of experts in the sciences are trustworthy. It seems that "experts in the sciences are highly reliable. Individual experts tend, to a very high degree, to form true, or at least approximately true, beliefs about the matters they investigate. It is for primarily this reason, of course, that opinions within the sciences tend to converge over time" (Kornblith 2013, p. 268). Recognizing these facts has important ramifications for the significance of disagreements concerning the sciences.

Considering the following simple case helps to illuminate the most basic lesson of the reliability of experts in the sciences and the way that disagreements in the sciences tend to go:

> Elisa is an expert in electrical engineering. Ned is not. He is a complete novice. Elisa believes some particular claim about electrical engineering, *p*. Ned does not believe that *p*, in fact, he believes not-*p*. Elisa and Ned have a conversation and discover that they disagree about *p*.

What happens to the justification that Elisa has for believing that *p* and the justification that Ned has for believing that not-*p* once they discover their disagreement? Assuming that Elisa and Ned are both aware of their comparative expertise, it seems clear that Elisa should simply continue to believe that *p* pretty much just as firmly as she did before she discovered that Ned disagreed. Ned, however, should drastically change his doxastic attitude—he should either now believe that *p*, as Elisa does, or at the very least he should greatly reduce his confidence in not-*p*.

The situation between Elisa and Ned is analogous to our position with respect to at least some areas of science—unless one is an expert in that particular scientific field, she will be in much the same position as Ned when it comes to disagreeing with the experts. When it comes to disagreement with experts in science a good general rule is that novices should simply defer to the opinion of experts. After all, as we have already noted, science has an excellent track record of converging on truth because experts in science tend to be very reliable at forming true beliefs on matters in their fields of specialty.

In fact, the advice to defer to the experts seems equally applicable to situations in which the subject is not a complete novice, but an informed layperson. For instance, if Ivan, who is reasonably well informed on matters of electrical engineering, discovers that he disagrees with Elisa about *p*, he should defer to her opinion. At a minimum Ivan should become considerably less confident of not-*p*. This becomes even clearer when we consider a situation where not only Elisa disagrees with Ivan, but the majority of experts in the field hold Elisa's position. In such a case it seems clear that Ivan should defer to the experts. This is true even if Ivan has examined the evidence concerning *p*/not-*p* himself and concluded that it supports not-*p*. Ivan is not an expert, so he knows that Elisa and the other experts are in a better position to evaluate the evidence than he is. In light of this, he should defer to their opinion.

The advice of deferring to the experts is applicable even to experts themselves. Of course, if Elisa, an expert electrical engineer, finds herself in a disagreement about astronomy with Alexis, an expert astronomer, she should defer to Alexis' opinion on the issue. It seems plausible that even if Elisa believes a particular claim about electrical engineering, *p*, but later discovers that the vast majority of electrical engineers disagree with her, she should defer to the opinion of the majority of experts (Kornblith 2013). After all, these other expert electrical engineers are each approximately as knowledgeable as Elisa is, so each one is about as likely to arrive at the truth concerning *p* as she is. The odds that the vast majority of experts are correct and Elisa is not in this case are very high. Consequently, the general rule of deferring to the majority opinion of the experts in the sciences when one disagrees (and even when one is simply coming to form an opinion about a topic in the sciences) seems to apply across the board to novices and experts alike.

Of course, there are exceptions and complications to this rule. For example, if Elisa knows that the majority of experts in her field have been bribed by a corporation to report what the corporation wants about *p*, then it seems that she should not defer to their opinion. Similarly, if Elisa knows that the experts are disagreeing with her while intoxicated at a party for experts in electrical engineering, then deferring to their opinion at that time is probably not reasonable. Further, if Elisa knows that she has made a new discovery which bears on the issue and has not yet shared the discovery with her colleagues, she has reason to resist deferring to the majority opinion of the experts.

In general, when one has good reason to think that the expert is not exhibiting her expertise in a particular instance or if there is reason to think that one has key information which the expert lacks, there is reason to be less deferential to expert opinion. Additionally, whether one should defer to the experts can be less clear in cases where the particular science contains a very small number of experts or it is not clear that the experts are independently arriving at their conclusions (perhaps because the experts are all students of one prominent scientist). Finally, there are cases where it is not clear that one should defer to the majority of expert opinion because it is not clear that the majority holds the lion's share of the expertise in the field. For instance, it could be that when one is considering whether to believe that *p* 95 % of the experts believe that *p* and 5 % believe not-*p*. In most cases the epistemically appropriate thing to do is to believe that *p*—even if one discovered the proportion of expert opinion after having already formed the belief that not-*p*. However, if the 5 % of experts who believe that not-*p* "are generally considered the epistemic superiors" of the 95 % of experts who believe that *p*, it is not clear that one should adopt the opinion of the majority because the minority is composed of the very best of the experts (Frances 2014, p. 72). Consequently, the advice to believe what the majority of experts believe does not always hold. Yet, despite the sorts of exceptions and complications that we have seen, more often than not we should defer to expert opinion in the sciences. Hilary Kornblith (2013, p. 267) quite reasonably suggests that when it comes to the sciences "even the experts should have opinions which are, *ceteris paribus*, dictated by the dominant opinion within the field. Where there is no dominant opinion, there is good reason to withhold belief."[11]

Before concluding this section it is worth noting that while the rule to defer to the opinion of the majority of experts in the sciences is an excellent epistemic rule, there may be important pragmatic reasons for scientists to not defer to other experts. It may be that scientists are better able to conduct research when they believe that their theories are true even when the majority of experts disagree. Perhaps believing that one's theory is true provides the necessary motivation to press through difficulties and setbacks. As we will see in the next chapter, science has the best prospects of discovering the truth when a variety of methods are employed and a variety of theories explored. As a result, it may be good for our overall progress in science for scientists to continue to believe that their theories are true in defiance of the opinion of the majority of experts. This means that there may be *pragmatic* reasons for continuing to believe a theory in the face of disagreement with the majority of experts. Nevertheless, when it comes to what we *epistemically* ought to do, we should to defer the majority of experts in the sciences.

[11]Whether the rule of deferring to the majority opinion of experts should be a more general epistemic rule which extends beyond the sciences is far from clear. See Carey and Matheson (2013) and Matheson (2015b) for some considerations for thinking that this rule should be fully general.

15.3 Conclusion

In this chapter we have explored some of the central issues related to the most prevalent form of social evidence: testimony. We have seen some of the best explanations of how it is that we come to have knowledge via the assertions of others. We have also examined the epistemic impact of another form of social evidence: evidence of disagreement. Understanding the nature and epistemic import of these forms of social evidence is important for understanding NOS because science is a social activity in which progress greatly depends upon sharing knowledge with one another and discovering how to epistemically best respond to evidence of disagreement. In the next chapter we will delve more fully into the social nature of scientific knowledge by exploring science as an epistemic system which generates much knowledge about the world.

References

Adler, J. (2014). Epistemological problems of testimony. In E. N. Zalta (Ed.), *The Stanford encyclopedia of philosophy* (Spring 2014 Edition). http://plato.stanford.edu/archives/spr2014/entries/testimony-episprob/

Audi, R. (2011). *Epistemology: A contemporary introduction to the theory of knowledge* (3rd ed.). New York: Routledge.

Bratman, M. (1993). Shared intention. *Ethics, 104*, 97–113.

Carey, B., & Matheson, J. (2013). How skeptical is the equal weight view? In D. E. Machuca (Ed.), *Disagreement and skepticism* (pp. 131–149). New York: Routledge.

Christensen, D. (2009). Disagreement as evidence: The epistemology of controversy. *Philosophy Compass, 4*, 756–767.

Christensen, D., & Lackey, J. (Eds.). (2012). *The epistemology of disagreement: New essays*. Oxford: Oxford University Press.

Conee, E. (2009). Peerage. *Episteme, 6*, 313–323.

Elga, A. (2007). Reflection and disagreement. *Nous, 41*, 478–502.

Faulkner, P. (2000). The social character of testimonial knowledge. *Journal of Philosophy, 97*, 581–601.

Faulkner, P. (2007). What is wrong with lying? *Philosophy and Phenomenological Research, 75*, 535–557.

Feldman, R., & Warfield, T. (Eds.). (2010). *Disagreement*. New York: Oxford University Press.

Frances, B. (2014). Disagreement. In S. Bernecker & D. Pritchard (Eds.), *The Routledge companion to epistemology* (pp. 68–74). New York: Routledge.

Fricker, E. (1994). Against gullibility. In B. K. Matilal & A. Chakrabarti (Eds.), *Knowing from words* (pp. 125–161). Dordrecht: Kluwer.

Fricker, E. (1995). Critical notice: Telling and trusting: Reductionism and anti-reductionism in the epistemology of testimony. *Mind, 104*, 393–411.

Gilbert, M. (1989). *On social facts*. London: Routledge.

Gleick, J. (2004). *Isaac Newton*. New York: Vintage Books.

Goldberg, S. (2008). Testimonial knowledge in early childhood, revisited. *Philosophy and Phenomenological Research, 76*, 1–36.

Goldman, A. I. (1999). *Knowledge in a social world*. Oxford: Oxford University Press.

Goldman, A. I. (2011). A guide to social epistemology. In A. Goldman & D. Whitcomb (Eds.), *Social epistemology: Essential readings* (pp. 11–37). New York: Oxford University Press.

Harman, G. (1965). The inference to the best explanation. *Philosophical Review, 74*, 88–95.
Harris, P., & Corriveau, K. (2011). Young children's selective trust in informants. *Philosophical Transactions of the Royal Society of London B (Biological Sciences), 366*, 1179–1187.
Hawking, S. (Ed.). (2002). *On the shoulders of giants: The great works of physics and astronomy*. Philadelphia: Running Press.
Hinchman, E. S. (2005). Telling as inviting trust. *Philosophy and Phenomenological Research, 70*, 562–587.
Kornblith, H. (2013). Is philosophical knowledge possible? In D. E. Machuca (Ed.), *Disagreement and skepticism* (pp. 260–276). New York: Routledge.
Lackey, J. (2005). Testimony and the infant/child objection. *Philosophical Studies, 126*, 163–190.
Lackey, J. (2008). *Learning from words*. Oxford: Oxford University Press.
Lackey, J. (2011). Testimony: Acquiring knowledge from others. In A. Goldman & D. Whitcomb (Eds.), *Social epistemology: Essential readings* (pp. 71–91). New York: Oxford University Press.
Lehrer, K. (2006). Testimony and trustworthiness. In J. Lackey & E. Sosa (Eds.), *The epistemology of testimony* (pp. 145–159). Oxford: Oxford University Press.
Lewis, D. (1969). *Convention*. Cambridge, MA: Harvard University Press.
Lipton, P. (2007). Alien abduction: Inference to the best explanation and the management of testimony. *Episteme, 4*, 238–251.
List, C. (2014). Three kinds of collective attitudes. *Erkenntnis, 79*, 1601–1622.
Longino, H. (2013). The social dimensions of scientific knowledge. In E. N. Zalta (Ed.), *The Stanford encyclopedia of philosophy* (Spring 2013 Edition). http://plato.stanford.edu/archives/spr2013/entries/scientific-knowledge-social/
Machuca, D. E. (Ed.). (2013). *Disagreement and skepticism*. New York: Routledge.
Malmgren, A.-S. (2006). Is there a priori knowledge by testimony? *Philosophical Review, 115*, 199–241.
Matheson, J. (2015a). Disagreement and epistemic peers. *Oxford Handbooks Online in Philosophy*. http://www.oxfordhandbooks.com/view/10.1093/oxfordhb/9780199935314.001.0001/oxfordhb-9780199935314-e-13?rskey=7HLZEJ&result=347
Matheson, J. (2015b). *The epistemic significance of disagreement*. New York: Palgrave-MacMillan.
Mill, J. S. (1859/2008). *On liberty*. Oxford: Oxford University Press.
Moran, R. (2006). Getting told and being believed. In J. Lackey & E. Sosa (Eds.), *The epistemology of testimony* (pp. 272–306). Oxford: Oxford University Press.
Peirce, C. S. (1868). Some consequences of four incapacities. *Journal of Speculative Philosophy, 2*, 140–157.
Peirce, C. S. (1878/1982). How to make our ideas clear. In H. S. Thayer (Ed.), *Pragmatism: The classical writings* (pp. 79–100). Indianapolis: Hackett.
Pettit, P. (2003). Groups with minds of their own. In F. Schmitt (Ed.), *Socializing metaphysics* (pp. 167–193). Lanham: Rowman & Littlefield.
Quinton, A. (1975). The presidential address: Social objects. *Proceedings of the Aristotelian Society, 76*, 1–27.
Rupert, R. (2011). Empirical arguments for group minds: A critical appraisal. *Philosophy Compass, 6*, 630–639.
Schiffer, S. (1972). *Meaning*. Oxford: Oxford University Press.
Schiffer, S. (2003). *The things we mean*. Oxford: Oxford University Press.
Searle, J. (1995). *The construction of social reality*. New York: Free Press.
Shatz, D. (2004). *Peer review: A critical inquiry*. Lanham: Rowman & Littlefield.
Wilson, R. A. (2004). *Boundaries of the mind: The individual in the fragile sciences*. Cambridge, UK: Cambridge University Press.

Chapter 16
Knowledge in a Scientific Community

Abstract In earlier chapters various social aspects of scientific knowledge have been explored. These have been aspects which allow for social evidence to provide scientific knowledge to an individual. The focus in this chapter, however, moves beyond the study of individualistic characteristics of scientific knowledge by looking at science itself as an epistemic system. The thoroughgoing social nature of science leads to some characteristics which make it an epistemic system particularly well suited for adding to the store of scientific knowledge. In particular, the social nature of science leads to a division of cognitive labor. This division of cognitive labor both makes it so that trust plays an integral role in the generation of scientific knowledge and so that scientific progress is enhanced by the scientific community hedging its bets by scientists pursuing a wide variety of research projects utilizing a variety of methods. Although the individual scientists who make up the scientific community are not perfect, various social institutions in science help to make good use of their baser motivations.

Scientists do not operate in a vacuum. As we noted in the previous chapter, science is a thoroughgoing social activity (Goldman 1999; Longino 2013). After all, scientists are not isolated from society-at-large, nor do they work in isolation from one another. They "read journals, go to conferences, establish collaborations, seek out grant money, win acclaim and prestige, and operate within a labor market" (Muldoon 2013, p. 117). In fact, when we talk of "science" we are not speaking of some abstract object that exists in an unchanging, eternal form. We are talking about a social activity in which many people are actively engaged, one from which many people beyond those actively engaged enjoy benefits. As Philip Kitcher (1993, p. 179) explains, "The science of a time is constituted by the collection of individuals engaged in doing science, their relations to one another and to the broader community, their cognitive propensities and their individual practices, the consensus practices of the various fields and subfields."

One of the most important facets of the social structure of science is the division of cognitive labor which it encourages. The fact that science incorporates a division of labor is relatively unsurprising. As any culture or society develops the members of the culture "divide up chores in ways that create different areas of expertise" (Keil 2006, p. 144). The idea that a division of labor is necessary for a society to be

K. McCain, *The Nature of Scientific Knowledge*, Springer Undergraduate Texts in Philosophy, DOI 10.1007/978-3-319-33405-9_16

sufficiently productive is not at all new—economists, philosophers, and sociologists have argued for this claim for centuries.[1] Importantly, "in most human cases, divisions of physical labor carry with them implications for divisions of cognitive labor" (Keil 2006, p. 144). Such divisions of cognitive labor seem to occur in just about every community (Lutz and Keil 2002). It seems to be clearly true that "the division of cognitive labor is an essential infrastructure that allows us to transcend the very limited understandings that exist in the mind of any one individual" (Keil 2006, p. 164). Even a mind as great as that of Isaac Newton is very limited in the grand scheme of things. As he himself is reported to have said when reflecting on his achievements as a scientist, "I don't know what I may seem to the world, but, as to myself, I seem to have been only like a boy playing on the sea-shore, and diverting myself in now and then finding a smoother pebble or a prettier shell than ordinary, whilst the ocean of truth lay all undiscovered before me" (Gleick 2004, p. 4). A division of cognitive labor is necessary if science is to move beyond the amount of knowledge that a single person can have at a particular time. Thus, it is no surprise that science is structured in a way that involves various divisions of labor.

One obvious way that science involves division of cognitive labor is the various disciplines and sub-disciplines into which scientists tend to group themselves (Muldoon 2013). Physicists tend to work on different issues than chemists, who in turn work on different issues than biologists, and so on. Even within broader disciplinary divides there are numerous divisions into sub-disciplines—astrophysicists tend to work on different problems than quantum physicists, for example.

Although the relatively macro-level division of cognitive labor into scientific disciplines and sub-disciplines is interesting, there is another more micro-level division of cognitive labor which is perhaps more central to better understanding the social character of NOS. There are two primary aspects of this micro-level division of cognitive labor: how scientists rely on the work of others in forming their own beliefs about which scientific theories and claims are true and how scientists go about making choices as to which research projects to pursue. The first issue concerns the role of trust in science—something which is unavoidable given the division of cognitive labor that is present, and something necessary for effectively making scientific progress. The second issue concerns how labor is divided when it comes to pursuing research projects in science. It is these two vital aspects of the epistemic system that is science which will be our focus in this chapter.

16.1 The Role of Trust in Science

In addition to displaying a healthy dose of intellectual humility, the quote from Newton above eloquently expresses a simple truth which holds for science as a whole—there is much more to the universe than what science has revealed so far.

[1]See Durkheim (1893/1997), Hume (1739–1740/1978), and Smith (1776/1904).

There is much scientific work to be done. Despite the fact that there are many truths yet to be uncovered by science, the extent of scientific knowledge we have already produced is daunting! In fact, there is so much scientific knowledge that no individual can reasonably claim to possess a significant share of it. As Michael Strevens (2010, p. 295) puts the point, "The expertise of even a professional scientist constitutes only a speck in the vast constellation of scientific knowledge... Our acceptance of the scientific image of the world is, then, based on trust in what scientists say, whether we are civilians or scientists ourselves."[2]

The trust that we must place in others when it comes to science is a result of our limitations as finite individuals. We simply do not have the resources to run every experiment for ourselves or to independently confirm every theory which is widely accepted by experts. It is because of our limitations that "only some results are so checked and many are simply accepted on trust" (Longino 2013, p. 7). We do not seem to have any other choice if we are to make much progress in science at all. This is especially clear when one considers the various divisions of labor which exist in science. No one can be an expert in all aspects of science, and no one has the time to independently verify every important scientific result in even a single scientific discipline let alone several of them. If scientists constantly have to check each and every result of other scientists, they will never get around to building on the scientific knowledge which already exists because there will be no time left for new experiments and the development of new theories. There is no getting around it—trust plays a critical role in the growth of scientific knowledge.

Given the vast importance of trust in science one might be inclined to worry that this puts science, and the knowledge which scientists seeks to generate, in a precarious position. After all, this great need for trust in science means that scientists and laypersons alike are largely dependent upon others for their scientific knowledge. Any time one is dependent on other people there is a chance that those other people will fail to live up to expectations. Sometimes people will be dishonest or incompetent or fall short of being a reliable source of information for some other reason(s). Unfortunately, "scientists cannot spend very much time checking the work of others if they are to make contributions" of their own (Hull 1988, p. 394). So, one might worry that this fact gives us reason to question how solid the knowledge produced by science is—or to question if our foundation for scientific beliefs is really sufficiently firm for knowledge at all.

Although there is always a chance that we will be led astray when we rely on others, as we have seen in the previous chapter, there are good reasons for thinking that we can, for the most part, trust the information we gain from others. This is particularly true in science where there are social structures set up to reward and punish in such a way that scientists have strong incentive to report honestly and competently (Hull 1988). Scientists, like people in general, are often greatly affected by the quality of their reputations. Hence, the potential risk to one's reputation by

[2]See Goldman (1999, 2001), Hardwig (1985), Hull (1988), Kitcher (1993), Longino (2013), and Shapin (1994) for considerations in support of thinking that trust is integral to science.

reporting information dishonestly or incompetently is a strong constraint on the quality of information that is shared among scientists and conveyed to laypeople (Adler 2014; Hull 1988). Additionally, things like the peer review process, the public availability of reports and findings, and the possibility of attempts to replicate experimental results help to ensure that the information circulated in the scientific community is trustworthy (Shatz 2004). Thus, it seems that we have good reason to trust that the foundation of knowledge arising from the scientific community is indeed solid.

Despite the constraints in place in science and our good reasons for trusting the information we receive from scientific experts, one might still worry about the reliance on trust which we find in science. When someone is a layperson, or even a scientist who is simply not an expert in the area of science in which she is relying on information from other scientists, one might worry that her reliance on the experts is "blind" (Hardwig 1985). That is to say, one might worry that the person, layperson or scientist, trusting an expert is in no position to evaluate the information she receives. In light of this, one might think that her position is the unenviable one of having to simply rely on information from an expert with no way of determining how reliable that expert really is. This is particularly disturbing since as we noted in the previous chapter, we should "always engage in some assessment of the speaker for trustworthiness" because "to believe what is asserted without doing so is to believe blindly, uncritically" (Fricker 1994, p. 145). If believing blindly is tantamount to believing uncritically, believing blindly does not seem to be very conducive to attaining knowledge. As Alvin Goldman (2001, p. 86) explains, if we really are "blind" in the way that some suggest we are when it comes to science, it "seems to imply that a layperson (or a scientist in a different field) cannot be *rationally justified* in trusting an expert". Of course, if we cannot be rationally justified in trusting experts, it seems likely that we cannot come to have knowledge on the basis of what they tell us. Thus, if the trust which is a necessary component of science is truly blind, science's ability to generate knowledge is dubious.

Fortunately, our trust in science (and in experts in general) is far from blind. We tend to be very good at determining who is an expert in what areas, and so, likely to provide us with reliable information on a given topic. "There are several distinct heuristics that can be used to figure out who knows what", and even "very young children are sensitive to many of these heuristics" (Keil 2006, p. 164). Given this sensitivity to heuristics for discerning expertise, it is unsurprising that it is widely assumed in educational settings that children are able to evaluate the expertise of others—this is a foundational point for the peer tutoring movement in educational practice.[3] In summarizing the results of their empirical research concerning the evaluation of expertise FrankKeil and his colleagues (2008, p. 298) claim:

> One of the most powerful ways of thinking about the organization of knowledge is based on the following idea: There are domain-specific patterns in the world that experts know and use to understand a wide range of phenomena that arise from those patterns. If one

[3]See Cohen et al. (1982) and Fuchs et al. (1997).

can also grasp those patterns in some coarse yet reliable manner, one knows which sort of expert to approach for further understanding. We have shown here that this appreciation of discipline-based ways of understanding shows its first signs quite early in childhood and develops substantially during the elementary school years.

It seems that we can determine who is an expert in which area from a very young age. This ability casts doubt on the idea that the trust so integral to science is truly blind.

In addition to our having the ability to determine who is an expert in a particular field, there is good reason to think that we have ways of determining which experts are more reliable than others. For one thing we have ways of "calibrating" experts (Kitcher 1993). A scientist who is an expert in a particular field can directly calibrate the reliability of other experts in that same field by using her own information about that field. A scientist can indirectly calibrate a particular expert's level of expertise by taking account of the opinions other scientists, who she has directly calibrated, have of the expert in question. As we noted above, reputation plays a significant role in science. One important function of reputation is to help with the calibration of expertise.

Finally, there are numerous additional sources of evidence which we (either as laypeople or as scientists) can consult when evaluating a particular expert. Alvin Goldman (2001, p. 93) lists five common forms of evidence we have when evaluating experts:

(A) Arguments presented by the contending experts to support their own views and critique their rivals' views.
(B) Agreement from additional putative experts on one side or other of the subject in question.
(C) Appraisals by "meta-experts" of the experts' expertise (including appraisals reflected in formal credentials earned by the experts).
(D) Evidence of the experts' interests and biases vis-a-vis the question at issue.
(E) Evidence of the experts' past "track-records"

In the majority of cases at least some of these sources of evidence will be readily available. Thus, it seems that our trust in experts in science is far from blind.

While trust plays a key role in science, and trust always carries the possibility of disappointment, it seems that this should not give us pause when considering the ability of science to generate knowledge. Understanding the importance of trust in science, and why this trust is not simply blind faith in experts, helps to illuminate the social nature of science.

16.2 The Division of Cognitive Labor

We have noted that trust is vital for scientific progress. A key reason for the necessity of trust is the division of cognitive labor which we find in the sciences. Such a division of cognitive labor is necessary for science to progress beyond the limitations of what individuals can know on their own. Although the division of cognitive labor plays a critical role in the discovery and distribution of scientific knowledge, it presents us with important questions. One of the most important questions, which will be our focus here, is how the division of cognitive labor leads scientists to choose various research projects to pursue.

In an ideal situation we would have an infinite amount of resources and a community of scientists who are each perfectly rational, purely altruistic, completely aware of what projects other scientists are pursuing, and motivated solely by the desire to make discoveries which increase the amount of scientific knowledge in the scientific community as a whole. The situation that we actually find ourselves in is far from this ideal though. Obviously, we do not have an infinite amount of resources. If we did, the need for a division of cognitive labor would be greatly diminished, if not entirely non-existent. Additionally, scientists, while typically rational, are like any other group of humans—they fall short of perfect rationality. They are also limited in their communication with one another, and so limited in their awareness of the projects that other scientists are pursuing. Despite being generally good people with a motivation to increase the total stock of scientific knowledge, scientists are not typically altruists, nor do they tend to be motivated solely by the desire to increase the amount of scientific knowledge possessed by the community. "Realistically... scientific research is driven by twin motives" the desire for scientific knowledge and for credit—"some amount of scientific credit, after all, is normally a precondition for job maintenance, and larger amounts of credit often conduce to a higher salary, a better job, and/or greater personal satisfaction" (Goldman 1999, p. 260). So, while scientists are probably, by and large, rational, good people with good intentions, like everyone else they are far from perfect.

It is because scientists are people too and our collective situation falls short of the ideal in the sorts of ways mentioned above that questions about the division of labor and how different research projects are chosen are worth considering. Our limitations are the source of tensions which can arise in science between what is rational for individual scientists and what is rational for the scientific community as a whole (Kuhn 1977). Interestingly, this sort of tension can even arise in cases where scientists are ideally motivated by a desire to attain scientific knowledge, and they are each behaving rationally as individuals. The reason for this is that it seems that what is rational for a group (such as the scientific community) is not merely a matter of each member of that group behaving in a way which is individually rational (Sarkar 1983).[4]

To see how group rationality can diverge from individual rationality, consider a simple case from Goldman (2011, p. 17). Assume that we have a group of three

[4]Thagard (1993) claims that it should be somewhat unsurprising that group rationality and individual rationality diverge in science. He points out that such divergences have long been recognized in decision theory in situations such as the "prisoners' dilemma" and the "tragedy of the commons".

individuals who are operating under the seemingly rational rule of majority vote. In other words, when at least two of the members agree on a judgment the group accepts that judgment. These three individuals are each evaluating the truth of three propositions: *p*, *q*, and <if *p*, then *q*>. Person A accepts all three propositions as true. Person B accepts *p*, but denies the other two propositions. Person C accepts <if *p*, then *q*>, but denies the other two propositions. Each individual accepts a set of propositions that is perfectly consistent, i.e. each individual makes a judgment which is plausibly rational. The group, however, does not end up with a rational set of judgments. The group ends up accepting *p* (A and B agree about it) and <if *p*, then *q*> (A and C agree about it) while denying *q* (B and C both deny it). Each individual made a consistent judgment about the three propositions and the group employed the reasonable rule of going with the majority opinion on each proposition in forming the group's judgment. And yet, the group ends up with an inconsistent set of commitments: *p* and <if *p*, then *q*> (which the group accepts) entail that *q* must be true, but the group denies *q*. Hence, just because each member of a group is acting rationally it does not follow that the group itself will act rationally.

This sort of divergence between individual and group rationality can occur in a scientific community. It can even occur in a scientific community composed entirely of "epistemically pure" scientists, who are only motivated by the desire to discover scientific truths (Kitcher 1993). For simplicity, assume that the scientific community is seeking to uncover the truth of whether a particular molecule exists, and there are only two possible research projects to pursue in order to find out the truth of the matter. Project 1 has a much better chance of success than Project 2. However, since science is not perfect, Project 2 also has some chance of success because the odds that Project 1 succeeds fall short of 100 %. In this case the rational thing for an individual motivated solely by the desire to discover truth to do is to pursue Project 1 because doing so has the best odds of leading to a successful discovery concerning the molecule. Nevertheless, if each scientist is motivated only by the desire to discover truth and does what is individually rational, then each scientist will pursue Project 1. This, however, means that the scientific community as a whole is not pursuing projects in the most rational fashion. The rational thing for the community to do is to expend at least some resources in pursuing Project 2. After all, the odds of discovering the truth concerning the molecule are much higher if both projects are pursued rather than just one. As Philip Kitcher (1993, p. 344) aptly notes, "a community that is prepared to hedge its bets when the situation is unclear is likely to do better than a community that moves quickly to a state of uniform opinion" about what projects to pursue.[5] Consequently, even a community

[5]It is worth noting that this situation does assume that scientists do not have perfect knowledge of what every other scientist is doing. If each individual scientist had this knowledge and was altruistic, then some might rationally accept the lower chances of their making the discovery by pursing Project 2 for the greater good of the community. However, the assumptions that scientists do not have perfect knowledge of what each other is doing and that they are not perfectly altruistic are exceedingly plausible.

of epistemically pure scientists might fail to behave rationally despite each member of the community behaving in a way which is individually rational.

One might worry that if tensions can arise between what is rational for the scientific community as a whole and what is rational for individual scientists when those individual scientists are motivated solely by the desire to increase the community's knowledge, they can surely arise in the situation we find ourselves in where scientists are motivated by a variety of factors only some of which are "pure". The desire for discovering truth is only one thing that motivates actual scientists because, like the rest of us, they are "epistemically sullied agents" (Kitcher 1993). They are also motivated by their desire to receive credit for making scientific discoveries and all of the goods which come along with that credit. As a result, one might worry that when we look at science as an epistemic system what we find is an irrationality—a system which falls far short of dividing its cognitive resources in a reasonable manner.

As we have seen time and time again, assuming that science fails to perform admirably well is a mistake. It simply does not follow that because a more "purely" motivated society of scientists fails to properly distribute cognitive labor that a more "sullied" community like we have will similarly fail. In fact, "from the community perspective, it is likely that sullied scientists will do better than the epistemically pure. This is because a pure community heads toward cognitive uniformity", but "the sullied community hedges its bets" by diversifying its cognitive labor (Kitcher 1993, pp. 310–311). Recognition of the fact that scientists are motivated by more than simply the desire to generate scientific knowledge has led to many economic models of the structure of science and a focus on the various incentive structures of science.[6] One of the chief results of examining such models is the recognition that "the very factors that are frequently thought of as interfering with the (epistemically well-designed) pursuit of science—the thirst for fame and fortune, for example—might actually play a constructive role in our community epistemic projects, enabling us, as a group, to do far better than we would have done had we behaved as independent epistemically pure individuals" (Kitcher 1993, p. 351). Quite simply,

[6]See, for example, Brock and Durlaf (1999), Goldman (1999), Goldman and Shaked (1991), Hagstrom (1965), Hull (1988), Kitcher (1993), Latour and Woolgar (1979), Merton (1973), Rescher (1990), and Strevens (2003). For criticisms of economic models of science see Hands (1995, 1997), Muldoon (2013), Muldoon and Weisberg (2011), Weisberg and Muldoon (2009), and Wray (2000). There are other ways of modeling science such as Thagard's (1993) distributed A.I. approach, the ecological model, which Muldoon and Weisberg (2011) and Weisberg and Muldoon (2009) draw from "hill climbing" models in computer science, and various consensus models such as those put forward by Hegselman and Kraus (2006), Lehrer (1975), Lehrer and Wagner (1981), Wagner (1985), and Zollman (2010). Although these other models are interesting and worthy of careful consideration, we will limit our focus to the economic approach. The reason for this is twofold. First, such models are by far the most widely accepted approaches to modeling science and its distribution of cognitive labor. Second, the other primary ways of modeling science and the distribution of cognitive labor seem to agree with the economic models on the crucial points that hedging its bets is the best way for science to proceed. They also agree with economic models that there are social structures in place which help ensure that science has a diversity of cognitive labor.

"a profit motive can *discipline* one's search for truth rather than distort it" (Goldman 1999, p. 260).

The central point illuminated by economic models of science (and other models) is that a beneficial division of cognitive labor is often achieved because "social institutions within science might take advantage of our personal foibles to channel our efforts toward community goals rather than the epistemic ends that we might set for ourselves as individuals" (Kitcher 1993, p. 351).[7] In order to see this it is worth taking a look at a particular social institution present in science which helps to use scientists' desire for credit as a way to ensure a useful division of cognitive labor.[8] We will examine what Robert Merton (1973) dubbed the "priority rule". The priority rule is the social institution that we find in science where all of the credit for a particular discovery goes to the first researchers to make that discovery. According to the priority rule, scientific discovery is a winner-takes-all matter where only first-place counts—everyone else loses when it comes to credit even when the independent discoveries of a particular fact made by different scientists are merely days apart. The priority rule is a very long-standing institution in science. Merton traces the priority rule back to such luminaries as Galileo and Newton—basically to the beginning of modern science. Interestingly, Merton maintains that the priority rule is one that most all scientists tend to view as a good thing. In fact, he claims that scientists believe that violations of this rule are *morally wrong*.

Michael Strevens (2003) argues that it is rational for credit to be awarded in accordance with the priority rule because society benefits from initial discoveries in science, but not from subsequent discoveries of the same fact. The recognition of this feature of how society as a whole benefits from scientific discoveries helps to explain why scientists view the priority rule as a just rule.[9] Nonetheless, one might wonder how this social institution helps to ensure a favorable distribution of cognitive labor.

The priority rule promotes a division of cognitive labor because when scientists realize that only the first to make a discovery will receive credit it will prompt some to pursue projects other than those pursued by other scientists. After all, "two equally talented, equally industrious scientists may receive different rewards under the priority system just because one of them, but not the other, is lucky enough to select a research program that achieves its goal (and does so before any rival)" (Strevens 2003, p. 74). So, the reasonable thing in many cases may be to pursue a different project than one's peers. Consider if Xenia knows that Xavier is

[7]Also see Hull (1988), Solomon (1992), and Thagard (1988).

[8]There are, of course, a number of such social institutions present in science. We will limit our focus here to just one of the most prominent ones because doing so is sufficient for our current illustrative purposes.

[9]Strevens (2003) suggests that the priority rule is perhaps simply a clear representation of a much more general reward scheme which we find in society as a whole. Namely, a reward scheme that rewards in proportion to the benefit provided to society as a whole. The priority rule exemplifies this rule because when it comes to scientific discovery only the first to make the discovery provides a benefit to society, so only the first to make a discovery gets rewarded.

investigating the cause of some phenomenon Y, then she knows that the only way she can receive credit for discovering the truth about Y is if she makes her discovery before Xavier. In some cases this may simply come down to luck—particularly if they are employing similar methods. Additionally, Xenia might be at a disadvantage if Xavier has been working on this project longer than she has. Xenia might decide that her best bet for getting credit for some important scientific discovery or other is to pursue a project related to phenomenon Z instead of investigating Y because there is less work being done related to Z. Thus, awareness of the priority rule can lead to a division of cognitive labor by capitalizing on scientists' desire for credit.

Of course, there are other ways of encouraging division of cognitive labor in science. Some of these ways will also appeal to our more "sullied" motivations; others will simply be a matter of resource allocation. Some of these social institutions are relatively indirect motivators like the priority rule; others are more direct methods for diversifying cognitive labor. For example, one way to directly encourage scientists to pursue a diversity of research projects is to simply offer increased funding for research projects which are underrepresented. "Very little of modern science can be conducted without funding, and as long as individual scientists at least propose to use different methods, funding agencies are in a position to encourage diversity by financially supporting it" (Goldman 1999, p. 257).

Admittedly, some of the motivators employed to encourage division of cognitive labor, and thus to increase the odds of science generating more and more knowledge, may be misguided.[10] Also, it is likely that the current set of social institutions providing incentives in science is not optimal (Kitcher 1993; Muldoon and Weisberg 2011). Despite all of this, exploration of the social institutions present in science for encouraging division of cognitive labor helps to demonstrate that the sort of division of cognitive labor which is necessary for significant scientific progress can be achieved even by, and perhaps especially by, epistemically "sullied" scientific communities like ours. In fact, what we have seen here should remind us that we should not think that we can "identify very general features of scientific life—reliance on authority, competition, desire for credit—as epistemically good or bad" because "particular kinds of social arrangements make good epistemic use of the grubbiest motives" (Kitcher 1993, p. 305).

[10]For example see Merton's (1973, 1988) discussion of the "Matthew effect" where more well-known scientists receive more credit than less well-known scientists do for the same achievements. The Matthew effect is a social institution in science which many believe is a misallocation of credit, and so, a negative side effect of credit-based motivational structures in science. Though see Strevens (2006) for persuasive arguments for thinking that the Matthew effect does in fact distribute credit fairly.

16.3 Conclusion

In this chapter we have moved beyond our study of the individualistic characteristics of scientific knowledge by looking at science as an epistemic system. We have seen that the thoroughgoing social nature of science leads to some characteristics which make it particularly well suited for adding to the store of scientific knowledge. In particular, the social nature of science leads to a division of cognitive labor. We have seen that this division of cognitive labor both makes it so that trust plays an integral role in the generation of scientific knowledge and so that scientific progress is enhanced by the scientific community hedging its bets by scientists pursuing a wide variety of research projects utilizing a variety of methods. Although the individual scientists who make up the scientific community are not perfect, various social institutions in science help to make good use of their baser motivations. We have seen that when it comes to the epistemic system that is science: "Flawed people, working in complex social environments, moved by all kinds of interests, have collectively achieved a vision of parts of nature that is broadly progressive and that rests on arguments meeting standards that have been refined and improved over centuries" (Kitcher 1993, p. 390). Science may not be perfect, but it is an epistemic system that has been tremendously successful at generating knowledge of the world around us.

References

Adler, J. (2014). Epistemological problems of testimony. In E. N. Zalta (Ed.), *The Stanford encyclopedia of philosophy* (Spring 2014 Edition). http://plato.stanford.edu/archives/spr2014/entries/testimony-episprob/

Brock, B., & Durlaf, S. (1999). A formal model of theory choice in science. *Economic Theory, 14*, 113–130.

Cohen, P. A., Kulik, J. A., & Kulik, C. C. (1982). Educational outcomes of tutoring: A meta-analysis of findings. *American Educational Research Journal, 19*, 237–248.

Durkheim, E. (1893/1997). *The division of labor in society*. New York: The Free Press.

Fricker, E. (1994). Against gullibility. In B. K. Matilal & A. Chakrabarti (Eds.), *Knowing from words* (pp. 125–161). Dordrecht: Kluwer.

Fuchs, D., Fuchs, L. S., Mathes, P. G., & Simmons, D. C. (1997). Peer-assisted learning strategies: Making classrooms more responsive to diversity. *American Educational Research Journal, 34*, 174–206.

Gleick, J. (2004). *Isaac Newton*. New York: Vintage Books.

Goldman, A. I. (1999). *Knowledge in a social world*. Oxford: Oxford University Press.

Goldman, A. I. (2001). Experts: Which ones should you trust? *Philosophy and Phenomenological Research, 63*, 85–110.

Goldman, A. I. (2011). A guide to social epistemology. In A. Goldman & D. Whitcomb (Eds.), *Social epistemology: Essential readings* (pp. 11–37). New York: Oxford University Press.

Goldman, A. I., & Shaked, M. (1991). An economic model of scientific activity and truth acquisition. *Philosophical Studies, 63*, 31–55.

Hagstrom, W. (1965). *The scientific community*. New York: Basic Books.

Hands, D. W. (1995). Social epistemology meets the invisible hand: Kitcher on the advancement of science. *Dialogue, 34*, 605–621.
Hands, D. W. (1997). Caveat emptor: Economics and contemporary philosophy of science. *Philosophy of Science, 64*, 107–116.
Hardwig, J. (1985). Epistemic dependence. *Journal of Philosophy, 82*, 335–349.
Hegselmann, R. & Krause, U. (2006). Truth and cognitive division of labour: First steps towards a computer aided social epistemology. *Journal of Artificial Societies and Social Stimulation, 9*. http://jasss.soc.surrey.ac.uk/9/3/10.html
Hull, D. (1988). *Science as a process*. Chicago: University of Chicago Press.
Hume, D. (1739–1740/1978). *A treatise of human nature*. Oxford: Clarendon Press.
Keil, F. C. (2006). Doubt, deference, and deliberation: Understanding and using the division of cognitive labor. In T. Z. Gendler & J. Hawthorne (Eds.), *Oxford studies in epistemology: Volume 1* (pp. 143–166). Oxford: Oxford University Press.
Keil, F. C., Stein, C., Webb, L., Billings, V. D., & Rozenblit, L. (2008). Discerning the division of cognitive labor: An emerging understanding of how knowledge is clustered in other minds. *Cognitive Science, 32*, 259–300.
Kitcher, P. (1993). *The advancement of science: Science without legend, objectivity without illusions*. New York: Oxford University Press.
Kuhn, T. S. (1977). *The essential tension: Selected studies in scientific tradition and change*. Chicago: University of Chicago Press.
Latour, B., & Woolgar, S. (1979). *Laboratory life: The social construction of scientific facts*. Beverly Hills: Sage.
Lehrer, K. (1975). Social consensus and rational agnoiology. *Synthese, 31*, 141–160.
Lehrer, K., & Wagner, C. (1981). *Rational consensus in science and society*. Dordrecht: Reidel.
Longino, H. (2013). The social dimensions of scientific knowledge. In E. N. Zalta (Ed.), *The Stanford encyclopedia of philosophy* (Spring 2013 Edition). http://plato.stanford.edu/archives/spr2013/entries/scientific-knowledge-social/
Lutz, D. J., & Keil, F. C. (2002). Early understanding of the division of cognitive labor. *Child Development, 73*, 1073–1084.
Merton, R. (1973). *The sociology of science*. Chicago: University of Chicago Press.
Merton, R. (1988). The Matthew effect in science, II: Cumulative advantage and the symbolism of intellectual property. *Isis, 79*, 607–623.
Muldoon, R. (2013). Diversity of the division of cognitive labor. *Philosophy Compass, 8*, 117–125.
Muldoon, R., & Weisberg, M. (2011). Robustness and idealization in models of cognitive labor. *Synthese, 183*, 161–174.
Rescher, N. (1990). *Cognitive economy: The economic dimension of the theory of knowledge*. Pittsburgh: University of Pittsburgh Press.
Sarkar, H. (1983). *A theory of method*. Berkeley: University of California Press.
Shapin, S. (1994). *A social history of truth: Civility and science in seventeenth-century England*. Chicago: University of Chicago Press.
Shatz, D. (2004). *Peer review: A critical inquiry*. Lanham: Rowman & Littlefield.
Smith, A. (1776/1904). *An inquiry into the nature and causes of the wealth of nations* (5th ed.). London: Methuen & Co.
Solomon, M. (1992). Scientific rationality and human reasoning. *Philosophy of Science, 59*, 439–455.
Strevens, M. (2003). The role of the priority rule in science. *Journal of Philosophy, 100*, 55–79.
Strevens, M. (2006). The role of the Matthew effect in science. *Studies in History and Philosophy of Science, 37*, 159–170.
Strevens, M. (2010). Reconsidering authority: Scientific expertise, bounded rationality, and epistemic backtracking. In T. Z. Gendler & J. Hawthorne (Eds.), *Oxford studies in epistemology: Volume 3* (pp. 294–330). Oxford: Oxford University Press.
Thagard, P. (1988). *Computational philosophy of science*. Cambridge, MA: MIT Press.
Thagard, P. (1993). Societies of minds: Science as distributed computing. *Studies in History and Philosophy of Science, 24*, 49–67.

Wagner, C. (1985). On the formal properties weighted averaging as a method of aggregation. *Synthese, 25*, 233–240.
Weisberg, M., & Muldoon, R. (2009). Epistemic landscapes and the division of cognitive labor. *Philosophy of Science, 76*, 225–252.
Wray, K. B. (2000). Invisible hands and the success of science. *Philosophy of Science, 67*, 163–175.
Zollman, K. J. S. (2010). The epistemic benefit of transient diversity. *Erkenntnis, 72*, 17–35.

Chapter 17
Looking Back and Looking Forward

Abstract This concluding chapter recaps some of the major insights of the earlier chapters of this book. It also points out how these insights can be used to supplement the science education literature on the nature of science discussed in the first chapter. The result is a more philosophically grounded science education literature. Such integration holds promise for strengthening both science education and philosophical approaches to the nature of science as well as providing a more in-depth understanding of scientific knowledge. Additionally, the chapter discusses some of the major areas where further research would be helpful. Although it is often a bit risky to do so, this chapter also makes some suggestions as to how some of the needed research might be fruitfully conducted and it speculates on what some of the results of such research might be. The goal of the chapter is not to offer precise predictions of how things will turn out, but rather, to encourage further research and helpfully gesture to good starting places for such research.

At the end of a book it is often helpful to take a look back at what has been accomplished before looking forward at what still remains to be done. We have worked to develop a philosophical foundation for better understanding debates in the science education literature which have an important epistemological dimension—particularly, but not limited to, the debate concerning NOS and debates about the goals for science education. The development of our philosophical foundation utilized an explanatory approach. As was fitting, we began with the heart of epistemology—an exploration of the general nature of knowledge. Next, we turned our attention to specific features of scientific knowledge before considering challenges to the possibility of our having scientific knowledge at all. As we saw, although many challenges to our having scientific knowledge are worth taking seriously, none provides grounds for thinking that we truly do not, or cannot, have scientific knowledge. Finally, we moved beyond the individualistic aspects of scientific knowledge to some of its social aspects. We discovered that science is a powerful epistemic system which is capable of generating a wealth of knowledge despite the limitations of the people who make up the scientific community. In some ways science is greater than the sum of its parts.

K. McCain, *The Nature of Scientific Knowledge*, Springer Undergraduate Texts in Philosophy, DOI 10.1007/978-3-319-33405-9_17

17.1 The Explanatory Approach and Shifting Focus

Throughout the development of our philosophical foundation for understanding important debates in science education our explanatory approach has yielded several insights into the nature of scientific knowledge. This explanatory approach has also suggested a shift in focus when it comes to understanding NOS.

First of all, we saw in several chapters that there are reasons for thinking that a shift from focusing on scientific knowledge to focusing on *evidence* and the *justification* evidence provides for scientific claims would be helpful. In many cases where we speak of "scientific knowledge" plausibly we really mean sufficient evidence for thinking that a particular set of scientific claims are true (or approximately true to a specific degree). Often, we are not careful to distinguish between knowledge and justified belief—as we saw in earlier chapters this could be because it is very difficult to say exactly what knowledge is. We have seen reasons for thinking that our focus in science is not really on knowledge in the strict sense at all, but rather on the sort of evidence, and methods of gaining that evidence, which can justify us in believing that particular scientific claims are true. Plausibly, this insight concerning the features of our scientific inquiry is helpful for better understanding NOS.

Additionally, this shift in focus helps to make the tentative nature of scientific knowledge clearer. It is considerably easier to understand how a theory that we are justified in believing to be true is tentative even though it is based on strong evidence than it is to understand how it is that a theory we *know* to be true is tentative.

The proposed shift in focus can also help to connect our understanding of NOS more clearly with the role that verisimilitude (truthlikeness) plays in science. As we noted in earlier chapters, science is often more concerned with verisimilitude than truth simpliciter. This suggests that what really matters for scientific inquiry is the evidence we have in support of particular claims and theories rather than possessing knowledge of them. Shifting our focus to evidence rather than knowledge would better account for this aspect of scientific inquiry.

Furthermore, earlier in the book we drew an important distinction between the attitude of acceptance as a working hypothesis and full acceptance/belief. Once this distinction has been appreciated the plausibility of shifting from a focus on scientific knowledge to a focus on evidence and justification in scientific inquiry becomes even clearer. Belief is a necessary condition for knowledge, however, in many cases some of the hypotheses which make up our current scientific theories are simply accepted as working hypotheses—they are not fully accepted/believed to be true. So, although we have ample evidence for accepting these theories we cannot count as actually knowing the theories because parts of our current theories are not believed to be true. With respect to these theories we fail to satisfy a necessary condition for knowledge—we lack the requisite belief. This is not a problem at all though, if we are concerned with evidence and justification rather than knowledge.

Moreover, the proposed shift in emphasis would not even require us to stop using the term "scientific knowledge". This is good because we have seen that there are practical reasons for continuing to use the *term* "scientific knowledge" even if our focus were to become more explicitly evidence-centric. One such practical reason is that talking about theories and claims which are justified or reasonable to believe in light of the evidence may lead to the mistaken thought that many of our best scientific theories and laws are "just theories". This sort of "just a theory" thinking may lead to misconceptions about strongly supported scientific theories.[1] Additionally, continuing to use the term "scientific knowledge" may help with the problems which arise when one fails to distinguish well-supported facts in science from things that one merely believes.[2] Hence, there are practical reasons for continuing to use the term "scientific knowledge" even if we shift our focus in the currently suggested way. A reasonable way of doing this is to make clear that the term "scientific knowledge" signifies scientific claims or theories for which we have sufficiently strong evidence for justifiedly believing they are true (or approximately true) whether or not the other conditions required for knowledge are met. Continuing to speak in terms of scientific knowledge is perfectly fine so long as we keep in mind that what we are really interested in (and talking about) is *evidence* for claims and whether we have sufficiently strong evidence to be *justified in believing those claims*.

17.2 Building on the Foundation of the Explanatory Approach

Although the explanatory approach to scientific knowledge developed here has helped elucidate some philosophical concepts and theories, the study of which may aid in facilitating improved understanding of NOS, much work remains to be done. Throughout this book we noted several philosophical debates that are still ongoing as well as numerous areas where additional research would be valuable. It would be cumbersome to recall each of these points here. Yet, it does seem that we should at least briefly consider a few key areas where further research would be especially pertinent before concluding our discussion. Some of this research is primarily in the field of education, some primarily in philosophy, and some primarily in psychology. However, as this book has hopefully made clearer, research in each of these disciplines can profit from dialog with the others. In fact, the first area where considerably more research is required is a straightforwardly interdisciplinary endeavor—models for optimizing the distribution of cognitive labor. As we have already seen, modeling cognitive labor and determining how to optimize the division of that labor already prominently draws on research in artificial intelligence, economics, computer science, and philosophy. It is plausible that with additional research we might come to better understand the best ways to organize

[1]See McCain and Weslake (2013) for discussion of this and other misconceptions which lead some to object to well supported scientific theories such as evolution.

[2]See Kampourakis (2014) for discussion of this problem.

our scientific practices so as to maximize the amount of scientific knowledge we can generate with our limited resources.

Related to this issue, further research needs to be done with respect to biases and illegitimate heuristics. We have seen that there are various errors of reasoning which humans tend to make fairly systematically. It would be worthwhile to explore the sorts of errors that we fall prey to more fully as well as strategies for how people can learn to better avoid these errors. Although this would not directly affect the division of cognitive labor, it holds the potential to greatly improve our chances of making the best use of our cognitive resources.

Another area worth exploring is the nature of understanding. Specifically, it would be particularly helpful to carefully examine what exactly is required in order to truly understand a theory and to use that theory to enhance understanding of phenomena. Going along with this, research into how best to facilitate increased understanding of scientific theories and improved skills in utilizing that understanding through education would be very useful.

Finally, an area of research that is of particular relevance to our discussion in this chapter is the educational benefits of implementing the recommended shift in focus when it comes to understanding NOS. It seems plausible that shifting our focus slightly to place more emphasis on evidence and justification instead of knowledge may help lead to less confusion when it comes to understanding scientific knowledge. Research into the effects of teaching students about science via the sort of evidence-centric approach advocated here could help illuminate the benefits (and costs) of such an approach. It is research that; hopefully, this book has shown is worth doing.

References

Kampourakis, K. (2014). *Understanding evolution*. Cambridge, UK: Cambridge University Press.

McCain, K., & Weslake, B. (2013). Evolutionary theory and the epistemology of science. In K. Kampourakis (Ed.), *The philosophy of biology: A companion for educators* (pp. 101–119). Dordrecht: Springer.

Index

K. McCain, *The Nature of Scientific Knowledge*, Springer Undergraduate Texts in Philosophy, DOI 10.1007/978-3-319-33405-9

F

G

H

I

CPSIA information can be obtained
at www.ICGtesting.com
Printed in the USA
LVOW01*1943181016
509278LV00012B/142/P

9 783319 334035